Spline-Interpolation Solution of One Elasticity Theory Problem

Authors & Editors:
Elena A. Shirokova
Pyotr N. Ivanshin

Contents

Chapters

Foreword

The main purpose of this work is to demonstrate the new methods of solution of some 3D elasticity theory problems. These methods are based on complex analysis application.

I have considered different types of approximate and precise solutions of different mechanical problems, throughout almost all of my scientific career. So I am always interested in getting acquainted with some new method of solution. It seems interesting to consider passing from thorough methods applied in the case of 2D problems to the case of 3D ones. So the book seems to be of certain value. I feel obliged to note that it is necessary to find not only some methods of problem solution but to make it possible to get actual numerical results. Since the book contains examples for almost all the considered questions, it becomes a nice piece of literature useful for almost immediate applications.

Dr.Sc. Damir F. Abzalilov

Preface

This book appeared as a result of the development of the Kolosov-Muskhelishvili plane elasticity theory solution methods. E.A.Shirokova applied the generalisation of these methods to the construction of the interpolation solution of the 3D problems of elasticity for cylinders and tubes in 2004.

The spline-interpolation solution was introduced as the development of the interpolation solution for solids different from cylinders. This solution is based on the interpolation solution but is more efficient computationally.

The second co-author began his investigations in this field recently. His achievement is construction of continuous and smooth spline-interpolation solutions for some types of cylindrical and non-cylindrical solids and approximation estimates.

Also the authors would like to express their thanks to A.N. Schermann for the construction of one important example and for proof-reading of the text.

The order of reading is the following:

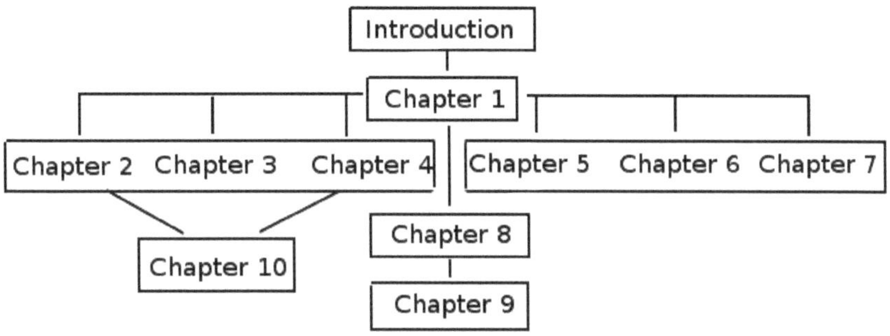

Elena A. Shirokova,

Pyotr N. Ivanshin

List of Contributors

Elena A. Shirokova, Kazan Federal University, Mechanics and Mathematics department

Pyotr N. Ivanshin, Kazan Federal University, Physics department

Introduction

We present the methods of approximate solution of the 3D basic problem of elasticity for the solids of the special types in this work. The classic formulation of the problem is the following: given the boundary displacements it should be possible to find the displacements in the whole elastic body which satisfy the equilibrium equations. This problem is named the second basic problem of elasticity [7]. There exist the well-known exact solutions of this problem in the symmetrical cases (e.g. the solids of revolution with the symmetrical stresses) [6]. The exact solution for the general case has not been found yet, so the engineers apply the approximate methods (Finite Element Method, Boundary Element Method). The application of these methods for the solids with the certain singularities (e.g. cones in the neighborhood of the vertex) or asymmetrical boundary conditions often fail to be correct.

Note that Kolosov-Muskhelishvili method based on the application of the complex variables and analytic functions yields the exact solutions for the wide range of the plane problems [7]. There were numerous attempts of the generalisation of Kolosov-Muskhelishvili method for the 3-dimensional solids, for example, by A.F. Tsalik, A. Alexandrov and F.A. Bogashov [14, 1, 3]. But their methods imply either solution of very large systems of symbol equations ([14, 3]) or transform the original problem to the different one (which is not equivalent to the given problem in the general case [1]). We might also recall the quaternion matrix representation method developed in [2], which also implies necessity of some complicated non-commutative calculations.

We construct the interpolation and the spline-interpolation solutions of the 3D second problem of the theory of elasticity in our work. Our solutions are based on the representation of the displacement coordinate functions in (x, y, h) space as polynomials in the variable h and solution of the plane boundary problems on the coefficients of the polynomials [10]:

$$u(x, y, h) = \sum_{k=0}^{n} u_k(x, y)h^k, \; v(x, y, h) = \sum_{k=0}^{n} v_k(x, y)h^k,$$

$$w(x, y, h) = \sum_{k=0}^{n} w_k(x, y)h^k. \qquad (1)$$

The spline-interpolation solution (SIS) may be considered as a variant of the finite element method. The differences between our spline-interpolation solution and the classical solution of finite element method are the following. 1) We consider the given boundary conditions not only at separated points but at the finite number of curves on the boundary of the solid, thus the basic element is the layer of the solid bounded by the planes $h = h_k$, $h = h_{k+1}$. 2) The initial system of equations is reduced to the system of partial differential equations, for which we present the exact solution and not linear one based on the information at the discrete set of points. 3) The method allows the body to possess singularities on the boundary. 4) The spline-interpolation solution can be arbitrary smooth on the boundary of the solid. The spline-interpolation solution is smooth almost everywhere in the solid. It can be discontinuous only at the interior points of the sections of the solid limited by the planes $h = h_k$, but the difference between values of the functions on the common plane can be made arbitrary small.

The main differences between our spline-interpolation solution and that of the boundary element method are as follows: 1) we consider the given boundary conditions not only at separated points but at the finite number of curves on the boundary of the solid, 2) our solution is continuously extendible to the boundary from the interior of the solid, 3) spline-interpolation solution can be arbitrary smooth at the boundary surface.

The interpolation solution (IS) of the problem for cylindrical solids is arbitrary smooth in the whole body. The drawback of this solution is the same as the drawback of the polynomial interpolation of the functions in one variable. So we must choose such boundary curves of interpolation that the corresponding polynomial interpolation of the boundary displacements over any of the generatrices of the cylinder have sufficiently adequate accuracy.

We change the boundary data of the classic problem in order to construct the interpolation solution of the first type: the boundary displacements are given only at the points of a finite number of the directrices of the cylinder. The special form of the solution given in equation (1) yields the exact particular solution of the formulated problem. This solution is not unique since we have no information about displacements at the points of the surface which differ from those on the given curves of interpolation. The classic formulation of the basic problem provides us with the displacements given on the complete external surface but the latter is often impossible in practical cases. The formulated problem with modified boundary data is close to the classic basic problem for adequate choice and sufficiently large number of the interpolation curves with the given displacements. Thus the particular solution of the formulated problem gives us a continuous extrapolation of the displacements to the complete cylindrical surface.

The interpolation solution of the second type is the solution of the equilibrium equations which coincides with the given displacements at one of the ends of the cylinder or approximates the displacements at the both ends. The interpolation solutions of both types are the base for construction of the spline-interpolation solution.

The first chapter is devoted to construction of the polynomial solution of the equilibrium equations. The coefficients of the polynomials turn out to be polyharmonic functions and therefore can be represented via the analytic functions of complex variables.

We construct the interpolation solutions for the circular cylinder, for the tube and for the certain type of non-circular cylinder in three following chapters. The data given at the surface provide the boundary value problems for the analytic functions which represent the polynomial coefficients. Note that we have to introduce some additional conditions at the inner surface for the tube.

We construct the spline-interpolation solutions of the 3D problem for the circular cylinder and for the pressurised tube in the fifth chapter on the base of interpolation solutions from the previous chapters.

We consider solids of revolution in the following chapter. The spline-interpolation solution for such bodies can be reduced to boundary value problems for the analytic functions in a disk with the boundary conditions at two concentric circles. We give the boundary data in the form of Fourier polynomials in order that these analytic functions be also polynomials.

The spline-interpolation solution for the conoids of the certain type is reduced to the boundary value problem for analytic functions with boundary conditions at two non-concentric circles in the seventh chapter.

The eighth chapter is devoted to the questions of approximation and convergence estimates of the interpolation and spline-interpolation solutions.

We consider the dynamic problems in the ninth chapter. The methods of interpolation and spline-interpolation solution of the 3D static problem are modified and applied to the 2D dynamic problem. We construct the interpolation solution of the 2D dynamic problem and generalise this method to the 3D case for the cylinder.

The tenth chapter contains the Appendices where we put the most bulky calculations for the spline-interpolation solutions.

The analytic form of both the interpolation solution and the spline-interpolation solution helps us to simulate static and dynamic stress distribution in elastic bodies. We demonstrate some examples where the stress components can be easily calculated and investigated.

The interpolation and spline-interpolation methods are based on the properties of analytic functions of complex variables which makes it possible to obtain the exact solutions of the plane boundary value problems when we restore the polynomial coefficients in plane domains. The technique of these methods is rather simple when we apply computer calculations. The analogy of processes of polynomial coefficients restoration leads to the possibility of computer programming application.

Spline-Interpolation Solution of One Elasticity Theory Problem

CHAPTER 1

Polynomial solution of the system of the equilibrium equations

Elena A. Shirokova,
Kazan Federal University
Elena.Shirokova@ksu.ru

ABSTRACT: This chapter is devoted to the solution of the system of equilibrium equations. We present the general form of the solution as a polynomial in the variable h.

Consider an elastic solid Ω in the coordinate (x,y,h) space. It is well known [7] that the boundary displacements define the inner displacements and stresses in the whole solid. Let

$$\vec{a} = (u(x,y,h), v(x,y,h), w(x,y,h)),\ (x,y,h) \in \Omega,$$

be the vector of displacements of the points of Ω. These displacements satisfy the equilibrium equations

$$(\lambda+\mu)\frac{\partial\theta}{\partial x}+\mu\Delta u=0,$$
$$(\lambda+\mu)\frac{\partial\theta}{\partial y}+\mu\Delta v=0, \qquad (1.1)$$
$$(\lambda+\mu)\frac{\partial\theta}{\partial h}+\mu\Delta w=0,$$

where

$$\theta=\frac{\partial u}{\partial x}+\frac{\partial v}{\partial y}+\frac{\partial w}{\partial h},$$

Δ is the Laplace operator and the factors λ and μ are Lamé coefficients. We search for the solution of the equilibrium equations given by formula (1.1) in the form of the polynomials in the variable h in order to reduce the given 3D boundary value problem to the set of plane boundary value problems.

Therefore

$$u(x,y,h)=\sum_{k=0}^{n}u_k(x,y)h^k, v(x,y,h)=\sum_{k=0}^{n}v_k(x,y)h^k,$$
$$w(x,y,h)=\sum_{k=0}^{n}w_k(x,y)h^k, \qquad (1.2)$$

the factors

$$u_k(x,y),\ v_k(x,y),\ w_k(x,y),$$

being the unknown functions at the projection D of the solid Ω on the plane XOY.

The equilibrium equations (1.1) imply according to (1.2) the relations

$$\lambda\{\sum_{k=0}^{n-1}[\frac{\partial^2 u_k}{\partial x^2}+\frac{\partial^2 v_k}{\partial x\partial y}+(k+1)\frac{\partial w_{k+1}}{\partial x}]h^k+(\frac{\partial^2 u_n}{\partial x^2}+\frac{\partial^2 v_n}{\partial x\partial y})h^n\}+$$
$$+\mu\sum_{k=0}^{n}(2\frac{\partial^2 u_k}{\partial x^2}+\frac{\partial^2 u_k}{\partial y^2}+\frac{\partial^2 v_k}{\partial x\partial y})h^k+\mu\{\sum_{k=1}^{n-1}[(k+1)u_{k+1}+ \quad (1.3)$$
$$+\frac{\partial w_k}{\partial x}]kh^{(k-1)}+n\frac{\partial w_n}{\partial x}h^{(n-1)}\}=0,$$

$$\lambda\{\sum_{k=0}^{n-1}[\frac{\partial^2 u_k}{\partial x\partial y}+\frac{\partial^2 v_k}{\partial y^2}+(k+1)\frac{\partial w_{k+1}}{\partial y}]h^k+(\frac{\partial^2 u_n}{\partial x\partial y}+\frac{\partial^2 v_n}{\partial y^2})h^n\}+$$
$$+\mu\sum_{k=0}^{n}(2\frac{\partial^2 v_k}{\partial y^2}+\frac{\partial^2 u_k}{\partial x\partial y}+\frac{\partial^2 v_k}{\partial x^2})h^k+\mu\{\sum_{k=1}^{n-1}[(k+1)v_{k+1}+ \quad (1.4)$$
$$+\frac{\partial w_k}{\partial y}]kh^{(k-1)}+n\frac{\partial w_n}{\partial y}h^{(n-1)}\}=0,$$

$$\lambda\{\sum_{k=1}^{n-1}[\frac{\partial u_k}{\partial x}+\frac{\partial v_k}{\partial y}+(k+1)w_{k+1}]kh^{(k-1)}+n(\frac{\partial u_n}{\partial x}+\frac{\partial v_n}{\partial y})h^{(n-1)}\}+$$
$$+\mu\{\sum_{k=0}^{n-1}[(k+1)(\frac{\partial u_{k+1}}{\partial x}+\frac{\partial v_{k+1}}{\partial y})+\frac{\partial^2 w_k}{\partial x^2}+\frac{\partial^2 w_k}{\partial y^2}]h^k+(\frac{\partial^2 w_n}{\partial x^2}+ \quad (1.5)$$
$$+\frac{\partial^2 w_n}{\partial y^2})h^n\}+2\mu\sum_{k=1}^{n-1}k(k+1)w_{k+1}h^{(k-1)}=0.$$

The displacement components satisfy equations (1.3),(1.4),(1.5) for every h from the projection of the solid Ω to the OZ axis if each factor with

$$h^k,\ k=0,1,...,n,$$

in (1.3), (1.4), (1.5) vanishes.

We see that the factors with the highest power of h give the conditions only on the functions

$$u_n(x,y), \quad v_n(x,y), \quad w_n(x,y).$$

The factors with $h^{(n-1)}$ give the relations between

$$u_n(x,y), \quad v_n(x,y), \quad w_n(x,y)$$

and

$$u_{n-1}(x,y), \quad v_{n-1}(x,y), \quad w_{n-1}(x,y).$$

Hence we must begin with the factors with the highest power of h and obtain the functions

$$u_n(x,y), \quad v_n(x,y), \quad w_n(x,y)$$

from the boundary data. Then we go to the lower power and obtain the functions

$$u_{n-1}(x,y), \quad v_{n-1}(x,y), \quad w_{n-1}(x,y)$$

with the help of the known functions

$$u_n(x,y), \quad v_n(x,y), \quad w_n(x,y)$$

from the boundary data.

The other factors

$$u_k(x,y), \quad v_k(x,y), \quad w_k(x,y)$$

can be obtained by the similar procedure consequently.

We use the complex variable $z = x + \imath y$ and some analytic on z functions in order to obtain the functions

$$u_k(x,y), \ v_k(x,y), \ w_k(x,y), \ k = 0, ..., n.$$

We consider these analytic functions in the projection D of the solid Ω

on the plane XOY. Recall that the derivatives over the complex variables z and $\overline{z}$ have the form

$$\frac{\partial}{\partial z}=\frac{1}{2}(\frac{\partial}{\partial x}-\imath\frac{\partial}{\partial y}),\ \frac{\partial}{\partial \overline{z}}=\frac{1}{2}(\frac{\partial}{\partial x}+\imath\frac{\partial}{\partial y}).$$

Let us consider the factors with h^n in (1.3) and (1.4). We obtain the system

$$\begin{cases} \frac{\partial}{\partial x}[(\lambda+2\mu)(\frac{\partial u_n}{\partial x}+\frac{\partial v_n}{\partial y})]-\frac{\partial}{\partial y}[\mu(\frac{\partial v_n}{\partial x}-\frac{\partial u_n}{\partial y})] & =0, \\ \frac{\partial}{\partial y}[(\lambda+2\mu)(\frac{\partial u_n}{\partial x}+\frac{\partial v_n}{\partial y})]+\frac{\partial}{\partial x}[\mu(\frac{\partial v_n}{\partial x}-\frac{\partial u_n}{\partial y})] & =0. \end{cases}$$

We multiply the first equation of this system by $\frac{1}{2}$ and add to the second one multiplied by $\frac{\imath}{2}$. Thus we get the equation

$$\frac{\partial}{\partial \overline{z}}[(\lambda+2\mu)(\frac{\partial u_n}{\partial x}+\frac{\partial v_n}{\partial y})+\imath\mu(\frac{\partial v_n}{\partial x}-\frac{\partial u_n}{\partial y})]=0.$$

So

$$(\lambda+2\mu)(\frac{\partial u_n}{\partial x}+\frac{\partial v_n}{\partial y})+\imath\mu(\frac{\partial v_n}{\partial x}-\frac{\partial u_n}{\partial y})=f_n(z),\ z=x+\imath y\in D,$$

where $f_n(z)$ is analytic in D.

Since

$$\frac{1}{2}[(\frac{\partial u_n}{\partial x}+\frac{\partial v_n}{\partial y})+\imath(\frac{\partial v_n}{\partial x}-\frac{\partial u_n}{\partial y})]=\frac{\partial}{\partial z}(u_n+\imath v_n),$$

we have

$$\frac{\partial}{\partial z}(u_n+\imath v_n)=\frac{\operatorname{Re} f_n(z)}{2(\lambda+2\mu)}+\frac{\imath\operatorname{Im} f_n(z)}{2\mu}=\frac{f_n+\overline{f_n}}{4(\lambda+2\mu)}+\frac{f_n-\overline{f_n}}{4\mu}.$$

Therefore

$$u_n+\imath v_n=\frac{\lambda+3\mu}{4\mu(\lambda+2\mu)}\int f_n(z)dz-\frac{\lambda+\mu}{4\mu(\lambda+2\mu)}z\overline{f_n(z)}+\overline{g_n(z)},$$

where $g_n(z)$ is analytic in D. Now if we introduce the following notation

$$\frac{\lambda+\mu}{2(\lambda+2\mu)}\int f_n(z)dz\equiv\phi_n(z),\ -2\mu g_n(z)\equiv\psi_n(z),\ \frac{\lambda+3\mu}{\lambda+\mu}\equiv\kappa,$$

we have the representation:

$$-2\mu(u_n + \imath v_n) = -\kappa\phi_n(z) + z\overline{\phi_n'(z)} + \overline{\psi_n(z)}.$$

Note that the representation of the function $(u_n + \imath v_n)$ is similar to the representation of the complex displacements for the second plane boundary value problem [7]. The analytic in D functions $\phi_n(z)$ and $\psi_n(z)$ can be restored, for example, via the boundary data of the function $(u_n + \imath v_n)$.

We obtain the Laplace equation

$$\Delta w_n = 0,$$

when we examine the factor with h^n in (1.5).

So $w_n(x,y) = \operatorname{Re}\rho_n(z)$, where the function $\rho_n(z)$ is analytic in D and can be restored via the given boundary data of the $w_n(x,y)$.

Now we examine the factors with $h^{(n-1)}$.

We equate the factors with $h^{(n-1)}$ in (1.3) and in (1.4) to zero and get the system of equations

$$\begin{cases} \frac{\partial}{\partial x}[(\lambda+2\mu)(\frac{\partial u_{n-1}}{\partial x}+\frac{\partial v_{n-1}}{\partial y})]-\frac{\partial}{\partial y}[\mu(\frac{\partial v_{n-1}}{\partial x}-\frac{\partial u_{n-1}}{\partial y})] & =-(\lambda+\mu)n\frac{\partial w_n}{\partial x}, \\ \frac{\partial}{\partial y}[(\lambda+2\mu)(\frac{u_{n-1}}{\partial x}+\frac{\partial v_{n-1}}{\partial y})]+\frac{\partial}{\partial x}[\mu(\frac{\partial v_{n-1}}{\partial x}-\frac{\partial u_{n-1}}{\partial y})] & =-(\lambda+\mu)n\frac{\partial w_n}{\partial y}, \end{cases}$$

which is equivalent to the equation

$$\frac{\partial}{\partial \overline{z}}[(\lambda+2\mu)(\frac{\partial u_{n-1}}{\partial x}+\frac{\partial v_{n-1}}{\partial y})+\imath\mu(\frac{\partial v_{n-1}}{\partial x}-\frac{\partial u_{n-1}}{\partial y})]=\frac{(\lambda+\mu)n}{2}\overline{\rho_n'(z)},$$

where $\rho_n(z)$ is known.

So

$$(\lambda+2\mu)(\frac{\partial u_{n-1}}{\partial x}+\frac{\partial v_{n-1}}{\partial y})+\imath\mu(\frac{\partial v_{n-1}}{\partial x}-\frac{\partial u_{n-1}}{\partial y})=\frac{(\lambda+\mu)n}{2}\overline{\rho_n(z)}+f_{n-1}(z),$$

where $f_{n-1}(z)$ is analytic in D. Now we have

$$\frac{\partial}{\partial z}(u_{n-1}+\imath v_{n-1}) = -\frac{(\lambda+\mu)n}{2}[\frac{\operatorname{Re}\rho_n(z)}{2(\lambda+2\mu)} - \imath\frac{\operatorname{Im}\rho_n(z)}{2\mu}] + \\ + \frac{\operatorname{Re} f_{n-1}(z)}{2(\lambda+2\mu)} + \frac{\imath \operatorname{Im} f_{n-1}(z)}{2\mu}.$$

So

$$-2\mu(u_{n-1}(x,y)+\imath v_{n-1}(x,y)) = -\kappa\phi_{n-1}(z) + z\overline{\phi'_{n-1}(z)} + \overline{\psi_{n-1}(z)} + \Phi_{n-1}(z,\overline{z}),$$

where

$$\phi_{n-1}(z) = \frac{\lambda+\mu}{2(\lambda+2\mu)}\int f_{n-1}(z)dz, \Phi_{n-1}(z,\overline{z}) = \\ = \mu(\lambda+\mu)n\int[\frac{\rho_n+\overline{\rho_n}}{4(\lambda+2\mu)} - \frac{\rho_n-\overline{\rho_n}}{4\mu}]dz.$$

The biharmonic function $\Phi_{n-1}(z,\overline{z})$ is the expression which contains the analytic function $\rho_n(z)$ introduced in the previous step; the analytic functions $\phi_{n-1}(z)$ and $\psi_{n-1}(z)$ are introduced in this step.

When we equate the factor with $h^{(n-1)}$ in (1.5) to zero, we obtain the equation

$$\mu\Delta w_{n-1}(x,y) = -(\lambda+\mu)n(\frac{\partial u_n}{\partial x} + \frac{\partial v_n}{\partial y}).$$

The right hand side of this equation is equal to $-2n\operatorname{Re}\phi'_n(z)$, with the analytic function $\phi_n(z)$ introduced in the previous step, so the last equation has the form

$$4\frac{\partial^2 w_{n-1}}{\partial z\partial\overline{z}} = -\frac{n}{\mu}[\phi'_n(z) + \overline{\phi'_n(z)}].$$

Therefore

$$4\frac{\partial w_{n-1}}{\partial z} = -\frac{n}{\mu}[\overline{z}\phi'_n(z) + \overline{\phi_n(z)}] + r_{n-1}(z),$$

where $r_{n-1}(z)$ is analytic in D. Now

$$w_{n-1}(x,y) = \operatorname{Re}\rho_{n-1}(z) + \Psi_{n-1}(z,\overline{z}),$$

where $\Psi_{n-1}(z,\overline{z}) = -n(z\overline{\phi_n(z)} + \overline{z}\phi_n(z))/(4\mu)$ is the biharmonic function which contains the analytic function $\phi_n(z)$ introduced in the previous step; the analytic function

$$\rho_{n-1}(z) = \frac{1}{4}\int r_{n-1}(z)dz, z \in D,$$

is introduced in this step.

We restore the coefficients

$$u_k(x,y), \quad v_k(x,y), \quad w_k(x,y)$$

of the displacement vector components (1.2) from $k = n$ till $k = 0$ successively using the same method.

The coefficients of the h^k satisfy the following equations

$$\begin{cases} \frac{\partial}{\partial x}[(\lambda+2\mu)(\frac{\partial u_k}{\partial x} + \frac{\partial v_k}{\partial y})] - & \frac{\partial}{\partial y}[\mu(\frac{\partial v_k}{\partial x} - \frac{\partial u_k}{\partial y})] = \\ = -\mu(k+2)(k+1)u_{k+2} & -(\lambda+\mu)(k+1)\frac{\partial w_{k+1}}{\partial x}, \\ \frac{\partial}{\partial y}[(\lambda+2\mu)(\frac{\partial u_k}{\partial x} + \frac{\partial v_k}{\partial y})] + & \frac{\partial}{\partial x}[\mu(\frac{\partial v_k}{\partial x} - \frac{\partial u_k}{\partial y})] = \\ = -\mu(k+2)(k+1)v_{k+2} & -(\lambda+\mu)(k+1)\frac{\partial w_{k+1}}{\partial y}, \end{cases}$$

$$\mu\Delta w_k = -(\lambda+\mu)(k+1)(\frac{\partial u_{k+1}}{\partial x} + \frac{\partial v_{k+1}}{\partial y}) - \\ -(k+1)(k+2)(\lambda+2\mu)w_{k+2}.$$

We obtain the coefficients in the following form:

$$-2\mu(u_k(x,y) + \imath v_k(x,y)) = -\kappa\phi_k(z) + z\overline{\phi_k'(z)} + \overline{\psi_k(z)} + \Phi_k(z,\overline{z}), \quad (1.6)$$

$$w_k(x,y) = \operatorname{Re}\rho_k(z) + \Psi_k(z,\overline{z}), \quad (1.7)$$

where the additional functions $\Phi_k(z,\overline{z})$ and $\Psi_k(z,\overline{z})$ should be obtained with the help of the recurrent relations

$$\Phi_k(z,\overline{z}) = \frac{(k+1)(k+2)}{8(\lambda+2\mu)}\left[\frac{2(\lambda^2+4\lambda\mu+5\mu^2)}{(\lambda+\mu)}\overline{z}\overline{\int \phi_{k+2}(z)dz} - \right.$$
$$-(\lambda+3\mu)z^2\overline{\phi_{k+2}(z)} - (\lambda+3\mu)z\int \psi_{k+2}(z)dz +$$
$$+(\lambda+\mu)\int dz\int \psi_{k+2}(z)dz - (\lambda+3\mu)\int dz\int \Phi_{k+2}(z,\overline{z})d\overline{z} +$$
$$\left. +(\lambda+\mu)\int dz\overline{\int \Phi_{k+2}(z,\overline{z})d\overline{z}}\right] + \tag{1.8}$$
$$+\frac{(\lambda+\mu)(k+1)}{4(\lambda+2\mu)}\left[4\mu\int \Psi_{k+1}(z,\overline{z})dz + \right.$$
$$\left. +(\lambda+3\mu)z\overline{\rho_{k+1}(z)} - (\lambda+\mu)\int \rho_{k+1}(z)dz\right],$$

$$\Psi_k(z,\overline{z}) = -\frac{(\lambda+\mu)(k+1)}{4\mu^2}\left[-\frac{2\mu}{(\lambda+\mu)}\mathrm{Re}(\overline{z}\phi_{k+1}(z)) + \right.$$
$$\left. +\mathrm{Re}(\int \Phi_{k+1}(z,\overline{z})d\overline{z})\right] - \tag{1.9}$$
$$-\frac{(k+1)(k+2)(\lambda+2\mu)}{4\mu}\left[\mathrm{Re}(\overline{z}\int \rho_{k+2}(z)dz) + \right.$$
$$\left. +\int dz\int \Psi_{k+2}(z,\overline{z})d\overline{z}\right].$$

Note that formula (1.8) must be changed if any of the functions $\phi_{k+2}(z)$ or $\rho_{k+1}(z)$ is a constant and formula (1.9) must be changed if the function $\phi_{k+1}(z)$ is a constant.

So, formula (1.8) has the form

$$\Phi_k(z,\overline{z}) = \frac{(k+1)(k+2)}{8(\lambda+2\mu)}\left[\frac{(\lambda+3\mu)^2}{(\lambda+\mu)}|z|^2\phi_{k+2} - \right.$$
$$-\frac{(\lambda+3\mu)}{2}z^2\overline{\phi_{k+2}} - (\lambda+3\mu)z\overline{\int \psi_{k+2}(z)dz} +$$
$$+(\lambda+\mu)\int dz\int \psi_{k+2}(z)dz - (\lambda+3\mu)\int dz\int \Phi_{k+2}(z,\overline{z})d\overline{z} +$$
$$\left. +(\lambda+\mu)\int dz\overline{\int \Phi_{k+2}(z,\overline{z})d\overline{z}}\right] + \tag{1.10}$$
$$+\frac{(\lambda+\mu)(k+1)}{4(\lambda+2\mu)}\left[4\mu\int \Psi_{k+1}(z,\overline{z})dz + \right.$$
$$\left. +(\lambda+3\mu)z\overline{\rho_{k+1}(z)} - (\lambda+\mu)\int \rho_{k+1}(z)dz\right],$$

if $\phi_{k+2}(z) \equiv \phi_{k+2}$,

$$\Phi_k(z,\overline{z}) = \frac{(k+1)(k+2)}{8(\lambda+2\mu)}\left[\frac{2(\lambda^2+4\lambda\mu+5\mu^2)}{(\lambda+\mu)}\overline{z}\int \phi_{k+2}(z)dz - \right.$$
$$-(\lambda+3\mu)z^2\overline{\phi_{k+2}(z)} -$$
$$-(\lambda+3\mu)z\overline{\int \psi_{k+2}(z)dz} + (\lambda+\mu)\int dz\int \psi_{k+2}(z)dz - \tag{1.11}$$
$$\left. -(\lambda+3\mu)\int dz\int \Phi_{k+2}(z,\overline{z})d\overline{z} + (\lambda+\mu)\int dz\overline{\int \Phi_{k+2}(z,\overline{z})d\overline{z}}\right] +$$
$$+\frac{(\lambda+\mu)(k+1)}{4(\lambda+2\mu)}\left[4\mu\int \Psi_{k+1}(z,\overline{z})dz\right],$$

if $\rho_{k+1}(z) \equiv \rho_{k+1}$.

Similarly, formula (1.9) has the form

$$\Psi_k(z,\overline{z}) = -\frac{(\lambda+\mu)(k+1)}{4\mu^2}\left[\mathrm{Re}(\int \Phi_{k+1}(z,\overline{z})d\overline{z})\right] - \\ -\frac{(k+1)(k+2)(\lambda+2\mu)}{4\mu}\left[\mathrm{Re}(\overline{z}\int \rho_{k+2}(z)dz) + \\ + \int dz \int \Psi_{k+2}(z,\overline{z})d\overline{z}\right], \tag{1.12}$$

if $\phi_{k+1}(z) \equiv \phi_{k+1}$.

Let us examine the order of harmonicity of the solution. The coefficients $u_n(x,y)$ and $v_n(x,y)$ are biharmonic functions and the coefficient $w_n(x,y)$ is a harmonic function in D. The coefficients $u_{n-1}(x,y)$ and $v_{n-1}(x,y)$ are biharmonic functions and the coefficient $w_{n-1}(x,y)$ is also a biharmonic function in D. The coefficients $u_{n-2}(x,y)$ and $v_{n-2}(x,y)$ are the three-harmonic functions and the coefficient $w_n(x,y)(x,y)$ is the biharmonic function in D....

The summand $-\kappa\phi_j(z)+z\overline{\phi_j'(z)}+\overline{\psi_j(z)}$ in the representation of $u_j(x,y)+$ $\imath v_j(x,y)$ is a biharmonic function, the summand $\mathrm{Re}\,\rho_j(z)$ in the representation of $w_j(x,y)$ is a harmonic function for all j. The order of harmonicity changes due to the presence of the functions $\Phi_j(z,\overline{z})$ and $\Psi_j(z,\overline{z})$ in the representations of $u_j(x,y)+$ $\imath v_j(x,y)$ and $w_j(x,y)$, respectively. Formulas (1.8) and (1.9) show that the harmonicity order of the polyharmonic coefficients of k-th power of h of the displacement vector components equals the harmonicity order of the polyharmonic coefficients of $(k-2)$-th power of h minus 1.

Indeed, the order of harmonicity of the function

$$\int dz \int \Phi_{j+2}(z,\overline{z})d\overline{z} \tag{1.13}$$

increases by 1 in comparison with the order of harmonicity of the function

$\Phi_{j+2}(z,\overline{z})$. And it is exactly the summand (1.13) we have in the representation of the function $\Phi_j(z,\overline{z})$. Similarly the summand

$$\int \Psi_{j+1}(z,\overline{z})dz$$

in the same representation raises by 1 the order of harmonicity in comparison with the order of harmonicity of the real function $\Psi_{j+1}(z,\overline{z})$.

The same with the function $\Psi_j(z,\overline{z})$. The order of harmonicity of the function

$$\int dz \int \Psi_{j+2}(z,\overline{z})d\overline{z} \tag{1.14}$$

increases by 1 in comparison with the order of harmonicity of the function $\Psi_{j+2}(z,\overline{z})$ and the summand given by formula (1.14) appears in the representation of $\Psi_j(z,\overline{z})$.

The representation of any $(m+1)$ harmonic function as the expansion with respect to the powers of $|z|^2$ has the form

$$\sum_{j=0}^{m} |z|^{2j}[F_j(z)+\overline{G_j(z)}],$$

where $F_j(z)$ and $G_j(z)$, $j=0,...,m,$ are analytic functions. Note that this representation for $u(x,y,h)+\ \imath v(x,y,h)$ contains only the anti-analytic summand with the highest power of $|z|^2$ that is $F_m(z)\equiv 0$. It follows from formulas (1.8) and (1.9) and can be proved by induction.

Finally, solution (1.2) of elasticity equations (1.1) is a polyharmonic in D function with the order of harmonicity dependent on the number n , it contains $3n$ analytic in D functions which are to be restored via boundary conditions.

CHAPTER 2

Interpolation solution of the second basic problem of elasticity for the circular cylinder

Elena A. Shirokova
Kazan Federal University
Elena.Shirokova@ksu.ru

Pyotr N. Ivanshin
Kazan Federal University
pivanshin@gmail.com

ABSTRACT: In this chapter we present the interpolation solution of the 3D second basic elasticity problem for the circular cylinder and explore some of its features. We present the solution of the first type which uses only the data on directrices of the cylindrical surface. Then we present the interpolation solution of the second type which uses the data at the ends of the circular cylinder.

The case of the circular cylinder is the most convenient case to restore the analytic functions $\phi_k(z)$, $\psi_k(z)$ and $\rho_k(z)$, which appear in Chapter 1, via the boundary conditions because the domain D is the unit disk and all boundary conditions are given at the same unit circle.

Let Ω be the circular cylinder in the (x, y, h) –coordinate space with the generatrix of the cylinder parallel to h -axis. Let the projection of the solid Ω on the XOY -plane be the unit disk D . Let the projection of the solid Ω on the h -axis be the interval $[A, B]$, D_A and D_B being the ends of the cylinder. We denote by C_A and C_B the boundaries of the disks D_A and D_B , respectively.

We distinguish two types of approximate interpolation solutions for the cylinder: the first is more accurate at the points distant from the ends of the cylinder than at the points close to the ends of the cylinder, the second is contrariwise more accurate at the points close to one of the ends or to the both ends of the cylinder than at the points distant from the ends of the cylinder.

We construct the interpolation solution of the first type — for the points distant from the ends of the cylinder — in section 2.1 using the given boundary displacements at the points of of the cylinder surface and ignoring the data at the points of the ends D_A and D_B of the cylinder. The procedure of construction of these interpolation solution is described in [10]. The example of such interpolation solution is given in section 2.2.

When we construct the interpolation solution of the second type we use the given displacements at the circles C_A and C_B and at the points of one of the ends D_A , D_B or of the both ends.

Sometimes it suffices to give the displacements not at all interior points of the end but at a finite number of interior points and to construct an interpolation solution which satisfies equilibrium equations (1.1) and the given displacements at the finite set of points of the end. Such solution interpolates the displacements at the corresponding end at the points where the displacements are not given. We construct such an interpolation solution of the second type in section 2.3 and give the example.

In section 2.4 we construct the interpolation solution of the second type for the case when this solution satisfies equilibrium equations (1.1) and takes the given displacements at C_A and C_B and at one of the ends D_A , D_B . The example of such construction is presented. Here we also consider the approximation of the data at the interior points of the ends.

2.1 Interpolation solution of the first type — for the points distant from the ends of the cylinder

Consider the first component of the displacement vector given on the cylindric boundary surface. We denote this continuous function by

$$U(\theta, h) = u(\cos\theta, \sin\theta, h),\ \ \theta \in [0, 2\pi],\ \ h \in [A, B].$$

Let us choose $(n-1)$ points in the each interval (A, B) for each value of $\theta \in [0, 2\pi]$ so that these points together with the points (θ, A) and (θ, B) are the nodes of "good" polynomial interpolation of the function $U(\theta, h)$ on the interval $[A, B]$.

Let these chosen points with different values $\theta \in [0, 2\pi]$ form $(n+1)$ simple closed curves $C_0 = C_A$, C_1 , $\ldots$, $C_{(n-1)}$, $C_n = C_B$, (see Fig.2.1) on the cylindric boundary surface:

$$C_j = \{(\cos\theta, \sin\theta, h_j(\theta)) | \theta \in [0, 2\pi]\}, j = 0, \ldots, n.$$

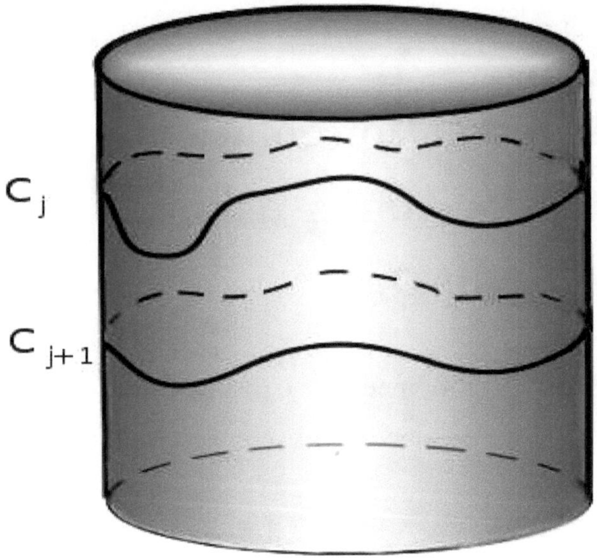

Figure 2.1: Directrices with given displacements

The function

$$U(\theta, h_j(\theta)), \quad \theta \in [0, 2\pi],$$

gives the boundary value of the first component of the displacement vector at all the points of the curve C_j , $j = 0, ..., n$.

We search for the first component of the displacement vector $u(x, y, h)$ in form (1.2). Now we obtain the boundary values of the unknown coefficients

$$u_k(cos\theta, \sin\theta) = \tilde{u}_k(\theta), \quad \theta \in [0, 2\pi], \quad k = 0, ..., n,$$

as the solution of the system

$$U(\theta, h_j(\theta)) = \sum_{k=0}^{n} u_k(\cos\theta, \sin\theta) h_j^k(\theta), j = 0, 1, \ldots, n.$$

This system has Vandermonde determinant for every $\theta \in [0, 2\pi]$ and therefore is resolvable. So we can regard the functions

$$u_k(\cos\theta, \sin\theta) = \tilde{u}_k(\theta), \quad \theta \in [0, 2\pi], \quad k = 0, \ldots, n,$$

as the given data.

We consider also the second and the third components of the vector of boundary displacements, choose the simple closed boundary curves on the cylinder surface which give "good" polynomial interpolations of the given boundary components and solve the corresponding systems. So we can also regard the values

$$v_k(\cos\theta, \sin\theta) = \tilde{v}_k(\theta), \quad w_k(\cos\theta, \sin\theta) = \tilde{w}_k(\theta), \quad \theta \in [0, 2\pi], \quad k = 0, ..., n,$$

as the given data.

The given boundary values of the unknown coefficients $u_k(x, y)$, $v_k(x, y)$ and $w_k(x, y)$, $k = 0, \ldots, n$, help us to obtain these coefficients $u_k(x, y)$, $v_k(x, y)$, $w_k(x, y)$ at every step beginning with the highest number $k = n$ as it was described at the previous chapter. Indeed, we obtain the analytic functions $\phi_n(z)$ and $\psi_n(z)$ via the boundary condition

$$-\kappa\phi_n(e^{\imath\theta}) + e^{\imath\theta}\overline{\phi_n'(e^{\imath\theta})} + \overline{\psi_n(e^{\imath\theta})} = -2\mu(\tilde{u}_n(\theta) + \imath\tilde{v}_n(\theta)),$$

and the analytic function $\rho_k(z)$ via the boundary condition

$$\operatorname{Re}\rho_n(e^{\imath\theta}) = \tilde{w}_n(\theta).$$

Now we construct the functions $\Phi_{n-1}(z,\overline{z})$ and $\Psi_{n-1}(z,\overline{z})$. So we can obtain the analytic functions $\phi_{n-1}(z)$ and $\psi_{n-1}(z)$ via the boundary condition

$$-\kappa\phi_{n-1}(e^{\imath\theta}) + e^{\imath\theta}\overline{\phi_{n-1}'(e^{\imath\theta})} + \overline{\psi_{n-1}(e^{\imath\theta})} = -2\mu(\tilde{u}_{n-1}(\theta) + \imath\tilde{v}_{n-1}(\theta)) - \\ -\Phi_{n-1}(e^{\imath\theta}, e^{-\imath\theta}),$$

and the analytic function $\rho_{n-1}(z)$ via the boundary condition

$$\operatorname{Re}\rho_{n-1}(e^{\imath\theta}) = \tilde{w}_{n-1}(\theta) - \Psi_{n-1}(e^{\imath\theta}, e^{-\imath\theta}).$$

Similarly the components of the functions $\Phi_k(z,\overline{z})$ and $\Psi_k(z,\overline{z})$ are already known when we find the coefficients $(u_k(x,y)+\imath v_k(x,y)$ and $w_k(x,y)$. Therefore the analytic functions $\phi_k(z)$ and $\psi_k(z)$ can be restored via the boundary condition

$$-\kappa\phi_k(e^{\imath\theta}) + e^{\imath\theta}\overline{\phi_k'(e^{\imath\theta})} + \overline{\psi_k(e^{\imath\theta})} = -2\mu(\tilde{u}_k(\theta) + \imath\tilde{v}_k(\theta)) - \Phi_k(e^{\imath\theta}, e^{-\imath\theta}),$$

and the analytic function $\rho_k(z)$ can be restored via the boundary condition

$$\operatorname{Re}\rho_k(e^{\imath\theta}) = \tilde{w}_k(\theta) - \Psi_k(e^{\imath\theta}, e^{-\imath\theta}).$$

Therefore we restore the analytic functions $\phi_k(z)$, $\psi_k(z)$ and $\rho_k(z)$, $k = 0, \ldots, n,$ at the same step as they are introduced.

It is convenient to restore the analytic functions via boundary conditions in the unit disk when the right hand side of the conditions are Fourier series or Fourier polynomials.

The boundary condition

$$-\kappa\phi(e^{\imath\theta}) + e^{\imath\theta}\overline{\phi'(e^{\imath\theta})} + \overline{\psi(e^{\imath\theta})} = \sum_{k=-\infty}^{+\infty} c_k e^{\imath k\theta} \tag{2.1}$$

leads to the solution

$$\phi(z) = \sum_{k=0}^{+\infty} a_k z^k, \ \psi(z) = \sum_{k=1}^{+\infty} b_k z^k, \tag{2.2}$$

where

$$a_0 = -\frac{2\overline{c_2} + c_0\kappa}{\kappa^2}, \ a_1 = \frac{\kappa c_1 + \overline{c_1}}{1-\kappa^2},$$

$$a_k = -\frac{c_k}{\kappa}, \ k = 2, 3, \ldots,$$

$$b_k = \overline{c_{-k}} + \frac{(k+2)c_{k+2}}{\kappa}, \ k = 1, 2, \ldots.$$

The boundary condition

$$\operatorname{Re}\rho(e^{\imath\theta}) = \frac{\alpha_0}{2} + \sum_{k=1}^{+\infty} \alpha_k \cos\theta + \beta_k \sin\theta \tag{2.3}$$

leads to the solution

$$\rho(z) = \frac{\alpha_0}{2} + \sum_{k=1}^{+\infty} (\alpha_k - \imath\beta_k) z^k. \tag{2.4}$$

So we can obtain all coefficients $u_k(x, y)$, $v_k(x, y)$ and $w_k(x, y)$, $k = 0, 1, \ldots, n$, and find the vector of displacements in the form (1.2).

We demonstrate the method of interpolation solution construction with the help of the simplest example of the cylinder subjected to bending and compression [10].

2.2 Example 2.1. The bent cylinder

Let us construct the interpolation solution for the case when $h \in [1,3]$, and the vectors of boundary displacements are known at three levels:

$$\vec{a}_{|h=1,x^2+y^2=1} = (0,0,\delta),$$
$$\vec{a}_{|h=2,x^2+y^2=1} = (\epsilon,0,0),$$
$$\vec{a}_{|h=3,x^2+y^2=1} = (0,0,-\delta).$$

We search for the coordinate displacement functions in the form of the polynomials:

$$u(x,y,h) = u_0(x,y) + u_1(x,y)h + u_2(x,y)h^2,$$
$$v(x,y,h) = v_0(x,y) + v_1(x,y)h + v_2(x,y)h^2,$$
$$w(x,y,h) = w_0(x,y) + w_1(x,y)h + w_2(x,y)h^2.$$

The boundary values of the coefficients can be easily obtained when we solve the system of three linear equations, so according to formulas (2.1)–(2.4) we have

$$\tilde{u}_0(\theta) = -3\epsilon, \tilde{u}_1(\theta) = 4\epsilon, \tilde{u}_2(\theta) = -\epsilon,$$
$$\tilde{v}_0(\theta) = \tilde{v}_1(\theta) = \tilde{v}_2(\theta) = 0,$$
$$\tilde{w}_0(\theta) = 2\delta, \tilde{w}_1(\theta) = -\delta, \tilde{w}_2(\theta) = 0.$$

We find the coefficients successively using the boundary conditions with the help of the procedure described above.

Finally we have the coordinates of the vector of displacements in the cylinder:

$$u(x,y,h) = -3\epsilon + (x^2+y^2-1)\frac{\epsilon\mu}{\lambda+3\mu} + 4\epsilon h - \epsilon h^2,$$
$$v(x,y,h) \equiv 0, \ \ w(x,y,h) = 2\delta - \delta h.$$

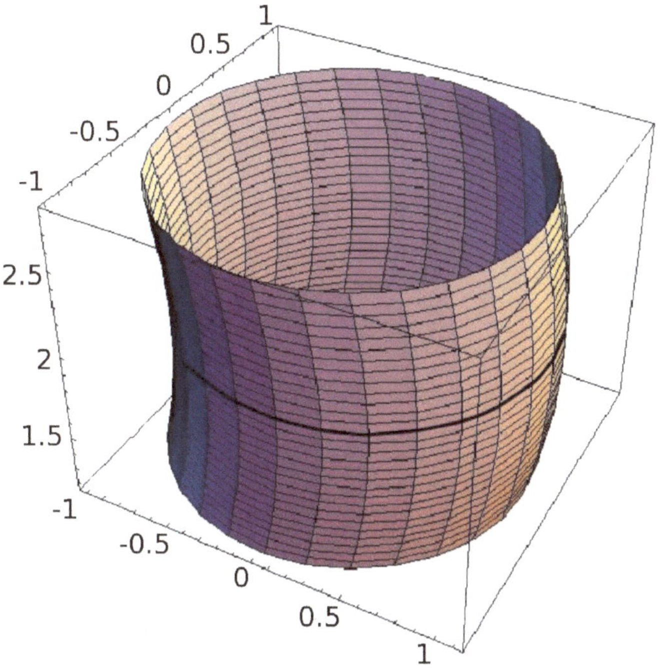

Figure 2.2: The bent cylinder side surface.

2.3 Interpolation solution of the second type when the data is given at a finite number of interior points of the ends

Let us formulate the problem. Given the vectors of displacements

$$(\tilde{u}_j(\theta), \tilde{v}_j(\theta), \tilde{w}_j(\theta)), \theta \in [0, 2\pi], \;\; j = 1, 2,$$

at the curves C_A and C_B and the vectors of displacements

$$\vec{a_j}(x_q^j, y_q^j) = (U_j(x_q^j, y_q^j), V_j(x_q^j, y_q^j), W_j(x_q^j, y_q^j)), \quad j = 1, 2,$$
$$(x_q^1, y_q^1) \in D_A, \quad q = 1, \ldots, Q_1; \;\; (x_q^2, y_q^2) \in D_B, \quad q = 1, \ldots, Q_2,$$

at the inner points of the ends D_A and D_B of the cylinder Ω it should be possible to find the displacements at any point of Ω close to the ends D_A and D_B .

We have only two curves (C_A and C_B) with given boundary data on the cylindrical surface, so it suffices to have linear over h coordinate functions of the vector of displacements. But we must introduce the additional constants in the solution in order to satisfy the given data at the points of the ends.

We can suppose every coordinate function of $\vec{a}$ to be polynomial of the second power in the variable h if we want to restore the displacements near one of the ends of the cylinder. So

$$u = \sum_{k=0}^{1} u_k(x,y)h^k + R^1(x,y)h^2, \; v = \sum_{k=0}^{1} v_k(x,y)h^k + R^2(x,y)h^2,$$

$$w = \sum_{k=0}^{1} w_k(x,y)h^k + R^3(x,y)h^2, \; (x,y) \in D, \; h \in [A,B], \quad (2.5)$$

where the functions

$$u_k(x,y), \;\; v_k(x,y) \;\; w_k(x,y), \;\; k = 0,1,\ldots,n, \;\; (x,y) \in D,$$

are to be found, and every polynomial

$$R^l(x,y), \;\; l = 1,2,3,$$

contains $Q = Q_1 + Q_2$ unknown coefficients. Therefore we have $3Q$ free parameters which will be found from the given displacements at Q points at the ends.

Note that the polynomials

$$R^l(x,y), \;\; l = 1,2,3,$$

are not arbitrary. According to the results of Chapter 1 we see that these polynomials must satisfy the following conditions:

$$R^1(x,y)+\imath R^2(x,y)=-\frac{\lambda+3\mu}{\lambda+\mu}\phi(z)+z\overline{\phi'(z)}+\overline{\psi(z)},$$

where the functions $\phi(z)$ and $\psi(z)$ are analytic with respect to $z=x+\imath y\in D$, hence, the polynomials $R^1(x,y)$ and $\imath R^2(x,y)$ must be biharmonic in D. The polynomial $R^3(x,y)$ must be harmonic in D.

Now we have from the boundary conditions

$$\sum_{k=0}^{1} u_k(\cos\theta,\sin\theta)h_j^k=\tilde{u}_j(\theta)-R^1(\cos\theta,\sin\theta)h_j^2,\ j=1,2. \qquad (2.6)$$

System (2.6) with the Vandermonde determinant gives the expressions of the boundary values $u_k(\cos\theta,\sin\theta)$, $k=0,1$. These expressions contain linear combinations of the unknown coefficients of the polynomial $R^1(x,y)$.

Similarly we get the expressions of boundary values

$$v_k(\cos\theta,\sin\theta),\quad k=0,1,$$

which contain linear combinations of the coefficients of the polynomial $R^2(x,y)$, and the expressions of boundary values

$$w_k(\cos\theta,\sin\theta),\quad k=0,1,$$

which contain linear combinations of the coefficients of the polynomial $R^3(x,y)$.

Now we find the unknown functions

$$u_k(x,y),\quad v_k(x,y),\quad w_k(x,y),\quad k=0,1,$$

by the same method as before. Beginning with $k=1$ we successively equate to zero every factor with z^k, $k=1,0$, at the left hand sides of equilibrium equations (1.1)

We solve the corresponding differential equations and apply the boundary conditions.

We have for every $k=1,0$, from the two first equilibrium equations (1.1) the following representation of $u_k(x,y)+\imath v_k(x,y)$:

$$-2\mu(u_k(x,y)+\imath v_k(x,y)) = -\kappa\phi_k(\zeta)+\zeta\overline{\phi'_k(\zeta)}+\overline{\psi_k(\zeta)}+\Phi_k(\zeta,\overline{\zeta}), \quad (2.7)$$

where $z=x+\imath y$, $\Phi_k(z,\overline{z})$ is the known function which contains linearly the unknown coefficients of the polynomials

$$R^l(x,y), \quad l=1,2,3,$$

and the functions

$$\phi_k(z), \quad \psi_k(z), \quad k=0,1,$$

are analytic in the unit disk D.

Now we apply the boundary conditions to relation (2.7)

$$[-\kappa\phi_k(z)+\zeta\overline{\phi'_k(z)}+\overline{\psi_k(z)}]|_{z=e^{\imath\theta}} =$$
$$= -2\mu[u_k(\cos\theta,\sin\theta)+\imath v_k(\cos\theta,\sin\theta)]-\Phi_k(e^{\imath\theta},e^{-\imath\theta}). \quad (2.8)$$

The boundary value problem (2.8) can be easily solved for the unit disk if the right hand side of (2.8) is represented as a Fourier series

$$-2\mu[P_0^1+\sum_{l=1}^{\infty}P_l^1\cos l\theta+Q_l^1\sin l\theta]-2\imath\mu[P_0^2+\sum_{l=1}^{\infty}P_l^2\cos l\theta+Q_l^2\sin l\theta].$$

We obtain then

$$-\kappa\phi_k(z)+z\overline{\phi'_k(z)}+\overline{\psi_k(z)} = -\mu\{2(P_0^1+\imath P_0^2)+\sum_{l=1}^{\infty}(P_l^1+Q_l^2-\imath Q_l^1+$$
$$+\imath P_l^2)z^l+\sum_{l=1}^{\infty}(P_l^1-Q_l^2+\imath Q_l^1+\imath P_l^2)\overline{z}^l+\frac{1}{\kappa}\sum_{l=2}^{\infty}l(P_l^1+Q_l^2+\imath Q_l^1-\imath P_l^2)\overline{z}^{l-1}\}.$$

We have for $k=1,0$, from the last equilibrium equation (1.1)

$$w_k(x,y)=\operatorname{Re}\rho_k(z)+\Psi_k(z,\overline{z}), \quad (2.9)$$

where $\Psi_k(z,\overline{z})$ is the known function which contains linearly the unknown coefficients of the polynomials

$$R^l(x,y), \quad l = 1,2,3,$$

and the function $\rho_k(z)$ is analytic in D. So using formula (2.9) we get the boundary condition

$$\mathrm{Re}\rho_k(e^{\imath\theta}) = w_k(\cos\theta, \sin\theta) - \Psi_k(e^{\imath\theta}, e^{-\imath\theta}). \tag{2.10}$$

The boundary value problem (2.10) can be easily solved in the unit disk. If the right hand side of (2.10) is

$$P_0^3 + \sum_{l=1}^{\infty} P_l^3 \cos l\theta + Q_l^3 \sin l\theta,$$

we have

$$\rho_k(z) = P_0^3 + \sum_{l=1}^{\infty}(P_l^3 - \imath Q_l^3)z^l,$$

$$w_k(x,y) = \frac{P_0^3}{2} + \sum_{l=1}^{\infty} \mathrm{Re}[(P_l^3 - \imath Q_l^3)(x + \imath y)^l].$$

After we find all the coefficients

$$u_k(x,y), \quad v_k(x,y), \quad w_k(x,y), \quad k = 0,1,\ldots,n,$$

we obtain the functions

$$u(x,y,h), \quad v_k(x,y,h), \quad w(x,y,h)$$

from (2.5). Each of these components of the vector of displacements contains linearly $3Q$ unknown constants. So, after we substitute the coordinates of the given Q points

$$(x_q^j, y_q^j), \quad j = 1,2, \quad q = 1,\ldots,Q_k, \quad k = 1,2,$$

at the ends of the cylinder to these components and compare them with the given values

$$U_1(x_q^j, y_q^j), \quad V_1(x_q^j, y_q^j), \quad W_1(x_q^j, y_q^j), \quad j = 1, 2,$$
$$(x_q^1, y_q^1) \in D_A, \quad q = 1, \ldots, Q_1, \quad (x_q^2, y_q^2) \in D_B, \quad q = 1, \ldots, Q_2,$$

we obtain a linear system over the unknown $3Q$ coefficients of the polynomials $R^l(x, y)$, $l = 1, 2, 3$, from (2.5). After we find all these constants we solve the problem.

Note that the presented form of the interpolation solution of the problem is not unique. One can also search for the solution in the form

$$u = \sum_{k=0}^{1} u_k(x, y)h^k + \sum_{t=1}^{l} R_t^1(x, y)h^{1+t}, \; v = \sum_{k=0}^{1} v_k(x, y)h^k + \sum_{t=1}^{l} R_t^2(x, y)h^{1+t},$$
$$w = \sum_{k=0}^{1} w_k(x, y)h^k + \sum_{t=1}^{l} R_t^3(x, y)h^{1+t}, (x, y) \in D, z \in [A, B],$$

where $R_t^j(x, y)$, $j = 1, 2, 3, \;\; t = 1, \ldots, l$, are polynomials with unknown coefficients. The number of these coefficients for each j must be $Q = Q_1 + Q_2$. The polynomials themselves must be polyharmonic functions and must have representations of the corresponding form in order that the solution satisfies equilibrium equations (1.1).

The following example was constructed by A.N. Schermann. Here the displacements are given at one point of each end, so the coordinate functions of the vector of displacements are the polynomials of the power 3 in the variable h .

2.3.1 Example 2. The bent cylinder with the displacements given at two points at the ends

This example demonstrates the interpolation solution of the problem of restoration of displacements in the circular cylinder $\Omega \; = \{(x, y, h) | x^2 + y^2 \leq$

$1, h \in [1,3]\}$ subjected to compression at the ends, and therefore to bending, with the given displacements at two points at the ends.

We use the interpolation method when the boundary displacements at two circles are given:

$$\vec{a}(\cos\theta, \sin\theta, 1) = (0,0,g), \; \vec{a}(\cos\theta, \sin\theta, 3) = (0,0,-g),$$

and also the displacements at two points at the ends of the cylinder are given:

$$\vec{a}(0,0,1) = (a,c,g+f), \; \vec{a}(0,0,3) = (b,d,-g-f).$$

We search for the interpolation solution in the form

$$u(x,y,h) = u_0(x,y) + u_1(x,y)h + u_2h^2 + u_3h^3,$$
$$v(x,y,h) = v_0(x,y) + v_1(x,y)h + v_2h^2 + v_3h^3,$$
$$w(x,y,h) = w_0(x,y) + w_1(x,y)h + (w_2 + \alpha x + \beta y)h^2 + w_3h^3,$$

where

$$\alpha = -3u_3\frac{\lambda+\mu}{\lambda+3\mu}, \; \beta = -3v_3\frac{\lambda+\mu}{\lambda+3\mu},$$

u_2 , v_2 , u_3 , v_3 , w_2 and w_3 are the constants to be found.

The coefficients of the powers of h^k , $k = 0,1,2,3$, can be found with the help of the method described in Chapter 1. The boundary conditions at the circles at the levels $h = 1$ and $h = 3$ give the boundary values of the unknown coefficients:

$$u_0(\cos\theta, \sin\theta) = 3u_2 + 12u_3, \; u_1(\cos\theta, \sin\theta) = -4u_2 - 13u_3,$$
$$v_0(\cos\theta, \sin\theta) = 3v_2 + 12v_3, \; v_1(\cos\theta, \sin\theta) = -4v_2 - 13v_3,$$
$$w_0(\cos\theta, \sin\theta) = 2g + 3w_2 + 3\alpha\cos\theta + 3\beta\sin\theta + 12w_3,$$
$$w_1(\cos\theta, \sin\theta) = -g - 4w_2 - 4\alpha\cos\theta - 4\beta\sin\theta - 13w_3.$$

These boundary values help us to restore the analytic in the unit disk functions $\phi_1(z)$, $\psi_1(z)$, $\rho_1(z)$, $\phi_0(z)$, $\psi_0(z)$, $\rho_0(z)$ which appear at every step when we get the coefficients $u_1(x,y)$, $v_1(x,y)$, $w_1(x,y)$, $u_0(x,y)$, $v_0(x,y)$, $w_0(x,y)$.

So the components of the vector of displacements u, v, w depend not only on the variables x, y, h but also on the introduced constants u_2, v_2, u_3, v_3, w_2 and w_3.

These introduced constants can be represented via the given constants a, b, c, d, f, g with the help of the following relations:

$$\begin{aligned}
u_0(0,0) + u_1(0,0) + u_2 + u_3 &= a, \\
v_0(0,0) + v_1(0,0) + v_2 + v_3 &= c, \\
w_0(0,0) + w_1(0,0) + w_2 + w_3 &= g + f, \\
u_0(0,0) + 3u_1(0,0) + 9u_2 + 27u_3 &= b, \\
v_0(0,0) + 3v_1(0,0) + 9v_2 + 27v_3 &= d, \\
w_0(0,0) + 3w_1(0,0) + 9w_2 + 27w_3 &= -g - f.
\end{aligned}$$

After we restore all analytic functions we obtain the interpolation solution:

$$\begin{aligned}
u(x,y,h) = \frac{1}{6\mu(\lambda+2\mu)}\{&a(-3+h)[3(-1+h)\lambda^2 + 3(-8+x^2+y^2+ \\
&+8h-h^2)\lambda\mu + (-34+6x^2+6y^2+33h-5h^2)\mu^2]+ \\
&+b(-1+h)[3(-3+h)\lambda^2 - \\
&-3(8+x^2+y^2-h^2)\lambda\mu - (18+6x^2+6y^2+7h-5h^2)\mu^2] - \\
&-6x(-1+h)(-3+h)(\lambda^2+3\lambda\mu+2\mu^2)f\},
\end{aligned}$$

$$v(x,y,h) = \frac{1}{6\mu(\lambda+2\mu)}\{c(-3+h)[3(-1+h)\lambda^2 + 3(-8+x^2+y^2+ \\ +8h-h^2)\lambda\mu + (-34+6x^2+6y^2+33h-5h^2)\mu^2] + \\ +d(-1+h)[3(-3+h)\lambda^2 - \\ -3(8+x^2+y^2-h^2)\lambda\mu - (18+6x^2+6y^2+7h-5h^2)\mu^2] - \\ -6y(-1+h)(-3+h)(\lambda^2+3\lambda\mu+2\mu^2)f\},$$

$$w(x,y,h) = \frac{1}{6\mu(\lambda+2\mu)}\{2f(-2+h)(\lambda+2\mu)[2(-3+h)(-1+h)\lambda + \\ +3(-1+x^2+y^2)\mu] - 3\mu[(a-b)x+(c-d)y](-3+h)(-1+h)(\lambda+\mu) - \\ -6g\mu(-2+h)(\lambda+2\mu)\}.$$

This solution can be extended to the entire cylinder if the newly-found displacements at the cylindrical surface happen to be close to those measured at the surface.

Note that for the case $a = b = c = d = 0$ the deformed cylinder is the axial-symmetric body with the generatrix

$$\{x(t), h(t), t \in [1,3]\},$$

where

$$x(t) = 1 - (1+\lambda/\mu)(t-3)(t-1)f,$$
$$h(t) = t + \frac{(t-2)[-3g\mu + 2f\lambda(t-1)(t-3)]}{3\mu},$$

the unit disk at the cross-section of Ω by the plane $h = 2$ being transformed to the disk of the radius $(1+(1+\lambda/\mu)f)$.

We give the example of the deformed side surface of the pressurised cylinder on Fig.2.3. The parameters for the drawing $\lambda = 15/13 * 10^{11}$, $\mu = 10/13 * 10^{11}$, $g = 0.1$, $f = 0.1$.

The upper surface of the deformed cylinder is the surface

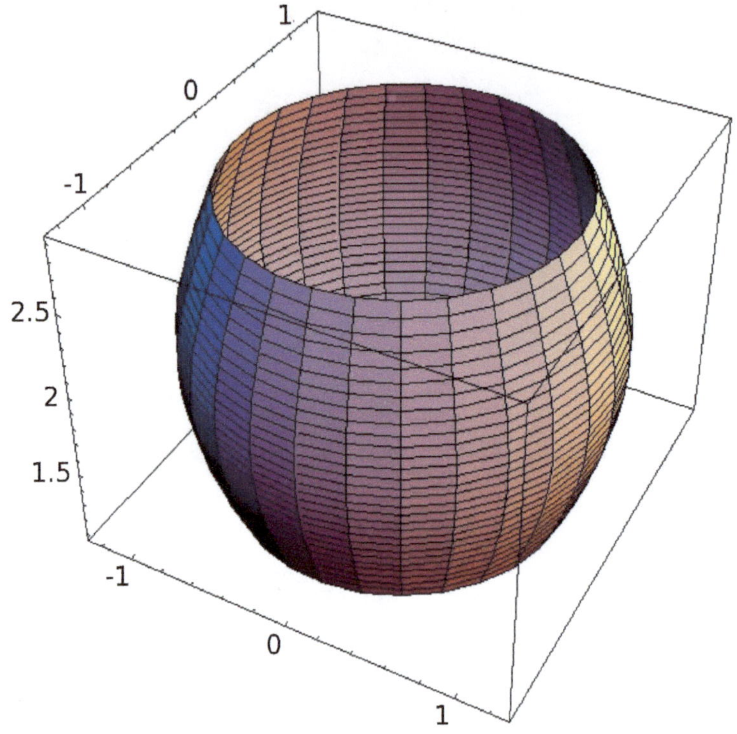

Figure 2.3: Symmetric deformation of the cylinder side surface.

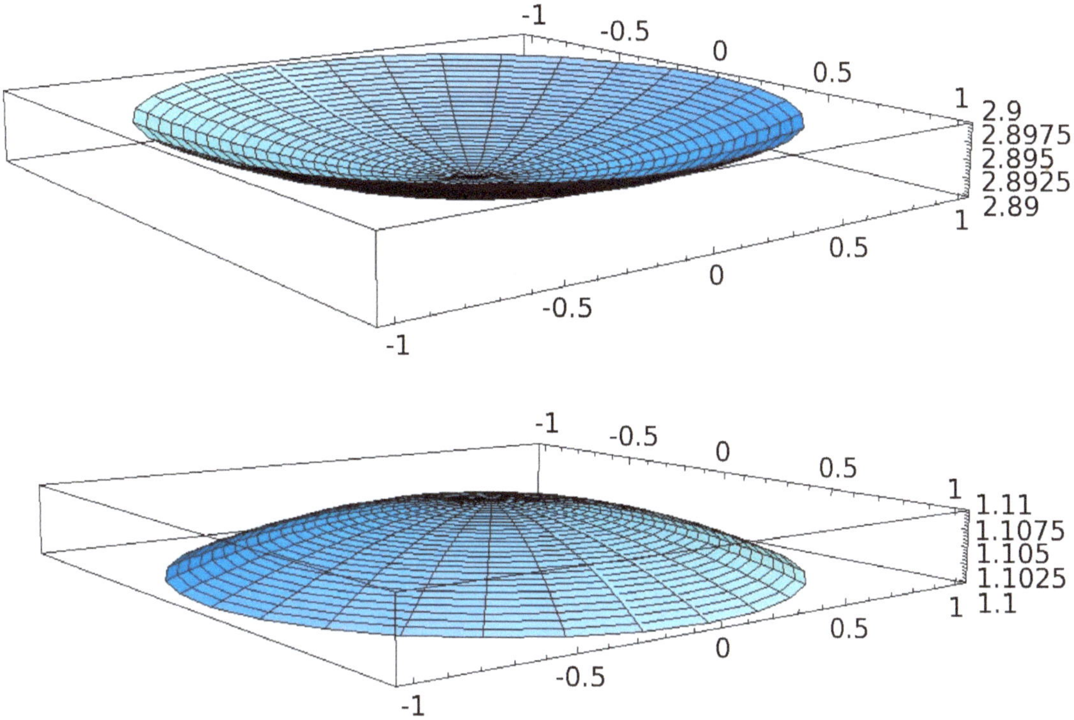

Figure 2.4: Deformations of the cylinder upper and lower ends by the planes $h = 3$ and $h = 1$, respectively.

$$\{(x + b(1 - x^2 - y^2), y + d(1 - x^2 - y^2),$$
$$3 - g - f(1 - x^2 - y^2))||x^2 + y^2 < 1\}.$$

The lower surface of the deformed cylinder is the surface

$$\{(x + a(1 - x^2 - y^2), y + c(1 - x^2 - y^2),$$
$$1 + g + f(1 - x^2 - y^2))|\ x^2 + y^2 < 1\}.$$

The drawings of these surfaces with $a = 0.03, b = 0.02, c = 0, d = 0, g = 0.1, f = 0.01, \mu = 80.5 \times 10^9, \lambda = 102 \times 10^9$ are given at Fig.2.4.

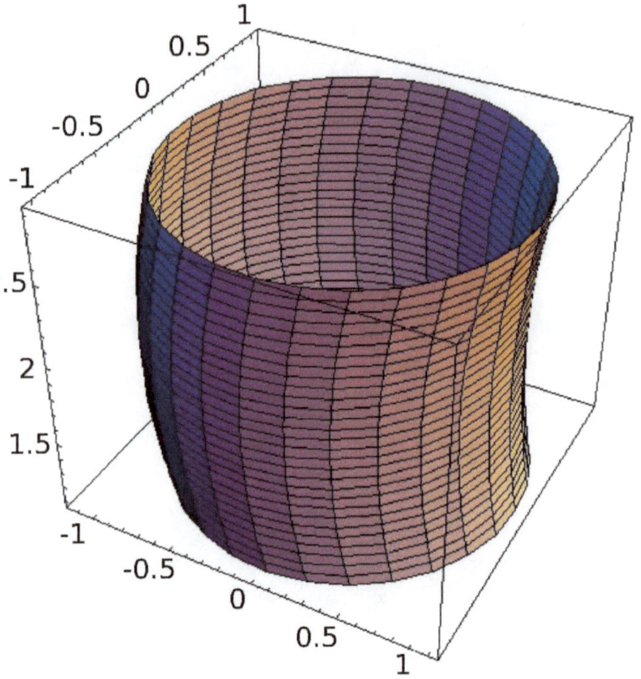

Figure 2.5: The form of the side surface of the deformed circular cylinder.

The side surface of the deformed cylinder with $a = 0.03, b = 0.02, c = 0, d = 0, g = 0.1, f = 0.01, \mu = 80.5 \times 10^9, \lambda = 102 \times 10^9$ is drawn on the Fig. 2.5.

Let us consider the boundary data of the cylinder mentioned above — with the side surface shown on Fig.2.5. It is rather interesting to examine the stresses at the points $(0, 0, 1)$, $(0, 0, 3)$ and also at the critical point on the generatrix $\{x = -1, y = 0, h \in [1, 3]\}$ where the displacement component u is minimal in order to predict possible breaking.

We find the principal stresses (critical values of the normal tear stress tensor components) at these points with the help of the well-known expression

[7] for a normal tear stress component:

$$\sigma_{nn} = \sigma_{11}\cos^2\alpha + \sigma_{22}\cos^2\beta + \sigma_{33}\cos^2\gamma + 2\sigma_{12}\cos\alpha\cos\beta + 2\sigma_{13}\cos\alpha\cos\gamma + 2\sigma_{23}\cos\beta\cos\gamma, \tag{2.11}$$

where $\vec{n} = (\cos\alpha, \cos\beta, \cos\gamma)$ is the direction vector.

The principle stresses at the point $(0,0,3)$ are -34.7052×10^9 with the direction vector $\vec{n} = (0.537064, 0, -0.843542)$, -9.49677×10^9, $\vec{n} = (0,1,0)$ and 0.721663×10^9, $\vec{n} = (0.843542, 0, 0.537064)$.

The same at $(0,0,1)$ are -36.7675×10^9, $\vec{n} = (0.557225, 0, 0.830361)$, -9.49677×10^9, $\vec{n} = (0,1,0)$, 2.78393×10^9, $\vec{n} = (-0.830361, 0, 0.557225)$.

The negative values of principle stresses lead to compression along the corresponding planes. The positive ones provide strain along the corresponding planes. The largest principal stress at the point where the displacement is smaller (the point $(0,0,3)$) is also smaller than the one at the point where the displacement is larger (the point $(0,0,1)$).

The critical point on the surface where the displacement component u is minimal is the point $(-1, 0, 1.97898)$. Note that the maximal value of the principal stress at this point equals 10.403×10^9, the corresponding direction vector is $\vec{n} = (0.999967, 0, 0.00808798)$.

The other two principal stresses equal to 2.31728×10^9 (with the direction vector $\vec{n} = (0,1,0)$) and -18.8147×10^9 (with $\vec{n} = (0.00808798, 0, -0.999967)$).

Evidently, neither of these positive tear tensions can break the body since the corresponding vectors $\vec{n}$ are directed into the body.

Let us now increase a and b so that the ratio $a : b$ stays constant in order that the critical point $(-1, 0, 1.97898)$ stays the same. We find the values of a and b such that the smallest principle stress equals 0. This happens for the approximate values $a = 0.1411926$ and $b = 0.0941284$.

Then the two larger stresses become 21.2443×10^9, 59.3175×10^9 with

Figure 2.6: The domain of large normal tear stress component

direction vectors $\vec{n} = (0,1,0)$ and $\vec{n} = (0.999727, 0, 0.0214329)$, respectively.

The stress value of $\sigma_{n_0 n_0} = 0$ from formula (2.11) corresponds to the direction vector $\vec{n}_0 = (0.0214329, 0, -0.999727)$.

The value of $\sigma_{n_0 n_0}$ at the points on the plane $0.0214329(x+1) - 0.999727(z - 1.97898) = 0$ inside the cylinder rapidly decreases as we pass from the point $(-1, 0, 1.97898)$ along the plane in the positive direction of x.

The points for which $\sigma_{n_0 n_0} = C < 0$ on this plane form the curve which is the intersection of the cylinder

$$23728.080436512533 + 2.548076847879 98y^2 - 22625.650280449892z + \\ +5374.282263108658z^2 = C$$

with the plane $0.0214329(x+1) - 0.999727(z - 1.97898) = 0$.

The situation is similar when we increase a and b so that all the principal stress values at the critical point $(-1, 0, 1.97898)$ become positive.

If we put, for example, $a = 0.1419$ $b = 0.0946$ the smallest principle stress is 0.119665×10^9, with the corresponding direction vector $\vec{n}_0 = (0.002, 0, -0.999769)$.

The points of the plane $0.002(x+1) - 0.999769(z - 1.97898) = 0$ for which $\sigma_{n_0 n_0} = C < 0.119665 \times 10^9$, $C > 0$ form the curve — ellipse which is the intersection of the ellipsoid

$$-10.334550117328218 - 16.712349156134113x + 2.5475086529249644x^2 + \\ +2.5475086529249644y^2 - 12.741837955812294z - 3.7141050669335125xz + \\ +2.3133867840044394z^2 = C$$

with the plane $0.002(x+1) - 0.999769(z - 1.97898) = 0$.

So one can conclude that the domain of possible rupture along the plane passing through the critical point $(-1, 0, 1.97898)$ with the normal vector equal to the direction vector correspondent to the smallest positive principle stress value at the critical point is a segment bounded by the cylinder boundary and an ellipse stretched on the plane along y axis (Fig. 2.6).

2.4 Interpolation solution of the second type when the data is given at all interior points of the end

Let the displacements (U_A, V_A, W_A) and (U_B, V_B, W_B) be given at the ends D_A and D_B respectively. We distinguish the solutions of the second type which take the given displacements at one of the ends D_A , D_B and the solutions which approximate the given data at the ends. Any of these solutions satisfies equilibrium equations (1.1) and takes given boundary displacements at the edges C_A and C_B . We suppose that the given displacements (U_A, V_A, W_A) and (U_B, V_B, W_B) are polyharmonic functions.

2.4.1 The exact solution at one end

Let us formulate the problem. Given the boundary displacements at the points of the circle C_B and also the displacements (U_A, V_A, W_A) at the points of the disc D_A , it should be possible to restore the displacements in the cylinder Ω which satisfy equilibrium equations (1.1) and take the given boundary displacements.

We suppose that the given functions $U_A + \imath V_A$ and W_A are three-harmonic in D_A and also the given three-harmonic function $U_A + \imath V_A$ has the following representation:

$$U_A + \imath V_A = F_0(z) + \overline{G_0(z)} + |z|^2[F_1(z) + \overline{G_1(z)}] + |z|^4[A_0^2 + \overline{G_0(z)}],$$

$$W_A = \frac{1}{2}[H_0(z) + \overline{H_0(z)}] + \frac{|z|^2}{2}[H_1(z) + \overline{H_1(z)}] + $$
$$+\frac{|z|^4}{2}[H_2(z) + \overline{H_2(z)}],$$

where

$$F_j(z) = \sum_{k=0}^{\infty} A_k^j z^k, \ G_j(z) = \sum_{k=1}^{\infty} B_k^j z^k,$$
$$H_j(z) = \sum_{k=0}^{\infty} D_k^j z^k,$$

A_0^2 is a complex constant.

Note that the coefficient of $|z|^4$ in the representation of the function $U_A + \imath V_A$ does not contain the analytic summand. The class of such functions is rather wide, it contains all biharmonic functions.

We assume that $A = 0$ and search for the solution in the form

$$u(x, y, h) = u_0(x, y) + u_1(x, y)h + u_2(x, y)h^2 + u_3(x, y)h^3,$$
$$v(x, y, h) = v_0(x, y) + v_1(x, y)h + v_2(x, y)h^2 + v_3(x, y)h^3,$$
$$w(x, y, h) = w_0(x, y) + w_1(x, y)h + w_2(x, y)h^2 + w_3(x, y)h^3.$$

After we substitute these displacement components in the equilibrium equation we get representations (1.6) and (1.7) for $u_k(x, y) + \imath v_k(x, y)$ and $w_k(x, y)$ which contain the functions $\Phi_k(z, \overline{z})$ and $\Psi_k(z, \overline{z})$ from (1.8) and (1.9), $k = 0, 1, 2$.

Due to these representations we have

$$
\begin{aligned}
u(x,y,h)+\imath v(x,y,h) &= -\kappa\sum_{k=0}^{\infty} a_k^0 z^k + z\sum_{k=1}^{\infty}\overline{a_k^0 z^{k-1}} + \sum_{k=1}^{\infty}\overline{b_k^0 z^k} + \Phi_0(z,\overline{z}) + \\
&+(-\kappa\sum_{k=0}^{\infty} a_k^1 z^k + z\sum_{k=1}^{\infty}\overline{a_k^1 z^{k-1}} + \sum_{k=1}^{\infty}\overline{b_k^1 z^k} + \Phi_1(z,\overline{z}))h + \\
&+(-\kappa\sum_{k=0}^{\infty} a_k^2 z^k + z\sum_{k=1}^{\infty}\overline{a_k^2 z^{k-1}} + \sum_{k=1}^{\infty}\overline{b_k^2 z^k} + \Phi_2(z,\overline{z}))h^2 + \\
&+(-\kappa\sum_{k=0}^{\infty} a_k^3 z^k + z\sum_{k=1}^{\infty}\overline{a_k^3 z^{k-1}} + \sum_{k=1}^{\infty}\overline{b_k^3 z^k})h^3,
\end{aligned}
$$

$$
\begin{aligned}
w(x,y,h) &= \frac{1}{2}(\sum_{k=0}^{\infty} c_k^0 z^k + \sum_{k=0}^{\infty}\overline{c_k^0 z^k}) + \Psi_0(z,\overline{z}) + \\
&+[\frac{1}{2}(\sum_{k=0}^{\infty} c_k^1 z^k + \sum_{k=0}^{\infty}\overline{c_k^1 z^k}) + \Psi_1(z,\overline{z})]h + \\
&+[\frac{1}{2}(\sum_{k=0}^{\infty} c_k^2 z^k + \sum_{k=0}^{\infty}\overline{c_k^2 z^k}) + \Psi_2(z,\overline{z})]h + \\
&+\frac{1}{2}(\sum_{k=0}^{\infty} c_k^3 z^k + \sum_{k=0}^{\infty}\overline{c_k^3 z^k})h^3,
\end{aligned}
$$

where the functions $\Phi_0(z,\overline{z})$, $\Phi_1(z,\overline{z})$, $\Phi_2(z,\overline{z})$, $\Psi_0(z,\overline{z})$, $\Psi_1(z,\overline{z})$, $\Psi_2(z,\overline{z})$ found via formulas (1.8) and (1.9) also contain the series with the introduced coefficients a_k^j , b_k^j , c_k^j , $j=1,2,3$, $k\in\mathbf{N}$.

We compare the functions $u(x,y,0)+\imath v(x,y,0)$ and $w(x,y,0)$ with the functions $U+\imath V$ and W correspondingly, equate the coefficients of the same powers of $z^m\overline{z}^n$ and get the linear system which consists of eight equations with respect to the coefficients a_k^j , b_k^j , c_k^j , $j=0,1,2,3$, $k\in\mathbf{N}\cup 0$. We add to this system the three equations of the coefficients of the same powers of $e^{\imath\theta}$ which appear when we compare the values of $u(\cos\theta,\sin\theta,B)+\imath v(\cos\theta,\sin\theta,B)$ and $w(\cos\theta,\sin\theta,B)$ with the given values of the displacements at the points of the circle C_B . Obviously, the number of the equations of the system is 11 and the number of the sets of the free coefficients is 12 , so the solution of the system depends on one of the sets of the coefficients, namely, on the coefficients b_k^3 , $k = 1, 2, \ldots$. The system itself is presented in Appendix 1.

The solution should be taken in the form of the polynomial of the odd power more than 3 in h if the order of harmonics of the given displacements at the end D_A is more than 3.

2.4.2 Example 2.3. The solution which coincides with the given displacements on one of the ends

Let us restore the displacements in the circular cylinder of the radius 1 and of the hight 1 via the boundary displacements given on the end on the level $h = 0$ and on the edge on the level $h = 1$. The given displacements are following:

$$(u + \imath v)(z, \overline{z}, 0) = [Q + \imath\hat{Q} + (Q_1 + \imath\hat{Q}_1)\overline{z} + (P_1 + \imath\hat{P}_1)z](|z|^2 - 1),$$

$$w(z, \overline{z}, 0) = [D_0 + (D_1 - \imath\hat{D}_1)\overline{z} +$$
$$+(D_2 - \imath\hat{D}_2)\overline{z}^2 + (D_1 + \imath\hat{D}_1)z + (D_2 + \imath\hat{D}_2)z^2](|z|^2 - 1);$$

$$(u + \imath v)(e^{\imath\phi}, e^{-\imath\phi}, 1) = 0, \ w(e^{\imath\phi}, e^{-\imath\phi}, 1) = 0.$$

We search for the displacements in the plane XOY in the form

$$u(x, y, h) + \imath v(x, y, h) = \sum_{j=0}^{3}(-\kappa\phi_j(z) + z\overline{\phi_j'(z)} + \overline{\psi_j(z)} + \Phi_j(z, \overline{z}))h^j,$$

and for the displacements in the direction of the axis OH in the form

$$w(x,y,h) = \sum_{j=0}^{3}(\operatorname{Re}\rho_j(z) + \Psi_j(z,\overline{z}))h^j,$$

where the functions $\phi_j(z)$, $\psi_j(z)$, $\rho_j(z)$ are analytic in the unit disk, the functions $\Phi_j(z,\overline{z})$, $\Psi_j(z,\overline{z})$ satisfy equalities (1.8), (1.9).

After we solve the system of equations with unknown constants a_k^j, b_k^j, c_k^j, $j = 0,1,2,3$, $k \in \mathbf{N} \cup 0$, we have the solution which depends on the free constants c_0^3, a_0^3, $\hat{a}_0^3$, b_1^3, $\hat{b}_0^3$.

The free variables c_0^3, a_0^3, $\hat{a}_0^3$, b_1^3, $\hat{b}_0^3$ can be used for better approximation of the function on the second circle, e.g. gluing of the derivatives but in the numerical example given below we make them vanish.

Numerical example

Put $\lambda = 15/13 \cdot 10^{11}$, $\mu = 10/13 \cdot 10^{11}$, $D_2 = 0.1$, $\hat{D}_2 = 0$, $D_1 = 0.2$, $\hat{D}_1 = 0$, $D_0 = 0.1$, $\hat{D}_0 = 0$, $c_0^3 = 0$, $b_1^3 = 0$, $\hat{b}_1^3 = 0$, $a_0^3 = 0$, $\hat{a}_0^3 = 0$, $\hat{a}_1^3 = 0$, $P_1 = -0.1$, $\hat{P}_1 = 0$, $Q = 0.1$, $\hat{Q} = 0$, $Q_1 = -0.2$, $\hat{Q}_1 = 0$.

Then

$$u(r,\phi,h) = -0.09999999999999999 + 0.9763157894736841h-$$

$$-1.3973684210526314h^2 + 0.3r\cos\phi - 0.4280701754385965hr\cos\phi+$$

$$+0.11228070175438587h^2r\cos\phi + 0.1r^2\cos^2\phi+$$

$$+0.42105263157894735hr^2\cos^2\phi - 0.30000000000000004r^3\cos^3\phi+$$

$$+0.3157894736842105hr^3\cos^3\phi + 0.1r^2\sin^2\phi+$$

$$+0.42105263157894735hr^2\sin^2\phi - 0.30000000000000004r^3\cos\phi\sin^2\phi+$$

$$+0.3157894736842105hr^3\cos\phi\sin^2\phi;$$

$$v(r,\phi,h) = -0.09999999999999998r\sin\phi + 0.9614035087719297hr\sin\phi-$$

$$-0.6456140350877191h^2 r\sin\phi + 0.1r^3\cos^2\phi\sin\phi-$$

$$-0.31578947368421056hr^3\cos^2(\phi)\sin\phi + 0.1r^3\sin^3\phi-$$

$$-0.31578947368421056hr^3\sin^3\phi;$$

$$w(r,\phi,h) = -0.1 - 0.41904761904761906h - 0.2476190476190476h^2-$$

$$-0.39999999999999997r\cos\phi + 0.7578947368421052hr\cos\phi-$$

$$-0.7578947368421053h^2 r\cos\phi - 0.1r^2\cos^2\phi+$$

$$+1.2350877192982455hr^2\cos^2\phi - 0.5684210526315789h^2 r^2\cos^2\phi+$$

$$+0.4r^3\cos^3\phi + 0.2r^4\cos^4\phi + 0.3r^2\sin^2\phi+$$

$$+0.09824561403508769hr^2\sin\phi + 0.5684210526315789h^2 r^2\sin^2\phi+$$

$$+0.4r^3\cos\phi\sin^2\phi - 0.19999999999999998r^4\sin^4\phi.$$

All these displacements can be visualized. It seems natural to consider the horizontal sections of the given cylinder. Note that in order to make the pictures more informative we already exaggerated the possible given deformations. On figure 2.7 we see that the given complicated deformation at the lower end of the cylinder simplifies as we approach the upper end. We have combinations of shifts and turns in different directions for different points on the lower end. The upper end deformation is then similar to the ordinary push of the central point. One may assume that this is due to intuitively clear vanishing of small deformations while they pass the solid.

So the displacement values on the upper disk of the first layer can be used as the given data on the lower disc of the second layer. Similarly we may construct the smooth solution. But in that case one must use polynomials of higher degrees in h than 3.

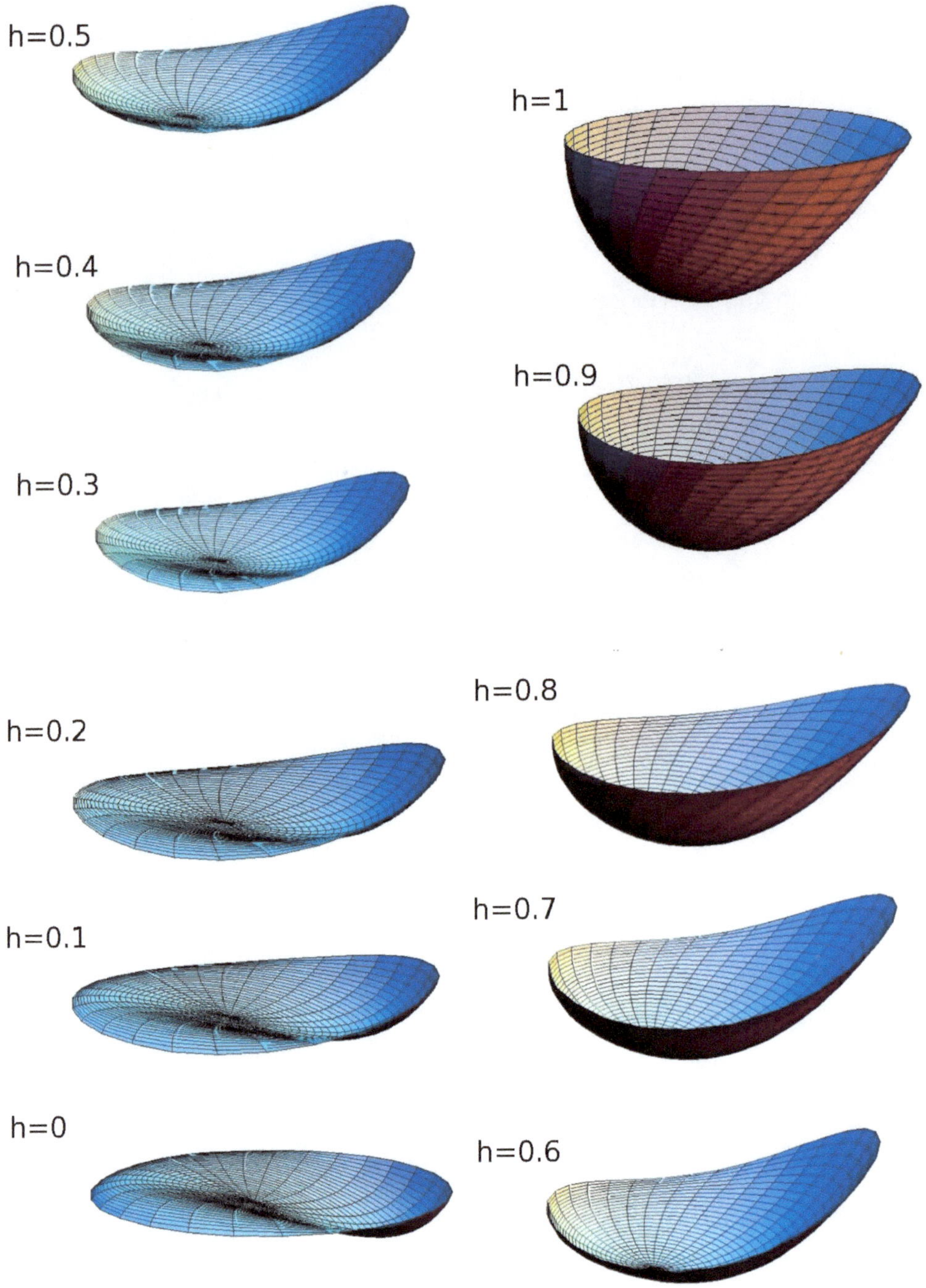

Figure 2.7: Deformations of the cylinder sections by the planes $h = 0$, 0.1, $\ldots$, 1.

2.4.3 The approximation solution near the ends

Suppose that we try to find the solution which satisfies equilibrium equations (1.1) and equals given displacements at both ends D_A and D_B. We would fail to get such solution in the form of the polynomial in the variable h. The matter is that the additional summands with the powers of h raise the order of harmonicity of the solution and the number of free analytic functions is less than the number of the analytic functions in the representations of the corresponding polyharmonic functions at the ends.

Therefore it is important to get the solution which satisfies equilibrium equations (1.1), vanishes at the edges C_A and C_B and approximates the data at the interior points of the ends.

We search for the solution in the traditional form of the polynomial in the variable h

$$u(x,y,h)+\imath v(x,y,h)=\sum_{j=0}^{m}(-\kappa\phi_j(z)+z\overline{\phi_j'(z)}+\overline{\psi_j(z)}+\Phi_j(z,\overline{z}))h^j,$$

$$w(x,y,h)=\sum_{j=0}^{m}(\mathrm{Re}\,\rho_j(z)+\Psi_j(z,\overline{z}))h^j.$$

where the functions $\phi_j(z)$, $\psi_j(z)$, $\rho_j(z)$ are analytic in the unit disk, the functions $\Phi_j(z,\overline{z})$, $\Psi_j(z,\overline{z})$ satisfy equalities (1.8), (1.9).

We represent six analytic functions $\phi_0(z)$, $\psi_0(z)$, $\rho_0(z)$, $\phi_1(z)$, $\psi_1(z)$, $\rho_1(z)$ via the other analytic functions $\phi_j(z)$, $\psi_j(z)$, $\rho_j(z)$, $j=2,...,m$, when we satisfy the null displacements at the edges C_A and C_B. Now we have the set of the free analytic functions which we must choose so that the polyharmonic functions $u(x,y,A)+\imath v(x,y,A)$, $u(x,y,B)+\imath v(x,y,B)$, $w(x,y,A)$ and $w(x,y,B)$ approximate the given polyharmonic functions $U_A+\imath V_A$, $U_B+\imath V_B$, W_A and W_B respectively.

Note that any n-harmonic in the unit disk D function $q(x,y)$ can be restored in D via the boundary values of

$$q, \frac{\partial q}{\partial r}, \frac{\partial^2 q}{\partial r^2}, ..., \frac{\partial^{n-1} q}{\partial r^{n-1}},$$

where $\partial/\partial r$ is the derivative over the polar radius of the boundary point [15].

Therefore we must minimise the values

$$|\frac{\partial^k[u(x,y,A) + \imath v(x,y,A) - U_A - \imath V_A]}{\partial r^k}|,$$

$$|\frac{\partial^k[u(x,y,B) + \imath v(x,y,B) - U_B - \imath V_B]}{\partial r^k}|,$$

$$|\frac{\partial^k[w(x,y,A) - W_A]}{\partial r^k}|, \ |\frac{\partial^k[w(x,y,B) - W_B]}{\partial r^k}|, \ k = 1, ..., l-1,$$

at the points of the unit circle, where l is the order of harmonicity of the difference of the end values of the solution and the given at the ends functions.

The easiest way of such minimisation is the minimisation in $L^2([0, 2\pi])$ metric.

Consider the functional

$$\sum_{k=1}^{l-1}[\int_0^{2\pi} |\frac{\partial^k[u(x,y,A) + \imath v(x,y,A) - U_A - \imath V_A]}{\partial r^k}|^2|_{x=\cos\theta, y=\sin\theta} d\theta +$$

$$+ \int_0^{2\pi} |\frac{\partial^k[u(x,y,B) + \imath v(x,y,B) - U_B - \imath V_B]}{\partial r^k}|^2|_{x=\cos\theta, y=\sin\theta} d\theta +$$

$$+ \int_0^{2\pi} [\frac{\partial^k[w(x,y,A) - W_A]}{\partial r^k}]^2|_{x=\cos\theta, y=\sin\theta} d\theta +$$

$$+ \int_0^{2\pi} [\frac{\partial^k[w(x,y,B) - W_B]}{\partial r^k}]^2|_{x=\cos\theta, y=\sin\theta} d\theta$$

on the space

$$\underbrace{C^\omega(D) \times C^\omega(D) \times ... \times C^\omega(D)}_{3m-3}.$$

The problem of minimisation of this functional due to Parseval's equality can be reduced to the problem of minimisation of the corresponding functional on the complete space

$$\underbrace{l^2 \times l^2 \times \dots \times l^2}_{3m-3},$$

where each space l^2 is the space of Taylor coefficients of the corresponding analytic function.

We can make the difference between the spline-interpolation solution at the end of the layer and the function given at this end arbitrary small if we glue not only the values of spline and the given function over the boundary but also the derivatives of the orders $1, \dots, m$ in the direction of the normal to the cylindrical surface at the boundary of the end. This gluing is possible if we increase the power of the spline with respect to h. In Chapter 8 we substantiate the spline-interpolation solution convergence to the exact one at the ends of the layer while increasing the power of the spline.

CHAPTER 3

Interpolation solution of the problems of elasticity for the pressurized tube

Elena A. Shirokova,
Kazan Federal University
Elena.Shirokova@ksu.ru

ABSTRACT: In this chapter we present the interpolation solution of two elasticity problems for the tube: with the given displacements and with the given stresses at the exterior surface. We introduce some additional conditions at the interior surface in order to obtain the solution.

We can apply the method of the interpolation solution construction to the hollow circular cylinder (or a tube). The displacements at the edges of a tube do not affect actively the displacements and stresses at the points distant from the edges. So it is rather reasonable to apply the interpolation solution to pressurised tubes.

We find the interpolation solution of the following problem for an elastic tube: given the displacements at $n+1$ levels of the exterior surface and a constant pressure and the "sliding condition" at the interior surface of the tube it should be possible to find the displacements at any point of the tube [11]. We name this problem the problem with given displacements. The initial problem then is reduced to a succession of plane boundary problems for the elliptic differential equations. We reduce each boundary value problem to a system of linear equations.

We also find the interpolation solution of the following problem for an elastic tube: given the stresses at $n+1$ levels of the exterior surface with the displacement and the rotation at one points of each level and a constant pressure and the "sliding condition" on the interior surface of the tube it should be possible to find the displacements and stresses at any point of the tube. We name this problem the problem with given stresses.

3.1 Formulation of the problem with given exterior displacements and analysis

Let Ω be a cylindrical tube in the coordinate space (x, y, h) where the element of the cylinder is parallel to the OH-axis, the exterior radius is 1, the interior radius is r, $A = h_0 \leq h \leq B = h_n$. The intersections of the exterior surface of Ω with the planes $h = h_j, j = 0, 1, \ldots, n,$ are denoted by C_j. We assume a constant pressure p acting at the interior surface of Ω. We also suppose that the vector of stresses which acts on the plane orthogonal to OH has the plane component which is orthogonal at the interior surface to the normal to the interior surface – we name this condition "the sliding condition".

Let $\vec{a} = (u(x, y, h), v(x, y, h), w(x, y, h))$ be the vector of displacements of the points of Ω. Given the vectors of displacements of the curves $C_j, j = 0, 1, \ldots, n,$ it should be possible to find the displacements at any point of Ω.

We suppose that every coordinate function of the desired vector of displacements $\vec{a}$ is a polynomial in h, so that

$$u = \sum_{k=0}^{n} u_k(x,y)h^k, \;\; v = \sum_{k=0}^{n} v_k(x,y)h^k,$$

$$w = \sum_{k=0}^{n} w_k(x,y)h^k, \;\; r^2 \leq x^2 + y^2 \leq 1, \;\; h \in [A, B]. \tag{3.1}$$

We put the displacements in form (3.1) into equilibrium equations (1.1) and obtain relations (1.3), (1.4) and (1.5) which give partial differential equations for the coefficients $u_k(x,y)$, $v_k(x,y)$, $w_k(x,y)$ in the same way as it was done in Chapter 1. We solve these equations and represent the coefficients of the displacement components with the help of analytic functions.

Let us introduce the following notation

$$u_{|h=h_j, x^2+y^2=1} = \hat{u}_j(\theta), \;\; v_{|h=h_j, x^2+y^2=1} = \hat{v}_j(\theta),$$

$$w_{|h=h_j, x^2+y^2=1} = \hat{w}_j(\theta), \theta \in [0, 2\pi], \tag{3.2}$$

where θ is the polar angle on the circle $x^2 + y^2 = 1$.

Note that we can find the values

$$u_k(x,y)_{|x^2+y^2=1}, \;\; v_k(x,y)_{|x^2+y^2=1},$$

$$w_k(x,y)_{|x^2+y^2=1}, \;\; k = 0, 1, \ldots, n.$$

Indeed, due to (3.2) the system

$$\hat{u}_j(\theta) = \sum_{k=0}^{n} \tilde{u}_k(\theta) h_j^k, \;\; j = 0, 1, \ldots, n,$$

with the Vandermonde determinant gives the values of $\tilde{u}_k(\theta)$. The values of $\tilde{v}_k(\theta)$ and $\tilde{w}_k(\theta)$ can be obtained in the same way.

Finally we have

$$u_k(x,y)_{|x^2+y^2=1} = \tilde{u}_k(\theta),\ v_k(x,y)_{|x^2+y^2=1} = \tilde{v}_k(\theta),$$
$$w_k(x,y)_{|x^2+y^2=1} = \tilde{w}_k(\theta),\ k = 0,..,n,\ \theta \in [0,2\pi] \quad . \tag{3.3}$$

Now we get the components of the stress tensor by using (3.1)

$$\sigma_{11} = \lambda\Lambda + 2\mu\sum_{k=0}^{n}\frac{\partial u_k}{\partial x}h^k, \sigma_{22} = \lambda\Lambda + 2\mu\sum_{k=0}^{n}\frac{\partial v_k}{\partial y}h^k,$$
$$\sigma_{33} = \lambda\Lambda + 2\mu\sum_{k=0}^{n-1}(k+1)w_{k+1}h^k, \sigma_{12} = \mu\sum_{k=0}^{n}(\frac{\partial u_k}{\partial y} + \frac{\partial v_k}{\partial x})h^k, \tag{3.4}$$
$$\sigma_{13} = \mu\{\sum_{k=0}^{n-1}[(k+1)u_{k+1} + \frac{\partial w_k}{\partial x}]h^k + \frac{\partial w_n}{\partial x}h^n\},$$
$$\sigma_{23} = \mu\{\sum_{k=0}^{n-1}[(k+1)v_{k+1} + \frac{\partial w_k}{\partial y}]h^k + \frac{\partial w_n}{\partial y}h^n\},$$

where

$$\Lambda \equiv \sum_{k=0}^{n-1}[\frac{\partial u_k}{\partial x} + \frac{\partial v_k}{\partial y} + (k+1)w_{k+1}]h^k + (\frac{\partial u_n}{\partial x} + \frac{\partial v_n}{\partial y})h^n. \tag{3.5}$$

The assumption of constant pressure acting at the interior surface has the following form:

$$\cos\theta\big[\lambda\{\sum_{k=0}^{n-1}[\frac{\partial u_k}{\partial x}+\frac{\partial v_k}{\partial y}+(k+1)w_{k+1}]h^k+$$
$$+(\frac{\partial u_n}{\partial x}+\frac{\partial v_n}{\partial y})h^n\}+2\mu\sum_{k=0}^{n}\frac{\partial u_k}{\partial x}h^k\big]_{|x^2+y^2=\rho^2}+$$
$$+\sin\theta\big[\mu\sum_{k=0}^{n}(\frac{\partial u_k}{\partial y}+\frac{\partial v_k}{\partial x})h^k\big]_{|x^2+y^2=\rho^2}=p\cos\theta, \tag{3.6}$$
$$\cos\theta\big[\mu\sum_{k=0}^{n}(\frac{\partial u_k}{\partial y}+\frac{\partial v_k}{\partial x})h^k\big]_{|x^2+y^2=\rho^2}+$$
$$+\sin\theta\big[\lambda\{\sum_{k=0}^{n-1}[\frac{\partial u_k}{\partial x}+\frac{\partial v_k}{\partial y}+(k+1)w_{k+1}]h^k+$$
$$+(\frac{\partial u_n}{\partial x}+\frac{\partial v_n}{\partial y})h^n\}+2\mu\sum_{k=0}^{n}\frac{\partial v_k}{\partial y}h^k\big]_{|x^2+y^2=\rho^2}=p\sin\theta.$$

The sliding condition can be written as

$$\sigma_{13_{|x^2+y^2=r^2}}\cos\theta+\sigma_{23_{|x^2+y^2=r^2}}\sin\theta=0.$$

So if we take equation (3.4) and (3.5) into account the condition has the form

$$\{\sum_{k=0}^{n-1}[(k+1)u_{k+1}+\frac{\partial w_k}{\partial x}]h^k+\frac{\partial w_n}{\partial x}h^n\}_{|x^2+y^2=r^2}\cos\theta+$$
$$+\{\sum_{k=0}^{n-1}[(k+1)v_{k+1}+\frac{\partial w_k}{\partial y}]h^k+\frac{\partial w_n}{\partial y}h^n\}_{|x^2+y^2=r^2}\sin\theta=0. \tag{3.7}$$

The polynomials in the left hand sides of equations (3.6) and (3.7) are equal to zero for every $h\in[A,B]$, $r^2<x^2+y^2<1$, so each factor with h^k, $k=0,1,\ldots,n,$ in equations (3.6), (3.7) vanishes.

3.2 Solution of the problem

3.2.1 Examination of coefficients of h^n

If we equate the coefficients of $\tilde{z}^n$ in equations (1.3) and (1.4) to zero we obtain the system which leads as in Chapter 1 to the following representation:

$$-2\mu(u_n + \imath v_n) = -\kappa\phi_n(z) + z\overline{\phi_n'(z)} + \overline{\psi_n(z)},$$

where

$$\kappa = \frac{\lambda + 3\mu}{\lambda + \mu},$$

the functions $\phi_n(z)$ and $\psi_n(z)$ are analytic in the ring $r < |z| < 1$.

We obtain the boundary condition at the outside of the ring $r < |z| < 1$

$$[-\kappa\phi_n(z) + z\overline{\phi_n'(z)} + \overline{\psi_n(z)}]_{|z=\exp\imath\theta} = $$
$$= -2\mu(\tilde{u}_n(\theta) + \imath\tilde{v}_n(\theta)), \theta \in [0, 2\pi], \tag{3.8}$$

if we take the first and the second relation in equation (3.3) with $k = n$.

The boundary condition at the inner circle of the ring $r < |z| < 1$ can be obtained if we equate the coefficients of $\tilde{z}^n$ in equation (3.6) to zero:

$$[-2\overline{z}\,\mathrm{Re}\,\phi_n'(z) + |z|^2\phi_n''(z) + z\psi_n'(z)]_{|z=r\exp\imath\theta} = 0. \tag{3.9}$$

We apply the Laurent series expansions of the unknown functions to obtain the exact solution of the problem. Let the functions $\phi_n(z)$ and $\psi_n(z)$ analytic in the ring $r < |z| < 1$ have the following representations:

$$\phi_n(z) = \sum_{m=-\infty}^{+\infty} \gamma_m^n z^m,$$

$$\psi_n(z) = \sum_{m=-\infty, m\neq 0}^{+\infty} c_m^n z^m,$$

the expansion of the known function

$$-2\mu(\tilde{u}_n(\theta) + \imath\tilde{v}_n(\theta))$$

in Fourier series being

$$-2\mu(\tilde{u}_n(\theta) + \imath\tilde{v}_n(\theta)) = \sum_{m=-\infty}^{+\infty} t_m^n e^{\imath m\theta}.$$

After we put the given expressions into equations (3.8), (3.9), we get the following relations:

$$-\kappa \sum_{m=-\infty}^{+\infty} \gamma_m^n e^{\imath m\theta} + e^{\imath\theta} \sum_{m=-\infty}^{+\infty} \overline{\gamma_m^n} m e^{-\imath(m-1)\theta} + \\ + \sum_{m=-\infty, m\neq 0}^{+\infty} \overline{c_m^n} e^{-\imath m\theta} = \sum_{m=-\infty}^{+\infty} t_m^n e^{\imath m\theta}, \tag{3.10}$$

$$-re^{-\imath\theta}\Big[\sum_{m=-\infty}^{+\infty} \gamma_m^n m r^{m-1} e^{\imath(m-1)\theta} + \sum_{m=-\infty}^{+\infty} \overline{\gamma_m^n} m r^{m-1} e^{-\imath(m-1)\theta}\Big] + \\ +r^2 \sum_{m=-\infty}^{+\infty} \gamma_m^n m(m-1) r^{m-2} e^{\imath(m-2)\theta} + \\ +re^{\imath\theta} \sum_{m=-\infty, m\neq 0}^{+\infty} m c_m^n r^{m-1} e^{\imath(m-1)\theta} = 0. \tag{3.11}$$

We compare constant terms of Fourier series in equation (3.10) and equation (3.11), the coefficients of $e^{2\imath\theta}$ in (3.10), the coefficients of $e^{-2\imath\theta}$ in equation (3.11) and obtain the system

$$\begin{cases} -\kappa\gamma_0^n + 2\overline{\gamma_2^n} = t_0^n, \\ -\kappa\gamma_2^n + \overline{c_{-2}^n} = t_2^n, \\ -2\overline{\gamma_2^n} r^2 - 2c_{-2}^n r^{-2} = 0, \end{cases}$$

which provides the values of

$$\gamma_0^n, \ \gamma_2^n, \ c_2^n.$$

After we compare the coefficients at $e^{\imath\theta}$ in equation (3.10) and the coefficients at $e^{-\imath\theta}$ in equation (3.11) we have the system which provides the values of γ_1^n and c_{-1}^n:

$$\begin{cases} -\kappa\gamma_1^n + \overline{\gamma_1^n} + \overline{c_{-1}^n} = t_1^n, \\ -r\gamma_1^n - r\overline{\gamma_1^n} - c_{-1}^n r^{-1} = 0. \end{cases}$$

The other coefficients of the Fourier series can be found via the systems of four equations which appear when we compare the coefficients at $e^{\imath k\theta}$ and $e^{(2-k)\imath\theta}$ in equation (3.10) and the coefficients at $e^{-k\imath\theta}$ and $e^{(k-2)\imath\theta}$ in equation (3.11):

$$\begin{cases} -\kappa\gamma_k^n + (2-k)\overline{\gamma_{2-k}^n} + \overline{c_{-k}^n} = t_k^n, \\ -k\overline{\gamma_k^n}r^k - k(2-k)\gamma_{2-k}^n r^{2-k} - kc_{-k}^n r^{-k} = 0, \\ -\kappa\gamma_{2-k}^n + k\overline{\gamma_k^n} + \overline{c_{k-2}^n} = t_{2-k}^n, \\ -(2-k)\overline{\gamma_{2-k}^n}r^{2-k} + k(k-2)\gamma_k^n r^k + (k-2)c_{k-2}^n r^{k-2} = 0. \end{cases}$$

This system provides the values of

$$\gamma_k^n, \ \gamma_{2-k}^n, \ c_{-k}^n, \ c_{k-2}^n$$

for all integers k, excluding $k = 0, 1, 2$ because the corresponding coefficients are found above. Therefore the functions $\phi(z)$ and $\psi(z)$ are obtained, so the values of the functions $u_n(x, y)$ and $v_n(x, y)$ are known.

Now we equate the coefficient of $\tilde{z}^n$ in equation (1.5) to zero and obtain the relation

$$\Delta w_n = 0.$$

We use the third relation in equation (3.3) with $k = n$ and equate the coefficient of $\bar{z}^n$ to zero in equation (3.7), so we have the boundary conditions

$$w_n(x,y)_{|x^2+y^2=1} = \tilde{w}(\theta),$$
$$\frac{\partial w_n}{\partial x}_{|x^2+y^2=r^2} \cos\theta + \frac{\partial w_n}{\partial y}_{|x^2+y^2=r^2} \sin\theta = 0.$$

We look for the function $\rho_n(z)$ analytic in the ring $r < |z| < 1$ and such that $\operatorname{Re}\rho_n(z) = w_n(x,y)$, $z = x + \imath y$, $r < |z| < 1$ using the boundary conditions

$$\operatorname{Re}\rho_n(e^{\imath\theta}) = \tilde{w}_n(\theta), \ \operatorname{Re}(z\rho_n'(z))_{|z=r\exp\imath\theta} = 0. \qquad (3.12)$$

Here we also use the expansion of the unknown analytic function

$$\rho_n(z) = \sum_{m=-\infty}^{+\infty} D_m^n z^m,$$

in equation (3.12) to compare the coefficients, the Fourier series expansion of the known function $\tilde{w}(\theta)$ being

$$\tilde{w}_n(\theta) = \frac{p_0^n}{2} + \sum_{k=1}^{\infty} p_k^n \cos k\theta + q_k^n \sin k\theta.$$

The coefficients of the function $\rho_n(z)$ can be obtained via the relation

$$D_0^n = \frac{p_0^n}{2},$$

and the systems

$$\begin{cases} \operatorname{Re} D_k^n + \operatorname{Re} D_{-k}^n & = p_k^n, \\ \operatorname{Re} D_k^n r^k - \operatorname{Re} D_{-k}^n r^{-k} & = 0, \end{cases}$$

$$\begin{cases} \operatorname{Im} D_k^n - \operatorname{Im} D_{-k}^n & = -q_k^n, \\ \operatorname{Im} D_k^n r^k + \operatorname{Im} D_{-k}^n r^{-k} & = 0. \end{cases}$$

Therefore the function $\rho_n(z)$ is found within an imaginary constant term, so we get the function $w_n(z,y)$.

3.2.2 Examination of coefficients of h^{n-1}

Now we equate the coefficients of $\bar{z}^{(n-1)}$ in equations (1.3), (1.4) to zero and have the system which leads to the relation

$$-2\mu(u_{n-1}+\imath v_{n-1}) = -\kappa\phi_{n-1}(z) + z\overline{\phi'_{n-1}(z)} + \overline{\psi_{n-1}(z)} + \Phi_{n-1}(z,\overline{z}),$$

where the functions $\phi_{n-1}(z)$ and $\psi_{n-1}(z)$ are analytic in the ring $r^2 < x^2+y^2 < 1$ and

$$\Phi_{n-1}(z,\overline{z}) = \mu(\lambda+\mu)n\int[\frac{\rho_n+\overline{\rho_n}}{4(\lambda+2\mu)} - \frac{\rho_n-\overline{\rho_n}}{4\mu}]dz.$$

The biharmonic function $\Phi_{n-1}(z,\overline{z})$ is known, the analytic functions $\phi_{n-1}(z)$ and $\psi_{n-1}(z)$ are to be found. We use the boundary conditions which are similar to equations (3.8), (3.9):

$$[-\kappa\phi_{n-1}(z) + z\overline{\phi'_{n-1}(z)} + \overline{\psi_{n-1}(z)}]_{|z=\exp\imath\theta} = -2\mu(\tilde{u}_{n-1}(\theta) + \\ +\imath\tilde{v}_{n-1}(\theta)) - \Phi_{n-1}(z,\overline{z})_{|z=\exp\imath\theta},\ \theta\in[0,2\pi], \tag{3.13}$$

$$[-2\overline{z}\,\mathrm{Re}\,\phi'_{n-1}(z) + |z|^2\phi''_{n-1}(z) + z\psi'_{n-1}(z)]_{|z=r\exp\imath\theta} = [-(1+ \\ +\frac{\lambda}{\mu})\overline{z}\,\mathrm{Re}\,\partial\Phi_{n-1}/\partial z - z\partial\overline{\Phi_{n-1}}/\partial z + \overline{z}\lambda n\,\mathrm{Re}\,h_n]_{|z=r\exp\imath\theta}, \tag{3.14}$$

in order to find $\phi_{n-1}(z)$, $\psi_{n-1}(z)$ with the same method as in the previous subsection.

Note that the right hand side of the boundary condition (3.13) contains the summand of the type $C_{n-1}\imath\theta$ and has the form

$$C_{n-1}\imath\theta + \sum_{m=-\infty}^{+\infty} t_m^{n-1}e^{\imath m\theta}$$

so we must take it into account when we apply the series expansions of the unknown functions. We add the additional summands with logarithms to the Laurent expansions of the functions $\phi(z)$ and $\psi(z)$ analytic in the ring

$r < |z| < 1$:

$$\phi_{n-1}(z) = \beta_{n-1} \ln z + \sum_{m=-\infty}^{+\infty} \gamma_m^{n-1} z^m,$$

$$\psi_{n-1}(z) = b_{n-1} \ln z + \sum_{m=-\infty, m\neq 0}^{+\infty} c_m^{n-1} z^m.$$

After we insert the given expansions in equations (3.8), (3.9) we have the following relations:

$$-\kappa[\beta_{n-1}\imath\theta + \sum_{m=-\infty}^{+\infty} \gamma_m^{n-1} e^{\imath m\theta}] + e^{\imath\theta}[\overline{\beta_{n-1}} e^{\imath\theta} + \sum_{m=-\infty}^{+\infty} \overline{\gamma_m^{n-1}} m e^{-\imath(m-1)\theta}] -$$

$$-\overline{b_{n-1}}\imath\theta + \sum_{m=-\infty}^{+\infty} \overline{c_m^{n-1}} e^{-\imath m\theta} = C_{n-1}\imath\theta + \sum_{m=-\infty}^{+\infty} t_m^{n-1} e^{\imath m\theta},$$

$$-re^{-\imath\theta}[\frac{\beta_{n-1}}{re^{\imath\theta}} + \sum_{m=-\infty}^{+\infty} \gamma_m^{n-1} m r^{m-1} e^{\imath(m-1)\theta} +$$

$$+\frac{\overline{\beta_{n-1}} e^{\imath\theta}}{r} + \sum_{m=-\infty}^{+\infty} \overline{\gamma_m^{n-1}} m r^{m-1} e^{-\imath(m-1)\theta}] +$$

$$+r^2[-\frac{\beta_{n-1}}{r^2} e^{-2\imath\theta} + \sum_{m=-\infty}^{+\infty} \gamma_m^{n-1} m(m-1) r^{m-2} e^{\imath(m-2)\theta}] +$$

$$+re^{\imath\theta}[\frac{b_{n-1}}{r} e^{-\imath\theta} + \sum_{m=-\infty}^{+\infty} m c_m^{n-1} r^{m-1} e^{\imath(m-1)\theta}] =$$

$$= \sum_{m=-\infty}^{+\infty} d_m^{n-1} e^{\imath m\theta}.$$

After we compare the summands with $\imath\theta$ we get

$$-\kappa\beta_{n-1} - \overline{b_{n-1}} = C_{n-1}.$$

The systems with respect to the coefficients are analogies of the ones from the previous subsection. Comparison of the constant terms in the both boundary conditions, the summands with $\imath\theta$ in the external boundary condition, the coefficients of $e^{2\imath\theta}$ in the external boundary condition and the coefficients of $e^{-2\imath\theta}$ in the internal boundary condition gives the system

$$\begin{cases} -\kappa\gamma_0^{n-1} + 2\overline{\gamma_2^{n-1}} = t_0^{n-1}, \\ -\overline{\beta_{n-1}} + b_{n-1} = d_0^{n-1}, \\ -\kappa\beta_{n-1} - \overline{b_{n-1}} = C_{n-1}, \\ -\kappa\gamma_2^{n-1} + \overline{\beta_{n-1}} + \overline{c_{-2}^{n-1}} = t_2^{n-1}, \\ -2\beta_{n-1} - 2\overline{\gamma_2^{n-1}}r^2 - 2c_{-2}^{n-1}r^{-2} = d_{-2}^{n-1}, \end{cases}$$

which provides the values of

$$\gamma_0^{n-1},\ \gamma_2^{n-1},\ \beta_{n-1},\ b_{n-1},\ c_2^{n-1}.$$

After we compare the coefficients at $e^{\imath\theta}$ in the external boundary condition and the coefficients at $e^{-\imath\theta}$ in the internal boundary condition we have the system which provides the values of γ_1^{n-1} and c_{-1}^{n-1}:

$$\begin{cases} -\kappa\gamma_1^{n-1} + \overline{\gamma_1^{n-1}} + \overline{c_{-1}^{n-1}} = t_1^{n-1}, \\ -r\gamma_1^{n-1} - r\overline{\gamma_1^{n-1}} - c_{-1}^{n-1}r^{-1} = d_{-1}^{n-1}. \end{cases}$$

The other coefficients of the Fourier series can be found as above via the systems of four equations which appear when we compare the coefficients at $e^{\imath k\theta}$ and $e^{(2-k)\imath\theta}$ in the external boundary condition and the coefficients at $e^{-k\imath\theta}$ and $e^{(k-2)\imath\theta}$ in the internal boundary condition:

$$\begin{cases} -\kappa\gamma_k^{n-1} + (2-k)\overline{\gamma_{2-k}^{n-1}} + \overline{c_{-k}^{n-1}} = t_k^{n-1}, \\ -k(2-k)r^{2-k}\gamma_{2-k}^{n-1} - k\overline{\gamma_k^{n-1}}r^k - kc_{-k}^{n-1}r^{-k} = d_{-k}^{n-1}, \\ -\kappa\gamma_{2-k}^{n-1} + k\overline{\gamma_k^{n-1}} + \overline{c_{k-2}^{n-1}} = t_{2-k}^{n-1}, \\ k(k-2)\gamma_k^{n-1}r^k - (2-k)\overline{\gamma_{2-k}^{n-1}}r^{2-k} + (k-2)c_{k-2}^{n-1}r^{k-2} = d_{k-2}^{n-1}. \end{cases}$$

This system provides the values of

$$\gamma_k^{n-1},\ \gamma_{2-k}^{n-1},\ c_{-k}^{n-1},\ c_{k-2}^{n-1}$$

for all integers k, excluding $k = 0, 1, 2$ because the corresponding coefficients are found above. Therefore the functions $\phi_{n-1}(z)$ and $\psi_{n-1}(z)$ are obtained, so the values of the functions $u_{n-1}(x, y)$ and $v_{n-1}(x, y)$ in the ring $r^2 < x^2 + y^2 < 1$ are known.

When we equate the coefficient at $\bar{z}^{(n-1)}$ in the last equilibrium equation to zero we obtain the relation

$$\mu \Delta w_{n-1} = -(\lambda + \mu) n (\frac{\partial u_n}{\partial x} + \frac{\partial v_n}{\partial y}).$$

The right hand side of this equation is equal to $-2n \operatorname{Re} \phi_n'(z)$, where $\phi_n(z)$ is just found, so the last equation has the form

$$4 \frac{\partial^2 w_{n-1}}{\partial z \partial \bar{z}} = -\frac{n}{\mu} [\phi_n'(z) + \overline{\phi_n'(z)}].$$

Therefore

$$4 \frac{\partial w_{n-1}}{\partial z} = -\frac{n}{\mu} [\bar{z} \phi_n'(z) + \overline{\phi_n(z)}] + H_{n-1}(z),$$

where $H_{n-1}(z)$ is analytic for $r < |z| < 1$. Now

$$w_{n-1}(x, y) = \operatorname{Re} \rho_{n-1}(z) + \Psi_{n-1}(z, \bar{z}),$$

where

$$\Psi_{n-1}(z, \bar{z}) = -n(z\overline{\phi_n} + \bar{z}\phi_n)/(4\mu)$$

is the known biharmonic function, the analytic function

$$\rho_{n-1}(z) = \frac{1}{4} \int H_{n-1}(z) dz,$$

is to be found. So we use the boundary conditions similar to equation (3.12):

$$\operatorname{Re} \rho_{n-1}(z)_{|z=\exp \imath\theta} = \tilde{w}_{n-1}(\theta) - \Psi_{n-1}(z, \bar{z})_{|z=\exp \imath\theta}, \theta \in [0, 2\pi],$$

$$\operatorname{Re}(z\rho_{n-1}'(z))_{|z=r\exp \imath\theta} = -[z\partial\Psi_{n-1}/\partial z + \bar{z}\partial\Psi_{n-1}/\partial\bar{z}]_{|z=r\exp \imath\theta} + \\ + \frac{n}{2\mu} \operatorname{Re}[\bar{z}(-\kappa\phi_n + z\overline{\phi_n'} + \overline{\psi_n})]_{|z=r\exp \imath\theta},$$

where $\phi_n(z)$ is the function, which was found in the previous subsection. We also can represent the function $\rho_{n-1}(z)$ as an the Laurent expansion with unknown coefficients of the logarithmic summand $B_{n-1}\ln z$ and find all coefficients via the boundary conditions. We have systems similar to the corresponding systems in the previous subsection with different right sides and the additional equations over B_{n-1}.

Thus the functions $u_{n-1}(x,y), v_{n-1}(x,y), w_{n-1}(x,y)$ are found. They are biharmonic in the ring $r < |z| < 1$.

3.2.3 Examination of coefficients of h^k

When we equate the coefficients at h^k, $k < n-1$, in equations (1.3) and (1.4) to zero we obtain the relation

$$-2\mu(u_k + \imath v_k) = -\kappa\phi_k(z) + z\overline{\phi_k'(z)} + \overline{\psi_k(z)} + \Phi_k(z,\overline{z}),$$

where the polyharmonic function $\Phi_k(z,\overline{z})$ is known, the analytic functions $\phi_k(z)$ and $\psi_k(z)$ are to be found.

We use the boundary conditions

$$[-\kappa\phi_k(z) + z\overline{\phi_k'(z)} + \overline{\psi_k(z)}]_{|z=\exp\imath\theta} =$$
$$= -2\mu(\tilde{u}_k(\theta) + \imath\tilde{v}_k(\theta)) - \Phi_k(z,\overline{z})_{|z=\exp\imath\theta},$$

$$[-2\overline{z}\operatorname{Re}\phi_k'(z) + |z|^2\phi_k''(z) + z\psi_k'(z)]_{|z=r\exp\imath\theta} = [-(1+$$
$$+\frac{\lambda}{\mu})\overline{z}\operatorname{Re}\partial\Phi_k/\partial z - z\partial\overline{\Phi_k}/\partial z +$$
$$+\overline{z}\lambda(k+1)(\operatorname{Re}\rho_{k+1} + \Psi_{k+1})]_{|z=r\exp\imath\theta},$$

in order to find $\phi_k(z), \psi_k(z)$, the first of these conditions being from equation (3.3), the second from equation (3.6).

We must note that the form of the second condition with $k = 0$ is different:

$$[-2\overline{z}\,\mathrm{Re}\,\phi_0'(z)+|z|^2\phi_0''(z)+z\psi_0'(z)]_{|z=r\exp i\theta}=[-p\overline{z}-(1+ \\ +\frac{\lambda}{\mu})\overline{z}\,\mathrm{Re}\,\partial\Phi_0/\partial z-z\partial\overline{\Phi_0}/\partial z+\overline{z}\lambda(\mathrm{Re}\,\rho_1+\Psi_1)]_{|z=r\exp i\theta}.$$

Note that the right-hand sides of the external and the internal boundary conditions contain the functions $\Phi_k(z,\overline{z})$ and $\Psi_k(z,\overline{z})$ with fail to be unique at the boundary. So the expansion of the functions $\phi_k(z)$, $\psi_k(z)$ should contain the summands of $z^m\ln z$ type. We expand the analytic functions $\phi_k(z)$, $\psi_k(z)$ in Laurent series in the ring $r<|z|<1$ with the summands of $z^m\ln z$ type, $0\ \le m\ \le n-k-1$, and find the coefficients with the help of the procedure described above.

Thus, the values of $u_k(x,y),v_k(x,y),r^2<x^2+y^2<1$ are found.

When we equate the coefficient at $\tilde{z}^{(k)}$ in the last equilibrium equation to zero we obtain the relation

$$\mu\Delta w_k=-(\lambda+\mu)(k+1)(\frac{\partial u_{k+1}}{\partial x}+\frac{\partial v_{k+1}}{\partial y})- \\ -(k+1)(k+2)(\lambda+2\mu)w_{k+2},$$

where the right-hand side of the equality is known.

Now

$$w_k(x,y)=\mathrm{Re}\,\rho_k(z)+\Psi_k(z,\overline{z}),$$

where the function $\Psi_k(z,\overline{z})$ is known. One should apply recurrent formulas (1.8), (1.9), (1.10), (1.11) and (1.12) in the order to find the functions $\Phi_k(z,\overline{z})$ and $\Psi_k(z,\overline{z})$.

The analytic function $\rho_k(z)$ can be found via the boundary conditions similar to equation (3.12):

$$\operatorname{Re}\rho_k(z)_{|z=\exp\imath\theta} = \tilde{w}_k(\theta) - \Psi_k(z,\overline{z})_{|z=\exp\imath\theta}, \theta \in [0, 2\pi],$$
$$\operatorname{Re}(z\rho_k'(z))_{|z=r\exp\imath\theta} = -[z\partial\Psi_k/\partial z + \overline{z}\partial\Psi_k/\partial\overline{z}]_{|z=r\exp\imath\theta} +$$
$$+\frac{k+1}{2\mu}\operatorname{Re}[\overline{z}(-\kappa\phi_{k+1} + \overline{z\phi_{k+1}'} + \overline{\psi_{k+1}} + \Phi_{k+1})]_{|z=r\exp\imath\theta}.$$

The first of the conditions is from equation (3.3), the second is from equation (3.7). We also apply the Laurent expansion of the function $\rho_k(z)$ with the logarithmic summands in order to find this function.

Thus the polyharmonic function $w_k(x,y)$ in the ring $r^2 < x^2 + y^2) < 1$ is found.

When all the functions $u_k(x,y), v_k(x,y), w_k(x,y)$ are obtained we use equation (3.1) to get the vector of displacements $\vec{a} = (u,v,w)$ at any point of Ω. This method gives the programmable algorithm for the numerical solution.

3.3 Example of a Pressurised Tube Subjected to Compression and Bending

We illustrate the described method with the following example. Let the tube Ω with the sliding condition and constant pressure action p at the inner surface be compressed and bent so that there are the following displacements at the three levels on the exterior surface:

$$\vec{a}_{|x^2+y^2=1,\tilde{z}=-1} = (0,0,\delta),$$
$$\vec{a}_{|x^2+y^2=1,\tilde{z}=0} = (\epsilon,0,0),$$
$$\vec{a}_{|x^2+y^2=1,\tilde{z}=1} = (0,0,-\delta).$$

We search for the coordinates of the vector of displacements in the form

$$u = \sum_{k=0}^{2} u_k(x,y)h^k, \; v = \sum_{k=0}^{2} v_k(x,y)h^k,$$
$$w = \sum_{k=0}^{2} w_k(x,y)h^k, \; r^2 \le x^2 + y^2 \le 1.$$

We follow the method described above for $n = 2$; finally we obtain

$$u(x,y,h) = \frac{\epsilon[\mu(3\lambda+5\mu) + r^2(\lambda^2 + 7\lambda\mu + 8\mu^2)]}{4\mu(\lambda+2\mu)(1+r^2)} + P -$$
$$-x\frac{r^2(p+\lambda\delta)}{(\lambda+\mu)(1-\kappa-2r^2)} +$$
$$+(x^2-y^2)[\frac{\epsilon(\lambda+\mu)(-\mu+\lambda r^2)}{8\mu(\lambda+2\mu)(1+r^2)} + \frac{\kappa}{2}P] +$$
$$+\frac{(x^2-y^2)}{(x^2+y^2)}\frac{\epsilon r^2(\lambda+\mu)}{(\lambda+2\mu)(1+r^2)} +$$
$$+(x^2+y^2)[\frac{\epsilon(\lambda+3\mu)(\mu-\lambda r^2)}{4\mu(\lambda+2\mu)(1+r^2)} - P] -$$
$$-\ln(x^2+y^2)\frac{\epsilon r^2\mu}{(\lambda+2\mu)(1+r^2)} +$$
$$+\frac{x}{(x^2+y^2)}\frac{r^2(p+\lambda\delta)}{(\lambda+\mu)(1-\kappa-2r^2)} - \frac{(x^2-y^2)}{(x^2+y^2)^2}Q - \epsilon h^2,$$

$$v(x,y,h) = -y\frac{r^2(p+\lambda\delta)}{(\lambda+\mu)(1-\kappa-2r^2)} + 2xy[\frac{\epsilon(\lambda+\mu)(-\mu+\lambda r^2)}{8\mu(\lambda+2\mu)(1+r^2)} + \frac{\kappa}{2}P] +$$
$$+\frac{2xy}{(x^2+y^2)}\frac{\epsilon r^2(\lambda+\mu)}{(\lambda+2\mu)(1+r^2)} + \frac{y}{(x^2+y^2)}\frac{r^2(p+\lambda\delta)}{(\lambda+\mu)(1-\kappa-2r^2)} -$$
$$-\frac{2xy}{(x^2+y^2)^2}Q,$$

$$w(x,y,h) = h[\frac{2r^2\epsilon x}{1+r^2} - \frac{2r^2\epsilon x}{(1+r^2)(x^2+y^2)} - \delta],$$

where

$$P = \frac{\epsilon}{4\mu(\kappa + r^4)(1 + r^2)(\lambda + 2\mu)}\{(\lambda + \mu)[\mu - r^2(\lambda + 8\mu)] + \\ + r^4\mu(5\lambda + 3\mu) - r^6\lambda(\lambda + 3\mu)\},$$

$$Q = -\frac{\epsilon(\lambda + \mu)[\mu - r^2(\lambda + 8\mu)]}{8\mu(1 + r^2)(\lambda + 2\mu)} + \frac{\kappa}{2}P.$$

The drawing of the pressurised tube subjected to compression and bending is shown on Fig.3.1.

It can be easily observed that the case when there is no bending (i.e. $\epsilon = 0$) brings to the displacements u, v symmetric with respect to the axis OH :

$$u(x, y, h) = \frac{x}{(x^2 + y^2)}\frac{r^2(p + \lambda\delta)}{(\lambda + \mu)(1 - \kappa - 2r^2)}[1 - (x^2 + y^2)],$$

$$v(x, y, h) = \frac{y}{(x^2 + y^2)}\frac{r^2(p + \lambda\delta)}{(\lambda + \mu)(1 - \kappa - 2r^2)}[1 - (x^2 + y^2)],$$

and the displacement $w(x, y, h) = -\delta h$ directly proportional to h .

The exterior surface bending on the level $h = 0$ in OX direction gives additional terms dependent on x, y, λ, μ, r, h but independent on p .

The formulated problem and the method of its solution can be applied to practical cases (e.g. a buried pipeline, a bent pressurised tube etc.).

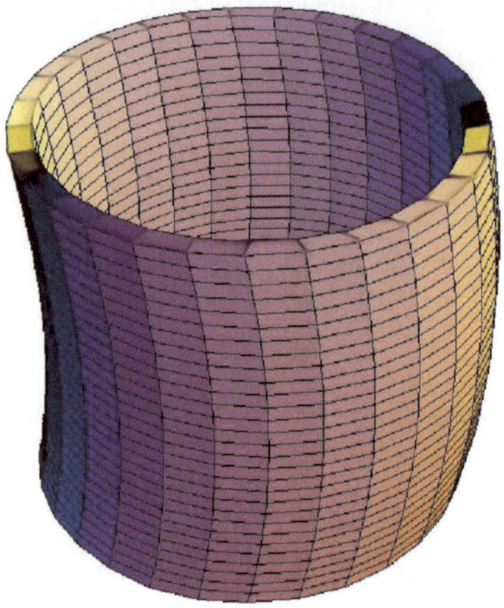

Figure 3.1: Deformed tube.

3.4 Formulation of the problem with given exterior stresses and analysis

Let Ω be a cylindrical tube in the coordinate space (x, y, h) where the element of the cylinder is parallel to the h-axis, the exterior radius is 1, the interior radius is r, $A \le h \le B$. The intersections of the exterior surface of Ω with the planes $h = h_j, j = 0, 1, \ldots, n,$ are denoted by C_j. We assume that a constant pressure p is acting at the interior surface of Ω. We also suppose that the sliding condition holds at the interior surface.

Now we formulate the following problem: given the vectors of exterior stresses

$$(F_1^j(\theta), F_2^j(\theta), F_3^j(\theta)),\ \theta \in [0, 2\pi],$$

at the points $(\cos\theta, \sin\theta, h_j)$ of the curves C^j, $j = 0, 1, \ldots, n,$ on the exterior cylindrical surface, given the vector of displacements $\vec{a}_0$ at the boundary point

$$(\cos\theta_0, \sin\theta_0, \widehat{h}_0),\ \widehat{h}_0 \in [A, B],$$

and the values $\tilde{r}^j$, $j = 1, 2$, of the rotation in XOY plane at two boundary points of Ω, it should be possible to find the displacements and stresses at any point of Ω.

The given boundary stresses can bring the solution to multi-valued displacements, so we add the following sufficient conditions of solvability:

$$\int_{-\pi}^{\pi} e^{\imath k\theta} F_m^j(\theta) d\theta = 0,\ m = 1, 2, 3;\ k = 0, 1, \ldots, n;\ j = 0, 1, \ldots, n.$$

3.4.1 Analysis

According to the boundary data we have the relations

$$[\sigma_{11}\cos(n,x)+\sigma_{12}\cos(n,y)+\sigma_{13}\cos(n,h)]_{C^j}=F_1^j(\theta),$$

$$[\sigma_{12}\cos(n,x)+\sigma_{22}\cos(n,y)+\sigma_{23}\cos(n,h)]_{C^j}=F_2^j(\theta),$$

$$[\sigma_{13}\cos(n,x)+\sigma_{23}\cos(n,y)+\sigma_{33}\cos(n,h)]_{C^j}=F_3^j(\theta), j=0,1,\dots,n.$$

Here the vector $\vec{n}$ is normal to the cylindrical surface, so $\cos(n,h)=0$ for $(x,y,h)\in C^j$. Finally we have the boundary conditions

$$[\sigma_{11}\frac{dy}{d\theta}-\sigma_{12}\frac{dx}{d\theta}]_{C^j}=F_1^j(\theta),$$

$$[\sigma_{12}\frac{dy}{d\theta}-\sigma_{22}\frac{dx}{d\theta}]_{C^j}=F_2^j(\theta), \tag{3.15}$$

$$[\sigma_{13}\frac{dy}{d\theta}-\sigma_{23}\frac{dx}{d\theta}]_{C^j}=F_3^j(\theta), j=0,1,\dots,n.$$

We suppose that every coordinate function of the desired vector of displacements $\vec{a}$ is a polynomial in h as in (3.1), so that

$$u(x,y,h)=\sum_{k=0}^{n}u_k(x,y)h^k,\ v(x,y,h)=\sum_{k=0}^{n}v_k(x,y)h^k,$$

$$w=\sum_{k=0}^{n}w_k(x,y)h^k,\ r^2\le x^2+y^2\le 1,\ h\in[A,B].$$

Now we obtain the components of the stress tensor as in (3.4), (3.5):

$$\sigma_{11} = \lambda\Lambda + 2\mu\sum_{k=0}^{n}\frac{\partial u_k}{\partial x}h^k, \sigma_{22} = \lambda\Lambda + 2\mu\sum_{k=0}^{n}\frac{\partial v_k}{\partial y}h^k,$$

$$\sigma_{33} = \lambda\Lambda + 2\mu\sum_{k=0}^{n-1}(k+1)w_{k+1}h^k, \sigma_{12} = \mu\sum_{k=0}^{n}(\frac{\partial u_k}{\partial y} + \frac{\partial v_k}{\partial x})h^k,$$

$$\sigma_{13} = \mu\{\sum_{k=0}^{n-1}[(k+1)u_{k+1} + \frac{\partial w_k}{\partial x}]h^k + \frac{\partial w_n}{\partial x}h^n\},$$

$$\sigma_{23} = \mu\{\sum_{k=0}^{n-1}[(k+1)v_{k+1} + \frac{\partial w_k}{\partial y}]h^k + \frac{\partial w_n}{\partial y}h^n\},$$

where

$$\Lambda \equiv \sum_{k=0}^{n-1}[\frac{\partial u_k}{\partial x} + \frac{\partial v_k}{\partial y} + (k+1)w_{k+1}]h^k + (\frac{\partial u_n}{\partial x} + \frac{\partial v_n}{\partial y})h^n.$$

After we put the expressions of stress tensor components in (3.15) we have

$$\sum_{k=0}^{n}\{\{\lambda[\frac{\partial u_k}{\partial x} + \frac{\partial v_k}{\partial y} + (k+1)w_{k+1}] + 2\mu\frac{\partial u_k}{\partial x}\}dy -$$
$$-\mu(\frac{\partial u_k}{\partial y} + \frac{\partial v_k}{\partial x})dx\}_{x=\cos\theta, y=\sin\theta}h_j^k = F_1^j(\theta)d\theta, \quad (3.16)$$

$$-\sum_{k=0}^{n}\{\{\lambda[\frac{\partial u_k}{\partial x} + \frac{\partial v_k}{\partial y} + (k+1)w_{k+1}] + 2\mu\frac{\partial v_k}{\partial y}\}dx +$$
$$+\mu(\frac{\partial u_k}{\partial y} + \frac{\partial v_k}{\partial x})dy\}_{x=\cos\theta, y=\sin\theta}h_j^k = F_2^j(\theta)d\theta, \quad (3.17)$$

$$\sum_{k=0}^{n}\mu\{[(k+1)u_{k+1} + \frac{\partial w_k}{\partial x}]dy -$$
$$-[(k+1)v_{k+1}) + \frac{\partial w_k}{\partial y}]dx\}_{x=\cos\theta, y=\sin\theta}h_j^k = F_3^j(\theta)d\theta. \quad (3.18)$$

Each of linear systems (3.16), (3.17), (3.18) with Vandermonde determinant can be easily solved, so we have the boundary relations

$$\{\{\lambda[\frac{\partial u_k}{\partial x} + \frac{\partial v_k}{\partial y} + (k+1)w_{k+1}] + 2\mu\frac{\partial u_k}{\partial x}\}dy -$$

$$-\mu(\frac{\partial u_k}{\partial y} + \frac{\partial v_k}{\partial x})dx\}_{x=\cos\theta,y=\sin\theta} = \tilde{F}_1^k(\theta)d\theta, k = 0,1,\ldots,n, \tag{3.19}$$

$$-\{\{\lambda[\frac{\partial u_k}{\partial x} + \frac{\partial v_k}{\partial y} + (k+1)w_{k+1}] + 2\mu\frac{\partial v_k}{\partial y}\}dx +$$

$$+\mu(\frac{\partial u_k}{\partial y} + \frac{\partial v_k}{\partial x})dy\}_{x=\cos\theta,y=\sin\theta} = \tilde{F}_2^k(\theta)d\theta, k = 0,1,\ldots,n, \tag{3.20}$$

$$\mu\{[(k+1)u_{k+1} + \frac{\partial w_k}{\partial x}]dy - [(k+1)v_{k+1} +$$

$$+\frac{\partial w_k}{\partial y}]dx\}_{x=\cos\theta,y=\sin\theta} = \tilde{F}_3^k(\theta)d\theta, k = 0,1,\ldots,n. \tag{3.21}$$

Note that the solvability condition gives the equalities

$$\int_{-\pi}^{+\pi} e^{ij\theta}\tilde{F}_m^k(\theta)d\theta = 0,\ \ m = 1,2,3;\ \ k = 0,1,\ldots,n,\ \ j = 0,1,\ldots,n.$$

The assumption of constant pressure acting at the interior surface (3.6) has the form:

$$\cos\theta\big[\lambda\{\sum_{k=0}^{n-1}[\frac{\partial u_k}{\partial x}+\frac{\partial v_k}{\partial y}+(k+1)w_{k+1}]h^k+$$
$$+(\frac{\partial u_n}{\partial x}+\frac{\partial v_n}{\partial y})h^n\}+2\mu\sum_{k=0}^{n}\frac{\partial u_k}{\partial x}h^k\big]_{|x^2+y^2=\rho^2}+$$
$$+\sin\theta\big[\mu\sum_{k=0}^{n}(\frac{\partial u_k}{\partial y}+\frac{\partial v_k}{\partial x})h^k\big]_{|x^2+y^2=\rho^2}=p\cos\theta,$$

$$\cos\theta\big[\mu\sum_{k=0}^{n}(\frac{\partial u_k}{\partial y}+\frac{\partial v_k}{\partial x})h^k\big]_{|x^2+y^2=\rho^2}+$$
$$+\sin\theta\big[\lambda\{\sum_{k=0}^{n-1}[\frac{\partial u_k}{\partial x}+\frac{\partial v_k}{\partial y}+(k+1)w_{k+1}]h^k+$$
$$+(\frac{\partial u_n}{\partial x}+\frac{\partial v_n}{\partial y})h^n\}+2\mu\sum_{k=0}^{n}\frac{\partial v_k}{\partial y}h^k\big]_{|x^2+y^2=\rho^2}=p\sin\theta.$$

Sliding condition (3.7) can be written as follows:

$$\{\sum_{k=0}^{n-1}[(k+1)u_{k+1}+\frac{\partial w_k}{\partial x}]h^k+\frac{\partial w_n}{\partial x}h^n\}_{|x^2+y^2=r^2}\cos\theta+$$
$$+\{\sum_{k=0}^{n-1}[(k+1)v_{k+1}+\frac{\partial w_k}{\partial y}]h^k+\frac{\partial w_n}{\partial y}h^n\}_{|x^2+y^2=r^2}\sin\theta=0.$$

The equilibrium equations give the representation of the coefficients $u_k(x,y)$, $v_k(x,y)$ and $w_k(x,y)$ with the help of the analytic functions $\phi_k(z)$, $\psi_k(z)$ and $\rho_k(z)$ in the ring $r<|z|<1$. Boundary relations (3.19), (3.20) and (3.21) give the boundary conditions at the points of the exterior circle $|z|=1$, the constant pressure and the sliding condition on the inner surface give the boundary conditions at the points of the inner circle $|z|=r$.

The solution contains $n+1$ steps when we examine the factors with h^t, $t=0,1,\ldots,n$, beginning with the factor with h^n. Here we describe the method of restoration of the coefficients $u_k(x,y)$, $v_k(x,y)$, $w_k(x,y)$ and analyse the solvability of the boundary problems.

3.4.2 Examination of coefficients of h^k

We have from the first two equilibrium equations

$$-2\mu(u_k(x,y)+\imath v_k(x,y)) = -\kappa\phi_k(z) + z\overline{\phi_k'(z)} + \overline{\psi_k(z)} + \Phi_k(z,\overline{z}),$$

where $\kappa = \frac{\lambda+3\mu}{\lambda+\mu}$, $\Phi_k(z,\overline{z})$ is the known function and the analytic functions $\phi_k(z)$ and $\psi_k(z)$ are to be found.

We use boundary condition (3.19):

$$\{\{\lambda[\frac{\partial u_k}{\partial x} + \frac{\partial v_k}{\partial y} + (k+1)w_{k+1}] + 2\mu\frac{\partial u_k}{\partial x}\}dy(\theta) -$$
$$-\mu(\frac{\partial u_k}{\partial y} + \frac{\partial v_k}{\partial x})dx(\theta)\}_{x=\cos\theta, y=\sin\theta} = \tilde{F}_1^k(\theta)d\theta,$$

which now has the form

$$\begin{aligned}2\operatorname{Re}\phi_k'(z(\theta))dy(\theta) + \operatorname{Im}[(z\overline{\phi_k''} + \overline{\psi_k'})_{z=z(\theta)}\overline{dz(\theta)}] = \\ = \tilde{F}_1^k(\theta)d\theta - \lambda(k+1)w_{k+1}(x(\theta),y(\theta))dy(\theta) + \\ +\frac{\lambda+\mu}{\mu}\operatorname{Re}(\frac{\partial\Phi_k}{\partial z})_{z=z(\theta)}dy(\theta) - \operatorname{Im}[(\overline{\frac{\partial\Phi_k}{\partial\overline{z}}})_{z=z(\theta)}\overline{dz(\theta)}],\end{aligned} \tag{3.22}$$

where $z(\theta) \equiv x(\theta) + \imath y(\theta) = \exp(\imath\theta)$.

In a similar way boundary condition (3.20) reduces to the equality

$$\begin{aligned}-2\operatorname{Re}\phi_k'(z(\theta))dx(\theta) - \operatorname{Re}[(z\overline{\phi_k''} + \overline{\psi_k'})_{z=z(\theta)}\overline{dz(\theta)}] = \\ = \tilde{F}_2^k(\theta)d\theta + \lambda(k+1)w_{k+1}(x(\theta),y(\theta))dx(\theta) - \\ -\frac{\lambda+\mu}{\mu}\operatorname{Re}(\frac{\partial\Phi_k}{\partial z})_{z=z(\theta)}dx(\theta) + \operatorname{Re}[(\overline{\frac{\partial\Phi_k}{\partial\overline{z}}})_{z=z(\theta)}\overline{dz(\theta)}].\end{aligned} \tag{3.23}$$

Now (3.22) and (3.23) together give the boundary condition which is similar to that for the first basic problem of the plane theory of elasticity [7]:

$$[\phi_k(z) + z\overline{\phi'_k} + \overline{\psi_k}]_{|z=\exp(\imath\theta)} = \imath\int[\tilde{F}_1^k(\theta) + \imath\tilde{F}_2^k(\theta)]d\theta -$$
$$-\lambda(k+1)\int[\operatorname{Re}\rho_{k+1}(z) + \Psi_{k+1}(z,\overline{z})]_{|z=\exp\imath\theta}de^{i\theta} +$$
$$+\frac{\lambda+\mu}{\mu}\int \operatorname{Re}(\frac{\partial\Phi_k}{\partial z})_{|z=\exp(\imath\theta)}de^{\imath\theta} - \int[(\frac{\partial\Phi_k}{\partial\overline{z}})_{|z=\exp(\imath\theta)}de^{-i\theta}].$$

In spite of this analogy with the plane problem we use the other boundary condition which can be got from the last one by its differentiation with respect to θ:

$$\begin{aligned}[-2\overline{z}\operatorname{Re}\phi'_k(z) + |z|^2\phi''_k(z) + z\psi'_k(z)]_{|z=\exp\imath\theta} = -\tilde{F}_1^k(\theta) + \imath\tilde{F}_2^k(\theta) +\\ +[-(1+\frac{\lambda}{\mu})\overline{z}\operatorname{Re}\partial\Phi_k/\partial z - z\partial\overline{\Phi_k}/\partial z +\\ +\overline{z}\lambda(k+1)(\operatorname{Re}\rho_{k+1}(z) + \Psi_{k+1}(z,\overline{z}))]_{|z=\exp\imath\theta},\end{aligned} \tag{3.24}$$

The boundary condition at the inner circle of the ring $r < |z| < 1$ is the same as in the previous problem:

$$\begin{aligned}[-2\overline{z}\operatorname{Re}\phi'_k(z) + |z|^2\phi''_k(z) + z\psi'_k(z)]_{|z=r\exp\imath\theta} = [-(1+\\ +\frac{\lambda}{\mu})\overline{z}\operatorname{Re}\partial\Phi_k/\partial z - z\partial\overline{\Phi_k}/\partial z +\\ +\overline{z}\lambda(k+1)(\operatorname{Re}\rho_{k+1}(z) + \Psi_{k+1}(z,\overline{z}))]_{|z=r\exp\imath\theta}.\end{aligned} \tag{3.25}$$

We must note that the form of the second condition with $k = 0$ is different from (3.25):

$$[-2\overline{z}\operatorname{Re}\phi'_0(z) + |z|^2\phi''_0(z) + z\psi'_0(z)]_{|z=r\exp i\theta} = [-p\overline{z} - (1+$$
$$+\frac{\lambda}{\mu})\overline{z}\operatorname{Re}\partial\Phi_0/\partial z - z\partial\overline{\Phi_0}/\partial z + \overline{z}\lambda(\operatorname{Re}\rho_1(z) + \Psi_1(z,\overline{z}))]_{|z=r\exp\imath\theta}.$$

Now we have to restore the analytic in the ring $r < |z| < 1$ functions $\phi_k(z)$ and $\psi_k(z)$ via the boundary conditions (3.24) and (3.25) where the recurrent formulas for the functions $\Phi_k(z,\overline{z})$ and $\Psi_k(z,\overline{z})$ are formulas

(1.8)–(1.12). We must note that the necessary condition of solvability of this boundary problem is the following: the coefficients of e^0 of Fourier expansions of the functions at the right-hand sides of the expressions (3.24) and (3.25) should vanish.

In order to obtain the single-valued displacements in the tube we have to get the single-valued in the ring $r < |z| < 1$ functions $\phi_k(z)$ and $\psi_k(z)$ and $\Phi_k(z, \overline{z})$. So the right-hand sides of the expressions (3.24) and (3.25) must also be single-valued.

We again apply the Laurent series expansions for the analytic in the ring $r < |z| < 1$ functions $\phi_k(z)$ and $\psi_k(z)$ as it is demonstrated in the solution of the problem with given exterior displacements. But there are some differences in the solution of the corresponding systems of coefficients. We see it while restoring coefficients of h^n.

Let

$$\phi_n(z) = \sum_{m=-\infty}^{+\infty} \gamma_m^n z^m,$$

$$\psi_n(z) = \sum_{m=-\infty, m\neq 0}^{+\infty} c_m^n z^m.$$

So the boundary conditions (3.24) and (3.25) with $k = n$ give

$$-e^{-\imath\theta}[\sum_{m=-\infty}^{+\infty} m\gamma_m^n e^{\imath(m-1)\theta} + \sum_{m=-\infty}^{+\infty} m\overline{\gamma_m^n} e^{-\imath(m-1)\theta}] +$$

$$+ \sum_{m=-\infty}^{+\infty} m(m-1)\gamma_m^n e^{\imath(m-2)\theta} + e^{\imath\theta} \sum_{m=-\infty, m\neq 0}^{+\infty} mc_m^n e^{\imath(m-1)\theta} = \quad (3.26)$$

$$= -\tilde{F}_1^n(\theta) + \imath\tilde{F}_2^n(\theta),$$

$$-re^{-\imath\theta}[\sum_{m=-\infty}^{+\infty} m\gamma_m^n r^{m-1}e^{\imath(m-1)\theta} + \sum_{m=-\infty}^{+\infty} m\overline{\gamma_m^n}r^{m-1}e^{-\imath(m-1)\theta}] +$$

$$+r^2\sum_{m=-\infty}^{+\infty} m(m-1)\gamma_m^n e^{\imath(m-2)\theta} + re^{\imath\theta}\sum_{m=-\infty,m\neq 0}^{+\infty} mc_m^n e^{\imath(m-1)\theta} = 0. \quad (3.27)$$

The solvability condition here is

$$\int_{-\pi}^{+\pi} \tilde{F}_m^n(\theta)d\theta = 0, \ m = 1, 2,$$

which holds due to the sufficient conditions.

We compare the coefficients of different powers of $e^{\imath\theta}$ in the left-hand sides and in the right-hand sides of (3.26) and (3.27) correspondingly.

The coefficients of e^0 in both parts of (3.26) and (3.27) vanish due to the solvability condition.

We compare the coefficients of $e^{-\imath\theta}$ and have the system

$$\begin{cases} -\gamma_1^n - \overline{\gamma_1^n} - c_{-1}^n = \frac{1}{2\pi}\int_{-\pi}^{\pi} e^{\imath\theta}[-\tilde{F}_1^n(\theta) + \imath\tilde{F}_2^n(\theta)]d\theta, \\ -r\gamma_1^n - r\overline{\gamma_1^n} - c_{-1}^n r^{-1} = 0. \end{cases}$$

This system provides us with $\operatorname{Re}\gamma_1^n$ and c_{-1}^n. These coefficients vanish due to the sufficient condition.

We compare the coefficients of $e^{-2\imath\theta}$ and have the system

$$\begin{cases} -2\overline{\gamma_2^n} - 2c_{-2}^n = \frac{1}{2\pi}\int_{-\pi}^{\pi} e^{2\imath\theta}[-\tilde{F}_1^n(\theta) + \imath\tilde{F}_2^n(\theta)]d\theta, \\ -2r^2\overline{\gamma_1^n} - 2c_{-2}^n r^{-2} = 0. \end{cases}$$

This system provides us with the coefficients γ_2^n and c_{-2}^n which vanish due to the sufficient condition.

We compare the coefficients of $e^{k\imath\theta}$ and the coefficients of $e^{(-k-2)\imath\theta}$,

$k \geq 1$, in (3.26) and (3,27). This comparison gives the system

$$\begin{cases} k[(k+2)\gamma^n_{k+2} + \overline{\gamma^n_{-k}} + c^n_k] = \frac{1}{2\pi}\int\limits_{-\pi}^{\pi} e^{-\imath k\theta}[-\tilde{F}^n_1(\theta) + \imath\tilde{F}^n_2(\theta)]d\theta, \\ k[(k+2)\gamma^n_{k+2}r^{k+2} + \overline{\gamma^n_{-k}}r^{-k} + c^n_k r^k] = 0, \\ (k+2)[k\gamma^n_{-k} - \overline{\gamma^n_{k+2}} - c^n_{-k-2}] = \frac{1}{2\pi}\int\limits_{-\pi}^{\pi} e^{\imath(k+2)\theta}[-\tilde{F}^n_1(\theta) + \imath\tilde{F}^n_2(\theta)]d\theta, \\ (k+2)[k\gamma^n_{-k}r^{-k} - \overline{\gamma^n_{k+2}}r^{k+2} - c^n_{-k-2}r^{-k-2}] = 0. \end{cases}$$

This system provides us with the values of the coefficients γ^n_{-k}, γ^n_{k+2}, c^n_k, c^n_{-k-2}. The coefficients vanish according to the sufficient conditions for $k = 1, ..., n$.

Note that the values of γ^n_0 and $\operatorname{Im}\gamma^n_1$ can not be found from the systems, so we restore the coefficient $u_n(x,y) + \imath v_n(x,y)$ within the summand $\nu_n \imath z + C_n$, where ν_n is an arbitrary real constant and C_n is an arbitrary complex constant. These constants are to be defined later.

The last equilibrium equation gives the relation

$$\mu\Delta w_k = -(\lambda+\mu)(k+1)(\frac{\partial u_{k+1}}{\partial x} + \frac{\partial v_{k+1}}{\partial y}) - \\ -(k+1)(k+2)(\lambda+2\mu)w_{k+2},$$

which can be presented in the form

$$w_k(x,y) = \operatorname{Re}\rho_k(z) + \Psi_k(z,\overline{z}),$$

where $\rho_k(z)$ is analytic in the ring $r < |z| < 1$, the function $\Psi_k(z,\overline{z})$ is known.

We have two boundary conditions for the function $\rho_k(z)$. The first one is relation (3.21) which now has the form

$$\operatorname{Re}(z\rho'_k(z))_{|z=\exp\imath\theta} = \frac{1}{\mu}\tilde{F}^k_3(\theta) - [z\partial\Psi_k/\partial z + \overline{z}\partial\Psi_k/\partial\overline{z}]_{|z=\exp\imath\theta} + \\ +\frac{k+1}{2\mu}\operatorname{Re}[\overline{z}(-\kappa\phi_{k+1} + z\overline{\phi'_{k+1}} + \overline{\psi_{k+1}} + \Phi_{k+1}(z,\overline{z}))]_{|z=\exp\imath\theta}. \quad (3.28)$$

The boundary condition at the inner circle for the analytic in the ring $1 > |z| > r$ function $\rho_k(z)$ is the sliding condition in the form of relation (3.7), so we have as for the previous problem

$$\mathrm{Re}(z\rho_k'(z))_{|z=r\exp\imath\theta} = -[z\partial\Psi_k/\partial z + \overline{z}\partial\Psi_k/\partial\overline{z}]_{|z=r\exp\imath\theta} +$$
$$+\frac{k+1}{2\mu}\mathrm{Re}[\overline{z}(-\kappa\phi_{k+1} + z\overline{\phi_{k+1}'} + \overline{\psi_{k+1}} + \Phi_{k+1}(z,\overline{z}))]_{|z=r\exp\imath\theta}. \qquad (3.29)$$

Note that the solvability condition here is the same as that for the previous boundary value problem: the coefficients of e^0 of the expansions of the expressions at the right-hand sides of (3.28) and (3.29) vanish. The right-hand sides of the expressions (3.24) and (3.25) must be single-valued in order that the displacements in the tube be single-valued.

We represent the functions $\rho_k(z)$ as the Laurent series expansions and put them in the external and internal boundary conditions. So we obtain the linear systems with respect to the coefficients of the expansions.

We present here the system which helps us to restore the coefficient $w_n(x,y)$. The analytic in the ring $r < |z| < 1$ function $\rho_n(z)$ is to be found in the form

$$\rho_n(z) = \sum_{m=-\infty}^{+\infty} D_m^n z^n.$$

The solvability condition holds due to the relation $\int_{-\pi}^{\pi} \tilde{F}_3^n(\theta)d\theta = 0$, which is the implication of the sufficient conditions.

After we put the expression of the function $\rho_n(z)$ into the boundary conditions (3.28), (3.29) with $k = n$ and compare the coefficients of $e^{\imath m\theta}$, $m \in \mathbf{N}$ at the both parts we have the system

$$\begin{cases} D_m^n - \overline{D_{-m}^n} = \frac{1}{m\mu\pi}\int_{-\pi}^{\pi} e^{-\imath m\theta}\tilde{F}_3^n(\theta)d\theta, \\ D_m^n r^m - \overline{D_{-m}^n} r^{-m} = 0, \end{cases}$$

which yields the values of D^n_m and D^n_{-m} for all $m \in \mathbf{N}$. Note that these coefficients vanish for $m = 1, ..., n$ due to the sufficient conditions.

We can not find D^n_0 from the last system, so we obtain the coefficient $w_n(x, y)$ within an arbitrary real summand.

Let $k = n - 1$. We have the following system of the boundary conditions for the analytic in the ring $r < |z| < 1$ functions $\phi_{n-1}(z)$ and $\psi_{n-1}(z)$:

$$\left[-2\overline{z}\operatorname{Re}\phi'_{n-1}(z) + |z|^2\phi''_{n-1}(z) + z\psi'_{n-1}(z)\right]_{|z=\exp\imath\theta} = -\tilde{F}_1^{n-1}(\theta) + \imath\tilde{F}_2^{n-1}(\theta) +$$

$$+\frac{n(\lambda^2 + 2\lambda\mu - \mu^2)}{4(\lambda + 2\mu)}e^{-\imath\theta}(\rho_n(e^{\imath\theta}) + \overline{\rho_n(e^{\imath\theta})}) - \frac{n(\lambda + \mu)(\lambda + 3\mu)}{4(\lambda + 2\mu)}\rho'_n(e^{\imath\theta}),$$

$$\left[-2\overline{z}\operatorname{Re}\phi'_{n-1}(z) + |z|^2\phi''_{n-1}(z) + z\psi'_{n-1}(z)\right]_{|z=r\exp\imath\theta} =$$

$$= \frac{n(\lambda^2 + 2\lambda\mu - \mu^2)}{4(\lambda + 2\mu)}re^{-\imath\theta}(\rho_n(re^{\imath\theta}) + \overline{\rho_n(re^{\imath\theta})}) - r^2\frac{n(\lambda + \mu)(\lambda + 3\mu)}{4(\lambda + 2\mu)}\rho'_n(re^{\imath\theta}).$$

The solvability condition holds due to the sufficient conditions and vanishing of some Laurent coefficients of the known functions. The latter also is the consequence of the sufficient conditions. We put $D^n_0 = 0$ which stays to be undetermined in order to make the coefficients γ^{n-1}_{-1}, γ^{n-1}_3, c^{n-1}_1 and c^{n-1}_{-3} vanish. This condition prevents the possible multi-valence of the coefficients of displacement components. So we find the Laurent coefficients of the functions $\phi_{n-1}(z)$ and $\psi_{n-1}(z)$ and restore the coefficient $u_{n-1} + \imath v_{n-1}$ within the summand $\nu_{n-1}\imath z + C_{n-1}$.

We have the following system of boundary conditions for the analytic in the ring $r < |z| < 1$ function $\rho_{n-1}(z)$:

$$\operatorname{Re}(z\rho'_{n-1}(z))_{|z=\exp\imath\theta} = \frac{1}{\mu}\tilde{F}_3^k(\theta) + \frac{n}{2\mu}(\phi'_n(e^{\imath\theta}) + \overline{\phi'_n(e^{\imath\theta})}) +$$

$$+\frac{n}{4\mu}(1 - \kappa)(e^{-\imath\theta}\phi_n(e^{\imath\theta}) + e^{\imath\theta}\overline{\phi_n(e^{\imath\theta})}) + \frac{n}{4\mu}(e^{\imath\theta}\psi_n(e^{\imath\theta}) + e^{-\imath\theta}\overline{\psi_n(e^{\imath\theta})}),$$

$$\operatorname{Re}(z\rho'_k(z))_{|z=r\exp\imath\theta} = \frac{nr^2}{2\mu}(\phi'_n(re^{\imath\theta}) + \overline{\phi'_n(re^{\imath\theta})}) +$$

$$+\frac{n}{4\mu}(1 - \kappa)(re^{-\imath\theta}\phi_n(re^{\imath\theta}) + re^{\imath\theta}\overline{\phi_n(re^{\imath\theta})}) + \frac{n}{4\mu}(re^{\imath\theta}\psi_n(re^{\imath\theta}) + re^{-\imath\theta}\overline{\psi_n(re^{\imath\theta})}).$$

The solvability condition holds here. We define the undetermined constant C_n so that the coefficients of Laurent expansion of $\rho_{n-1}(z)$ with z and z^{-1} vanish, and restore all coefficients of Laurent expansion of the function $\rho_{n-1}(z)$. The coefficient of displacement component w_{n-1} is restored within a real summand D_0^{n-1}.

Therefore we restore the single-valued coefficients of the displacement components consequently and simultaneously define the unknown constants C_k, D_0^k and ν_k. The latter constant we define when we search for $\phi_{k-2}(z)$ and $\psi_{k-2}(z)$. Hence after we restore all the coefficients we still have undetermined complex constant C_0 and undetermined real constants D_0^0, ν_1 and ν_2. These parameters can be defined via the additional data: the given displacements at one boundary point of the tube and the given rotation in XOY plane at two boundary points. We put our solution into the relations

$$(u(\cos\theta_0, \sin\theta_0, \widehat{h}_0), v(\cos\theta_0, \sin\theta_0, \widehat{h}_0), w(\cos\theta_0, \sin\theta_0, \widehat{h}_0) = \vec{a}_0,$$

$$(\frac{\partial v}{\partial x} - \frac{\partial u}{\partial y})|_{x=\cos\theta_k, y=\sin\theta_k, \widehat{h}_k} = 2\tilde{r}^k, \ k = 1, 2,$$

and restore the undetermined constants.

CHAPTER 4

Interpolation solution of the problem of elasticity for the non-circular cylinder

Elena A. Shirokova,
Kazan Federal University
Elena.Shirokova@ksu.ru

ABSTRACT: In this chapter we present the interpolation solution of the 3D second basic elasticity problem for non-circular cylinder. We apply the rational conform mapping in order to reduce the boundary value problem in a non-circular domain to the boundary value problem in the unit disk.

Let Ω be a cylinder in (xyh) space with generatrices parallel to OH axis. We construct the interpolation solution of the second basic problem of elasticity for the case when the directrix C of the cylinder surface of the elastic solid Ω is not a circle and the section D of Ω by the plane normal to the generatrix is not a disk.

We construct the interpolation solution which is accurate either for the points distant from the ends of the cylinder or for the points near the ends of the cylinder.

We begin with construction of the interpolation solution accurate for the points distant from the ends using the displacements given at the cylindrical surface. We can formulate the problem as follows: given the displacements at $n+1$ directrices C_j , $j=0,...,n$, parallel to C of the cylinder surface, it should be possible to find the vector of displacements of the points of Ω which meets the equilibrium equations and the given data at the boundary curves C_j .

The method of the interpolation solution construction for a non-circular cylinder is similar to the method presented in Chapter 2. We search for the solution as the polynomial of the power n in the third variable h with the unknown coefficients $u_k(x,y)$, $v_k(x,y)$ and $w_k(x,y)$, $k=0,1,...,n$. The given data at $n+1$ directrices provide the boundary values $\tilde{u}_k(s)$, $\tilde{v}_k(s)$ and $\tilde{w}_k(s)$ of the unknown coefficients. We require that the displacements components satisfy the equilibrium equations and represent the unknown coefficients with the help of the analytic functions as it was done in Chapter 1.

The main difficulty now is to solve the boundary value problem — to find the functions $\phi(z)$ and $\psi(z)$ analytic in the domain D via the boundary condition

$$[-\kappa\phi(z)+z\overline{\phi'(z)}+\overline{\psi(z)}]_{|z=z(s)\in C}=f(s), \qquad (4.1)$$

where $f(s)$ is the complex valued function given on C , $\kappa=(\lambda+3\mu)/(\lambda+\mu)$.

We consider only the class R of the domains D being the rational maps of the unit disk E . There is the well-known method of solution of the corresponding boundary value problems presented in [7] which applies to them Cauchy type integral. We use the method based on Schwartz problem solution for meromorphic functions in the unit disk.

The interpolation solution accurate at the points near the ends of the cylinder also needs solution of this boundary value problem.

4.1 Reduction of the boundary value problem solution to Schwartz problems

Let D be the domain from the class R, the function $z(\zeta)$ be the rational function which maps the unit disk E to the domain D. We consider the following notation: $F(\zeta) = \phi(z(\zeta))$, $G(\zeta) = \psi(z(\zeta))$. Now the boundary condition (4.1) has the form

$$[-\kappa F(\zeta) + z(\zeta)\overline{\frac{F'(\zeta)}{z'(\zeta)}} + \overline{G(\zeta)}]_{|\zeta=e^{\imath\theta}} = f(s(\theta)). \tag{4.2}$$

We search for the functions $F(\zeta)$ and $G(\zeta)$ analytic in the unit disk E which satisfy the boundary condition (4.2).

We denote by $q(\zeta)$ the meromorphic in E function which coincides with the function $\overline{z(\zeta)}$ on the boundary of E, so we can write $q(\zeta) = \overline{z}(1/\zeta)$.

Now we separate the real and imaginary parts of the boundary condition (4.2), recall that $\operatorname{Re} Q = \operatorname{Re} \overline{Q}$, $\operatorname{Im} Q = -\operatorname{Im} \overline{Q}$ and rewrite (4.2) as two equations:

$$\operatorname{Re}[-\kappa F(\zeta) + q(\zeta)\frac{F'(\zeta)}{z'(\zeta)} + G(\zeta)]|_{\zeta=e^{\imath\theta}} = f_1(\theta), \tag{4.3}$$

and

$$\operatorname{Im}[-\kappa F(\zeta) - q(\zeta)\frac{F'(\zeta)}{z'(\zeta)} - G(\zeta)]|_{\zeta=e^{\imath\theta}} = f_2(\theta), \tag{4.4}$$

where $f_1(\theta) + \imath f_2(\theta) = f(s(\theta)$.

The functions

$$-\kappa F(\zeta) \pm q(\zeta)\frac{F'(\zeta)}{z'(\zeta)} \pm G(\zeta)$$

are meromorphic in E and have the poles at the same points of E where the function $q(\zeta)$ has the poles. Note that the principle part of Laurent expansion of the function

$$-\kappa F(\zeta) + q(\zeta)\frac{F'(\zeta)}{z'(\zeta)} + G(\zeta)$$

differs from the principle part of Laurent expansion of the function

$$-\kappa F(\zeta) - q(\zeta)\frac{F'(\zeta)}{z'(\zeta)} - G(\zeta)$$

only by the sign.

Suppose that the function $q(\zeta)$ has the poles of the order k_j at the points $a_j \in E$, $j = 1, ..., n$. Now we have from (4.3) according to [4]

$$-\kappa F(\zeta) + q(\zeta)\frac{F'(\zeta)}{z'(\zeta)} + G(\zeta) = \frac{1}{2\pi}\int_{-\pi}^{\pi} f_1(\theta)\frac{e^{\imath\theta}+\zeta}{e^{\imath\theta}-\zeta}d\theta +$$

$$+\sum_{j=i}^{n}\sum_{t=1}^{k_j}[C_t^j(\frac{1-\overline{a_j}\zeta}{\zeta - a_j})^t - \overline{C_t^j}(\frac{\zeta - a_j}{1-\overline{a_j}\zeta})^t] + \imath B, \tag{4.5}$$

where B is arbitrary real constant, C_t^j are arbitrary complex constants.

We have from (4.4)

$$-\kappa F(\zeta) - q(\zeta)\frac{F'(\zeta)}{z'(\zeta)} - G(\zeta) = \frac{1}{2\pi}\int_{-\pi}^{\pi} f_2(\theta)\frac{e^{\imath\theta}+\zeta}{e^{\imath\theta}-\zeta}d\theta +$$

$$+\sum_{j=i}^{n}\sum_{t=1}^{k_j}[-C_t^j(\frac{1-\overline{a_j}\zeta}{\zeta - a_j})^t - \overline{C_t^j}(\frac{\zeta - a_j}{1-\overline{a_j}\zeta})^t] + A, \tag{4.6}$$

where A is arbitrary real constant, C_t^j are the same as in (4.5).

Now we find the analytic function $F(\zeta)$ from (4.5) and (4.6). The expression of this function contains arbitrary constants A, B, C_t^j. We put

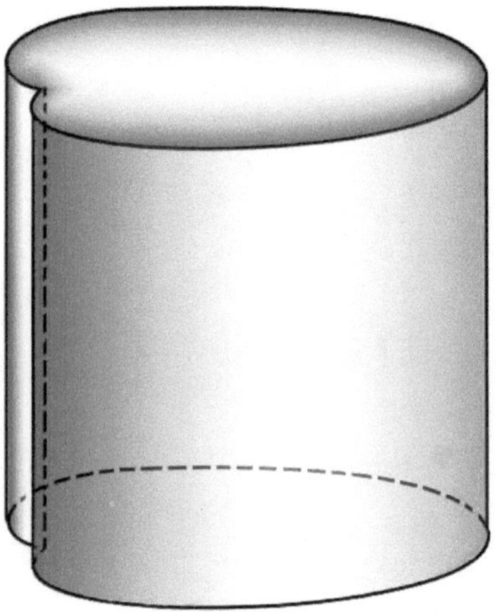

Figure 4.1: Cylinder with furrow.

the expression of $F(\zeta)$ in (4.5) and find the expression of the function $G(\zeta)$. The last expression also contains the arbitrary constants A, B, C_t^j and the summands with singularities of $(\zeta - a_j)^{-m}$ type. The function $G(\zeta)$ must be analytic in E, so we equate the coefficients of all singularities to null and obtain the linear system over the arbitrary constants which gives the values of all these constants.

So we have the functions $G(\zeta)$ and $F(\zeta)$ analytic in E.

4.2 Example 4.1. The bent cylinder with the furrow along the generatrix

Let the domain D be the image of the unit disk E mapped by the function

$$z(\zeta) = (4 - 2\zeta - \zeta^2)^{-1}, |\zeta| < 1.$$

This domain D is close to the disk with the radius approximately equal to 0.412, with the center at $(0.588, 0)$ and with the cusp at the boundary point $(0.2, 0)$. So there is the furrow over the generatrix $\{(0.2, 0, h)|h \in [1, 3]\}$ of the corresponding cylinder (Fig. 4.1).

Let us construct the interpolation solution for the case when $h \in [1, 3]$, and the vectors of boundary displacements given at three levels:

$$\vec{a}_{|h=1,(x,y)\in C} = (0, 0, \delta), \vec{a}_{|h=2,(x,y)\in C} = (\epsilon, 0, 0), \vec{a}_{|h=3,(x,y)\in C} = (0, 0, -\delta).$$

We search for the coordinate displacement functions in the form of the polynomials:

$$u(x, y, h) = u_0(x, y) + u_1(x, y)h + u_2(x, y)h^2,$$
$$v(x, y, h) = v_0(x, y) + v_1(x, y)h + v_2(x, y)h^2,$$
$$w(x, y, h) = w_0(x, y) + w_1(x, y)h + w_2(x, y)h^2, (x, y) \in D.$$

The boundary values of the coefficients are the same as in Example 1 from Chapter 2, so we have

$$\tilde{u}_0(s) = -3\epsilon, \tilde{u}_1(\theta) = 4\epsilon, \tilde{u}_2(s) = -\epsilon,$$
$$\tilde{v}_0(s) = \tilde{v}_1(\theta) = \tilde{v}_2(s) = 0,$$
$$\tilde{w}_0(s) = 2\delta, \tilde{w}_1(\theta) = -\delta, \tilde{w}_2(s) = 0.$$

We find the coefficients $u_k(x, y)$, $v_k(x, y)$, $w_k(x, y)$ successively beginning with $k = 2$.

Finally we have the expressions for $u(x, y, h) + \imath v(x, y, h)$ and $w(x, y, h)$

in terms of h and $z = x + \imath y$:

$$
\begin{aligned}
u(x,y,h) + \imath v(x,y,h) = -\Big[\Big[\epsilon\mu\,\big(16(-1+4z)\mu(\lambda^2+3\lambda\mu+2\mu^2)+ \\
\overline{z}\,\big(-275A(\overline{z})(-24+z^2)\lambda^3 + (-32+48856A(\overline{z})+ \\
+128z(-5+10A(\overline{z})+|A(z)|^2) - 1775z^2A(\overline{z})\big)\,\lambda^2\mu+ \\
+(16(-22+7449A(\overline{z}))+640z(-3+8A(\overline{z})+|A(z)|^2)-3525z^2A(\overline{z}))\lambda\mu^2+ \\
+\mu^3(64(-9+1496A(\overline{z}))+ \\
+256z(-5+20A(\overline{z})+3|A(z)|^2)-2025z^2|A(z)|^2)\big)+ \\
+10\overline{z}^2(11\lambda+27\mu)(5A(\overline{z})z(\lambda+\mu)^2+ \\
+8(1+A(\overline{z}))\mu(\lambda+2\mu)))\big]\,\big(100A(\overline{z})\overline{z}\,\big(11\lambda^3+ \\
+82\lambda^2\mu+201\lambda\mu^2+162\mu^3)\big)^{(-1)} + \frac{z^2\epsilon\mu(\lambda+\mu)}{4(\lambda+2\mu)} - \frac{|z|^2\epsilon\mu(\lambda+3\mu)}{2(\lambda+2\mu)}\Big]\,(2\mu)^{-1}+ \\
+4\epsilon h - \epsilon h^2,
\end{aligned}
$$

where $A(t) = \sqrt{5 - t^{-1}}$,

$$
w(x,y,h) = 2\delta - \delta h.
$$

4.3 Example 4.2. The bent cylinder with the furrow with the displacements given at two points at the ends

Here we construct the interpolation solution more accurate at the points at the ends of the non-circular cylinder. Now the boundary displacements are given at the edge of the cylinder and at two points at the ends of it.

We restore with the help of the interpolation solution method the displacements in the cylinder $\Omega = \{(x, y, h)|(x, y) \in D, z \in [1, 3]\}$ subjected to compression at the ends and bending when the domain D is the same as in the previous section — it is the image of the unit disk mapped by the function

$$z(\zeta) = (4 - 2\zeta - \zeta^2)^{-1}, |\zeta| < 1.$$

Recall that this domain D is close to the disk with the radius approximately equal to 0.412, with the center at $(0.588, 0)$ and with the cusp at the boundary point $(0.2, 0)$. So the cylinder Ω has the furrow over the generatrix $\{(0.2, 0, h)|h \in [1, 3]\}$ (Fig.3). We denote by

$$\{(x(s), y(s))|s \in [0, l]\}$$

the points of the boundary of D.

The boundary conditions at the edges of the cylinder are

$$\vec{a}(x(s), y(s), 1) = (0, 0, \epsilon),\ \vec{a}(x(s), y(s), 3) = (0, 0, -\epsilon).$$

There are given also the displacements at two points at the ends of the cylinder:

$$\vec{a}(0.588, 0, 1) = (\alpha, 0, \epsilon - \nu),\ \vec{a}(0.588, 0, 3) = (\gamma, 0, -\epsilon - \nu).$$

We search for the solution in the form

$$u(x,y,h)=u_0(x,y)+u_1(x,y)h+ah^2+bh^3,$$
$$v(x,y,h)=v_0(x,y)+v_1(x,y)h,$$
$$w(x,y,h)=w_0(x,y)+w_1(x,y)h+eh^2,$$

where a , b and e are the constants to obtain.

The boundary conditions at the edge give the boundary conditions for the unknown coefficients $u_0(x,y)$, $u_1(x,y)$, $v_0(x,y)$, $v_1(x,y)$, $w_0(x,y)$, $w_1(x,y)$:

$$u_0(x(s),y(s))=3a+12b,\ u_1(x(s),y(s))=-4a-13b,$$
$$v_0(x(s),y(s))=0,\ v_1(x(s),y(s))=0,$$
$$w_0(x(s),y(s))=2\epsilon+3e,\ w_1(x(s),y(s))=-\epsilon-4e.$$

Now we have to solve the boundary value problems in order to get the unknown coefficients $u_0(x,y)$, $u_1(x,y)$, $v_0(x,y)$, $v_1(x,y)$, $w_0(x,y)$ and $w_1(x,y)$. We describe only the restoration of the the coefficients $u_1(x,y)$, $v_1(x,y)$ and $w_1(x,y)$. The other coefficients can be found in the same way.

According to the results of Chapter 1 we have

$$u_1(x,y)+\imath v(x,y)=-\frac{1}{2\mu}[-\kappa\phi_1(z)+z\overline{\phi_1(z)}+\overline{\psi_1(z)}+\Phi_1(z,\overline{z})],$$

where $\phi_1(z)$ and $\psi_1(z)$ are analytic in D functions, $z=x+\imath y$,

$$\Phi_1(z,\overline{z})=\frac{3\mu b}{4(\lambda+2\mu)}[2|z|^2(\lambda+3\mu)-z^2(\lambda+\mu)].$$

So we have the boundary condition

$$[-\kappa\phi_1(z)+z\overline{\phi_1(z)}+\overline{\psi_1(z)}]|_{z=x(s)+\imath y(s)}=2\mu(4a+13b)-$$
$$-\frac{3\mu b}{4(\lambda+2\mu)}[2|z|^2(\lambda+3\mu)-z^2(\lambda+\mu)]|_{z=x(s)+\imath y(s)}.$$

We introduce the functions $F_1(\zeta) = \phi_1(z(\zeta))$, $G_1(\zeta) = \psi_1(z(\zeta))$, where

$$z(\zeta) = \frac{1}{(4 - 2\zeta - \zeta^2)}, \ \zeta \in E.$$

Now we restore the analytic in E functions $F_1(\zeta)$ and $G_1(\zeta)$ according to the previous section via the boundary conditions

$$\mathrm{Re}[-\kappa F_1(\zeta) + \frac{\zeta^2(4 - 2\zeta - \zeta^2)^2}{2(1+\zeta)(4\zeta^2 - 2\zeta - 1)} F_1'(\zeta) + G_1(\zeta)]|_{\zeta=e^{\imath\theta}} =$$

$$= 6\mu\epsilon - \frac{3\mu b}{4(\lambda + 2\mu)} \mathrm{Re}[\frac{2(\lambda + 3\mu)}{|4 - 2\zeta - \zeta^2|^2} - \frac{(\lambda + \mu)}{(4 - 2\zeta - \zeta^2)^2}]|_{\zeta=e^{\imath\theta}},$$

and

$$\mathrm{Im}[-\kappa F_1(\zeta) - \frac{\zeta^2(4 - 2\zeta - \zeta^2)^2}{2(1+\zeta)(4\zeta^2 - 2\zeta - 1)} F_1'(\zeta) - G_1(\zeta)]|_{\zeta=e^{\imath\theta}} =$$

$$= \frac{\mu\epsilon}{4(\lambda + 2\mu)} \mathrm{Im}[\frac{(\lambda + \mu)}{(4 - 2\zeta - \zeta^2)^2}]|_{\zeta=e^{\imath\theta}}.$$

After we solve the Schwartz problems for the meromorphic functions

$$-\kappa F_1(\zeta) \pm \frac{\zeta^2(4 - 2\zeta - \zeta^2)^2}{2(1+\zeta)(4\zeta^2 - 2\zeta - 1)} F_1'(\zeta) \pm G_1(\zeta)$$

we have two additional constants C_1 and C_2 . The expressions of the functions $F_1(\zeta)$ and $G_1(\zeta)$ contain these unknown constants. After we find the residues of the function $G_1(\zeta)$ and equate them to null we find these constants C_1 and C_2 and get the analytic functions $F_1(\zeta)$ and $G_1(\zeta)$. We get the expression of the coefficient $u_1(x, y) + \imath v_1(x, y)$ in terms of the variable z after we apply the inverse mapping

$$\zeta = -1 + \sqrt{5 - \frac{1}{z}}.$$

The third coefficient $w_1(x, y)$ satisfies the equation

$$\Delta w_1 = 0,$$

so we have taking in account the boundary values

$$w_1(x, y) \equiv -\epsilon - 4e.$$

We present the final expressions of the complex displacements $u(x, y, h)+ \imath v(x, y, h)$ in XOY -plane and the displacements $w(x, y)$ over OH . These expressions contain the parameters a , b , e , which can be expressed through the given data α , γ and ν .

Finally the displacements which satisfy the given conditions and the equilibrium equations have the form

$$u(x, y, h) + \imath v(x, y, h) = (a\,(-8(-1 + 4z)\mu(\lambda + \mu)+$$
$$250\bar{z}^3(-1 + 5z)\mu(11\lambda + 27\mu) -$$
$$25\bar{z}^{5/2}\sqrt{-1 + 5\bar{z}}(-6 + 25z)\mu(11\lambda + 27\mu) -$$
$$4\sqrt{\bar{z}}\sqrt{-1 + 5\bar{z}}\,(275(-3 + h)(-1 + h)\lambda^2+$$
$$4\left(1113 + 375(-4 + h)h + 45z + 4\sqrt{z}\sqrt{-1 + 5z}\right)\lambda\mu+$$
$$\left(5979 + 2025(-4 + h)h + 340z + 48\sqrt{z}\sqrt{-1 + 5z}\right)\mu^2) +$$
$$2\bar{z}\left(1375(-3 + h)(-1 + h)\lambda^2 + 4\,(5567+\right.$$
$$\left.+1875(-4 + h)h + 265z + 20\sqrt{z}\sqrt{-1 + 5z}\right)\lambda\mu +$$
$$3\left(9989 + 3375(-4 + h)h + 620z + 80\sqrt{z}\sqrt{-1 + 5z}\right)\mu^2) +$$
$$10\bar{z}^2\left(-1375(-3 + h)(-1 + h)\lambda^2 - (22229+\right.$$
$$\left.7500(-4 + h)h + 1275z + 80\sqrt{z}\sqrt{-1 + 5z}\right)\lambda\mu -$$

$$\left(29848+10125(-4+h)h+2475z+240\sqrt{z}\sqrt{-1+5z}\right)\mu^2\big)+$$
$$5\overline{z}^{3/2}\sqrt{-1+5\overline{z}}\,\big(1375(-3+h)(-1+h)\lambda^2+$$
$$5\left(4441+1500(-4+h)h+236z+16\sqrt{z}\sqrt{-1+5z}\right)\lambda\mu+$$
$$\left(10125(-4+h)h+4\left(7444+575z+60\sqrt{z}\sqrt{-1+5z}\right)\right)\mu^2\big)\big)+$$
$$b\,\big(-24h(-1+4z)\mu(\lambda+\mu)+750h\overline{z}^3(-1+5z)\mu(11\lambda+27\mu)-$$
$$75h\overline{z}^{5/2}\sqrt{-1+5\overline{z}}(-6+25z)\mu(11\lambda+27\mu)-$$
$$4\sqrt{\overline{z}}\sqrt{-1+5\overline{z}}\,\big(275\left(12-13h+h^3\right)\lambda^2+$$
$$12\left(1500+h\left(-1637+125h^2+45z+4\sqrt{z}\sqrt{-1+5z}\right)\right)\lambda\mu+$$
$$3\left(8100+h\left(-8871+675h^2+340z+48\sqrt{z}\sqrt{-1+5z}\right)\right)\mu^2\big)+$$
$$2\overline{z}\,\big(1375\left(12-13h+h^3\right)\lambda^2+12\,\big(7500+$$
$$h\left(-8183+625h^2+265z+20\sqrt{z}\sqrt{-1+5z}\right)\big)\lambda\mu+$$
$$9\left(13500+h\left(-14761+1125h^2+620z+80\sqrt{z}\sqrt{-1+5z}\right)\right)\mu^2\big)+$$
$$5\overline{z}^{3/2}\sqrt{-1+5\overline{z}}\,\big(1375\left(12-13h+h^3\right)\lambda^2+$$
$$15\left(6000+h\left(-6559+500h^2+236z+16\sqrt{z}\sqrt{-1+5z}\right)\right)\lambda\mu+$$
$$3\left(40500+h\left(-44474+3375h^2+2300z+240\sqrt{z}\sqrt{-1+5z}\right)\right)\mu^2\big)+$$
$$10\overline{z}^2\,\big(-1375\left(12-13h+h^3\right)\lambda^2-$$
$$3\left(30000+h\left(-32771+2500h^2+1275z+80\sqrt{z}\sqrt{-1+5z}\right)\right)\lambda\mu-$$
$$3\left(40500+h\left(-44402+3375h^2+2475z+240\sqrt{z}\sqrt{-1+5z}\right)\right)\mu^2\big)\big)\big)\big/$$
$$\left(25\overline{z}\left(10-50\overline{z}-\frac{4\sqrt{-1+5\overline{z}}}{\sqrt{\overline{z}}}+25\sqrt{\overline{z}}\sqrt{-1+5\overline{z}}\right)(\lambda+3\mu)(11\lambda+27\mu)\right).$$

$$w(x,y,h)=\left(20e\left(\left(-3+\overline{z}\left(2\left(5+\sqrt{5-\frac{1}{\overline{z}}}\right)-25z\right)+\right.\right.\right.$$
$$\left.2\left(5+\sqrt{5-\frac{1}{z}}\right)z\right)\lambda+2\left(-100h+25h^2+\right.$$
$$\left.\left.\overline{z}\left(2\left(5+\sqrt{5-\frac{1}{\overline{z}}}\right)-25z\right)+2\left(36+\left(5+\sqrt{5-\frac{1}{z}}\right)z\right)\right)\mu\right)$$
$$\left(11\lambda^2+60\lambda\mu+81\mu^2\right)-\mu\left(3b(\lambda+\mu)\right.$$
$$\left(\left(208+55\overline{z}^2\left(2\left(5+\sqrt{5-\frac{1}{\overline{z}}}\right)-25z\right)-\left(895+102\sqrt{5-\frac{1}{z}}\right)z+\right.\right.$$
$$\left.\left.110\left(5+\sqrt{5-\frac{1}{z}}\right)z^2-\overline{z}\left(895+102\sqrt{5-\frac{1}{\overline{z}}}-3200z+1375z^2\right)\right)\lambda+$$
$$\left(464+135\overline{z}^2\left(2\left(5+\sqrt{5-\frac{1}{\overline{z}}}\right)-25z\right)-\left(1935+326\sqrt{5-\frac{1}{z}}\right)z+\right.$$
$$270\left(5+\sqrt{5-\frac{1}{z}}\right)z^2+\overline{z}\left(-1935-\right.$$
$$326\sqrt{5-\frac{1}{\overline{z}}}+320\left(20+\sqrt{5-\frac{1}{\overline{z}}}+\sqrt{5-\frac{1}{z}}\right)z-$$
$$\left.\left.\left.3375z^2\right)\right)\mu\right)+1000(-2+h)\epsilon\left(11\lambda^2+60\lambda\mu+81\mu^2\right)\right)\Big)/$$
$$\left(1000\mu\left(11\lambda^2+60\lambda\mu+81\mu^2\right)\right).$$

Note that the values of the displacements in the plane XOY of the points of the furrow can be found by the limit application and equal to

$$u(0.2,0,h)+\imath v(0.2,0,h)=(h-3)(h-1)[b(4+h)+a].$$

It seems rather interesting to examine the conditions of crack growth at the points of the furrow. The boundary of the domain D has the cusp at the point $(0.2,0)$, so the stress tensor components have the singularities at any point $(0.2,0,h_0), h_0 \in [1,3]$, namely,

$$\sigma_{ij} = \frac{K_{ij}}{\sqrt{(x-0.2)^2+y^2}},$$

where

$$K_{11} = \frac{\mu^2 0.32(a+3bh_0)[\cos\frac{5\tau}{2}(\lambda+\mu)+\cos\frac{\tau}{2}(7\lambda+11\mu)]}{(\lambda+3\mu)(11\lambda+27\mu)},$$

$$K_{22} = \frac{\mu^2 0.32(a+3bh_0)[-\cos\frac{5\tau}{2}(\lambda+\mu)+\cos\frac{\tau}{2}(\lambda-3\mu)]}{(\lambda+3\mu)(11\lambda+27\mu)},$$

$$K_{12} = \frac{\mu^2 0.32(a+3bh_0)[\sin\frac{5\tau}{2}(\lambda+\mu)+\cos\frac{\tau}{2}(3\lambda+7\mu)]}{(\lambda+3\mu)(11\lambda+27\mu)},$$

$$K_{33} = \frac{\cos\frac{\tau}{2}\,\lambda\mu^2 1.28(a+3bh_0)}{(\lambda+3\mu)(11\lambda+27\mu)},$$

$$K_{13} = \frac{2\cos\frac{\tau}{2}}{(\lambda+3\mu)(11\lambda+27\mu)}\left[b\mu(0.12\lambda^2+0.432\lambda\mu+0.312\mu^2)+\right.$$
$$\left.+e(0.22\lambda^3+1.64\lambda 2\mu+4.02\lambda\mu^2+3.24\mu^3)\right],$$

$$K_{23} = \frac{2\sin\frac{\tau}{2}}{(\lambda+3\mu)(11\lambda+27\mu)}\left[b\mu(0.12\lambda^2+0.432\lambda\mu+0.312\mu^2)+\right.$$
$$\left.+e(0.22\lambda^3+1.64\lambda 2\mu+4.02\lambda\mu^2+3.24\mu^3)\right],$$

and τ is the polar angle:

$$(x-0.2)+\imath y = \sqrt{(x-0.2)^2+y^2}e^{\imath\tau}.$$

The normal stress tensor component can be obtained by the well-known expression

$$\sigma_{nn} = \sigma_{11}\cos\alpha+\sigma_{22}\cos\beta+\sigma_{33}\cos\gamma+2\sigma_{12}\cos\alpha\cos\beta+$$
$$2\sigma_{13}\cos\alpha\cos\gamma+2\sigma_{23}\cos\beta\cos\gamma, \qquad (4.7)$$

where $\vec{n} = (\cos\alpha,\cos\beta,\cos\gamma)$ is the direction vector.

Here the normal stress tensor component σ_{nn} also has singularities at the points $(0.2, 0, h_0)$ and can be presented as

$$\sigma_{nn} = \frac{K_{nn}}{\sqrt{(x-0.2)^2+y^2}}. \tag{4.8}$$

We can not apply expression (4.7) in order to find the directions of the initial crack growth from the points of furrow, so we search for the plane along which the cylinder may crack from any point $(0.2, 0, h_0)$, $h_0 \in [1,3]$, with the help of the following considerations.

Let $\vec{r}$ be the vector with the origin at $(0.2, 0, h_0)$ which lies in the plane along which the cylinder can tear, the spherical coordinates of this vector being r, τ, ω, so

$$\vec{r} = (r\cos\tau\cos\omega, r\sin\tau\cos\omega, r\sin\omega)\ , \tau \in [-\pi, \pi], \omega \in [-\pi/2, \pi/2].$$

Then the normal to this plane vector $\vec{n} = (l, m, k)$ has the coordinates

$$l(\tau,\omega,\sigma) = \frac{\cos\sigma\tan\omega}{\sqrt{(\tan\omega)^2+(\cos(\sigma-\tau))^2}},$$
$$m(\tau,\omega,\sigma) = \frac{\sin\sigma\tan\omega}{\sqrt{(\tan\omega)^2+(\cos(\sigma-\tau))^2}},$$
$$k(\tau,\omega,\sigma) = \frac{-\cos(\sigma-\tau)}{\sqrt{(\tan\omega)^2+(\cos(\sigma-\tau))^2}},\ \sigma \in [-\pi, \pi].$$

We obtain the plane along which the bent cylinder can be torn if we know the values of τ, ω and σ. These values must be the coordinates of the critical points of the factor K_{nn} from expression (4.8). Since

$$K_{nn} = l^2K_{11} + m^2K_{22} + k^2K_{33} + 2lmK_{12} + 2lkK_{13} + 2mkK_{23},$$

here K_{nn} has the following form in terms of τ, ω and σ:

$$K_{nn}(\tau,\omega,\sigma)=\frac{2(0.64a+1.92bh_0)\lambda\mu^2\cos^2(\sigma-\tau)\cos\frac{\tau}{2}}{(\lambda+3\mu)(11\lambda+27\mu)\left(\cos^2(\sigma-\tau)+\tan^2\omega\right)}-$$
$$-4\left(b\mu\left(0.12\lambda^2+0.432\lambda\mu+0.312\mu^2\right)+e\left(0.22\lambda^3+1.64\lambda^2\mu+\right.\right.$$
$$\left.\left.+4.02\lambda\mu^2+3.24\mu^3\right)\right)\frac{\cos(\sigma-\tau/2)\cos(\sigma-\tau)\tan\omega}{(\lambda+3\mu)(11\lambda+27\mu)\left(\cos^2(\sigma-\tau)+\tan^2\omega\right)}+$$
$$+\frac{0.32(a+3bh_0)\mu^2\tan^2\omega}{(\lambda+3\mu)(11\lambda+27\mu)\left(\cos^2(\sigma-\tau)+\tan^2\omega\right)}\left[\cos^2\sigma\left((7\lambda+11\mu)\cos\frac{\tau}{2}+\right.\right.$$
$$\left.+(\lambda+\mu)\cos\frac{5\tau}{2}\right)+\sin^2\sigma\left((\lambda-3\mu)\cos\frac{\tau}{2}-(\lambda+\mu)\cos\frac{5\tau}{2}\right)+$$
$$\left.+\sin2\sigma\left((3\lambda+7\mu)\sin\frac{\tau}{2}+(\lambda+\mu)\sin\frac{5\tau}{2}\right)\right]$$

It can be easily verified that the values $\tau=0,\sigma=\pi/2$ and arbitrary ω provide nulls for the first derivatives of $K_{nn}(\tau,\omega,\sigma)$. Note that

$$K_{nn}(0,\omega,\pi/2)=-\frac{1.28(a+3bh_0)\mu^3}{(\lambda+3\mu)(11\lambda+27\mu)},$$

hence the cylinder does not destroy if $a+3bh_0>0$ at the point $(0.2,0,h_0)$ of the furrow along the plane $y=0$. Since

$$a+3bh_0=k[\gamma(h_0-1)+\alpha(3-h_0)],$$

where α and γ are the given displacements and $k\approx2.876(\lambda+3\mu)/\mu$ we arrive to the following conclusion: the crack over the plane $y=0$ can not grow from the points $(0.2,0,h_0)$ for which

$$\gamma(h_0-1)+\alpha(3-h_0)>0.$$

CHAPTER 5

Spline-interpolation solutions for the circular cylinder and for the tube

Elena A. Shirokova,
Kazan Federal University
Elena.Shirokova@ksu.ru

Pyotr N. Ivanshin,
Kazan Federal University
pivanshin@gmail.com

ABSTRACT: In this chapter we construct the spline-interpolation solution for the cylindrical solids.

5.1 Spline-interpolation solution for the circular cylinder

Suppose that we have to restore the displacements and the stresses at the interior points of the elastic circular cylinder via the given displacements at the ends of the cylinder and at the points of an arbitrary finite number of circles (directrices) on the surface of the cylinder. The interpolation solution would be rather awkward in this case if the number of the given directrices is large. We should apply the spline-interpolation solution for this problem.

5.1.1 Gluing of the inner points of adjacent sections

The spline-interpolation solution (SIS) is based on the method of interpolation solution. The gist of this method is following: we divide the cylinder into the finite number of cylindrical fragments by the planes which contain the given directrices. We begin to construct our solution with the fragment which borders on one of the ends of the cylinder. So the displacements at one end and at the edge of the other end of this fragment are given and we construct the solution of the equilibrium equations which coincides with the given displacements at one end and at the edge of the other end as it is done in 2.5.3.

Then we pass to the next fragment. We construct SIS for this fragment by the same method, the given displacements at the end common with the previous fragment being those which have been found at the previous step. So we glue the adjacent fragments over the common section....

Here the problem is how to construct SIS for the final fragment. We have to restore for this final fragment the displacements which satisfy the equilibrium equations and the given data at two ends of the fragment. Our

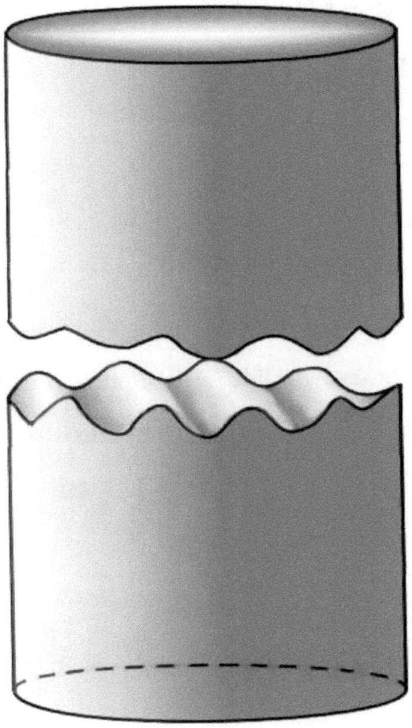

Figure 5.1: Gluing the displacements of the cylindrical fragments over the common end

method fails to construct such solution as it is mentioned in Chapter 2. So we can satisfy the data at one end, at the edge of the other end and at a finite number of points of the second end. Otherwise we can approximate the given data at the interior points of the both ends of the final fragment as it is described in 2.4.3.

We construct the solution in the form of a polynomial in h . The power of this polynomial is dependent on the order of harmonicity of the data at the ends of the cylinder. We get the additional set of 3 functions (u_k, v_k, w_k) for each additional degree of h . The harmonicity of the solution increases by 1 with every 2 additional degrees of h .

For example if we want to construct the solution that equals the given biharmonic function at the end of the solid and meets the boundary conditions at l circles on the boundary surface of it, we must consider the solution as polynomial of degree not less than 3 in the variable h. So we consider the solution in the form

$$(u + \imath v)(z, \overline{z}, h) = \sum_{k=0}^{3} (u_k + \imath v_k)(z, \overline{z}) h^k,$$

$$w(z, \overline{z}, h) = \sum_{k=0}^{3} w_k(z, \overline{z}) h^k.$$

All the coefficients in the series representations of $u + \imath v$ and w then are restored from the system of Appendix 1.

If the boundary data at the end of the cylinder is the function of a high order of harmonicity we must search for the solution in the form of a polynomial of higher power in h. Then we have to solve more difficult system with respect to the coefficients of the expansions of the functions $u_k + \imath v_k$ and w_k.

We have to raise the power of the SIS for the fragments beginning with the second one if we search not only for a continuous but also for a smooth at the given surface directrices solution. Indeed, the solution must now additionally to the data at one end and at the edge of the other end also satisfy the derivative condition at the edge of the first end, the derivative being taken with respect to h. So the power of the polynomial in h increases by 1.

The order of the smoothness of the solution on the boundary surface influences the power of the polynomial. If this order increases by 1 then the power of the polynomial increases by 1.

5.1.2 Smooth spline-interpolation

In this section we construct the spline-interpolation solution which is smooth at the boundary points. The spline must glue not only the displacements at the points of the edge of the common section of two adjacent layers but also the values of their first n derivatives with respect to h at this edge.

We consider a layer in the circular cylinder M bounded by two planes parallel to XOY. Without loss of generality we assume that the lower level is given by the section of the cylinder by the plane $h = 0$ and the upper level is the section of M by $h = h_0$. We must construct the solution of system (1.1) which meets the given displacements at the edges at the levels $h = 0$ and $h = h_0$ and also the given values of the first n derivatives with respect to h at the edge at the level $h = 0$. So we have the following system of boundary conditions:

$$u(x, y, h_0)|_{x^2+y^2=1} = \sum_{j=0}^{m} (\alpha_{0,j} \cos j\theta + \dot{\alpha}_{0,j} \sin j\theta),$$

$$v(x, y, h_0)|_{x^2+y^2=1} = \sum_{j=0}^{m} (\beta_{0,j} \cos j\theta + \dot{\beta}_{0,j} \sin j\theta),$$

$$w(x, y, h_0)|_{x^2+y^2=1} = \sum_{j=0}^{m} (\gamma_{0,j} \cos j\theta + \dot{\gamma}_{0,j} \sin j\theta);$$

$$u(x, y, 0)|_{x^2+y^2=1} = \sum_{j=0}^{m} (a_{0,j} \cos j\theta + A_{0,j} \sin j\theta),$$

$$v(x, y, 0)|_{x^2+y^2=1} = \sum_{j=0}^{m} (b_{0,j} \cos j\theta + B_{0,j} \sin j\theta),$$

$$w(x, y, 0)|_{x^2+y^2=1} = \sum_{j=0}^{m} (c_{0,j} \cos j\theta + C_{0,j} \sin j\theta);$$

$$\frac{\partial^k}{\partial h^k} u(x, y, h)|_{x^2+y^2=1, h=0} = \sum_{j=0}^{m} (a_{k,j} \cos j\theta + A_{k,j} \sin j\theta),$$

$$\frac{\partial^k}{\partial h^k} v(x, y, h)|_{x^2+y^2=1, h=0} = \sum_{j=0}^{m} (b_{k,j} \cos j\theta + B_{k,j} \sin j\theta),$$

$$\frac{\partial^k}{\partial h^k} w(x, y, h)|_{x^2+y^2=1, h=0} = \sum_{j=0}^{m} (c_{k,j} \cos j\theta + C_{k,j} \sin j\theta), \ k = 1, \ldots, n$$

Let us search for the solution in the form

$$u(x,y,h)=\sum_{j=0}^{n+1}u_j(x,y)h^j,\ v(x,y,h)=\sum_{j=0}^{n+1}v_j(x,y)h^j,$$

$$w(x,y,h)=\sum_{j=0}^{n+1}w_j(x,y)h^j.$$

Then we get three systems with respect to the boundary values of the unknown functions u_j , v_j and w_j , $j=0,\dots,n+1$:

$$A\begin{pmatrix}u_0\\ \vdots\\ u_{n+1}\end{pmatrix}=\begin{pmatrix}\sum\limits_{j=0}^{m}(\alpha_{0,j}\cos j\theta+\dot{\alpha}_{0,j}\sin j\theta)\\ \sum\limits_{j=0}^{m}(a_{0,j}\cos j\theta+A_{0,j}\sin j\theta)\\ \vdots\\ \sum\limits_{j=0}^{m}(a_{n,j}\cos j\theta+A_{n,j}\sin j\theta)\end{pmatrix};$$

$$A\begin{pmatrix}v_0\\ \vdots\\ v_{n+1}\end{pmatrix}=\begin{pmatrix}\sum\limits_{j=0}^{m}(\beta_{0,j}\cos j\theta+\dot{\beta}_{0,j}\sin j\theta)\\ \sum\limits_{j=0}^{m}(b_{0,j}\cos j\theta+B_{0,j}\sin j\theta)\\ \vdots\\ \sum\limits_{j=0}^{m}(b_{1,j}\cos j\theta+B_{1,j}\sin j\theta)\end{pmatrix};$$

$$A\begin{pmatrix}w_0\\ \vdots\\ w_{n+1}\end{pmatrix}=\begin{pmatrix}\sum\limits_{j=0}^{m}(\gamma_{0,j}\cos j\theta+\dot{\gamma}_{0,j}\sin j\theta)\\ \sum\limits_{j=0}^{m}(c_{0,j}\cos j\theta+C_{0,j}\sin j\theta)\\ \vdots\\ \sum\limits_{j=0}^{m}(c_{n,j}\cos j\theta+C_{n,j}\sin j\theta)\end{pmatrix}.$$

Here all these systems have the same matrix:

$$A = \begin{pmatrix} 1 & h_0 & h_0^2 & \dots & h_0^n & h_0^{n+1} \\ 1 & 0 & 0 & \dots & 0 & 0 \\ 0 & 2 & 0 & \dots & 0 & 0 \\ \dots & \dots & \dots & \dots & \dots & \dots \\ 0 & 0 & 0 & \dots & n! & 0 \end{pmatrix}$$

Clearly this matrix is non-degenerate since

$$|\mathrm{det} A| = h_0^{n+1} \prod_{k=1}^{n} k!,$$

hence all three systems of boundary equations have unique solutions.

We construct the spline using the same procedure as in Section 2.1. We reconstruct $(u_k + \imath v_k)(z, \overline{z})$, $w_k(z, \overline{z})$, $k = 0, \dots, n+1$ by their boundary values beginning with the highest numbers.

5.1.3 Gluing the inner points within the smooth spline-interpolation

Suppose that we need to find the solution of static elasticity theory problem which meets the boundary conditions given as m -harmonic functions at the ends of the solid and also whose first n derivatives with respect to h are continuous. So we need to construct the combination of the splines introduced before. Then we must consider solution in the form of the polynomials in h of degree $m + n$.

The matrix of the corresponding system then consists of two sets of relations. The first set is similar to one given in the previous section:

$$A = \begin{pmatrix} 1 & h_0 & h_0^2 & \dots & h_0^{n+m-1} & h_0^{n+m} \\ m! & 0 & 0 & \dots & 0 & 0 \\ 0 & (m+1)! & 0 & \dots & 0 & 0 \\ \dots & \dots & \dots & \dots & \dots & \dots \\ 0 & 0 & 0 & \dots & (n+m)! & 0 \end{pmatrix}$$

Also we have the other set of equations similar to one presented in Appendix 1. Assume that

$$(u + \imath v)(z, \overline{z}, 0) = \sum_{j=0}^{m} (F_j(z) + G_j(\overline{z}))|z|^{2j},$$

$$w(z, \overline{z}, 0) = \sum_{j=0}^{m} (H_j(z) + \overline{H_j}(\overline{z}))|z|^{2j}.$$

Let us assume that the displacements functions have the form

$$(u + \imath v)(x, y, 0) = \sum_{j=0}^{m} (u_j + \imath v_j)|z|^{2j}$$

and

$$w(x, y, 0) = \sum_{j=0}^{m} w_j |z|^{2j}$$

So in order to get the second set of equations we simply equate the multiples of the common degrees of $|z|^{2j}$

$$(u_j + \imath v_j) = F_j(z) + G_j(\overline{z}),$$

$$w_j(z) = H_j(z) + \overline{H_j}(\overline{z}).$$

5.2 Spline-interpolation solution for the tube

5.2.1 The scheme of the solution

The interpolation solution with given exterior displacements for the elastic tube presented in Chapter 3 is the base of the spline-interpolation solution presented in this section. The scheme of spline-interpolation solution was published in [12]. Here we also present the relations and the linear systems necessary for computer calculations.

Let the segment $[A, B]$ be the projection of the tube Ω on the coordinate axis OH. Let this segment be the union of the finite number of segments: $[A, B] = \cup_{j=0}^{n}[h_j, h_{j+1}]$. Let us find the components of the vector of displacements (u, v, w) which satisfy the equilibrium equations in the solid $\Omega_j = \{(x, y, h)|r < x^2 + y^2 < 1, \; h \in (h_j, h_{j+1})\}$, the displacements at the points of the circles

$$C_j = \{x^2 + y^2 = 1, \; h = h_j\}, \; C_{j+1} = \{x^2 + y^2 = 1, \; h = h_{j+1}\},$$
$$l_j = \{x^2 + y^2 = r, \; h = h_j\}$$

being given (the displacements at the points of the circle l_j being obtained after solution of the same problem for the previous solid fragment Ω_{j-1}). We assume a constant pressure p acting at the interior surface of Ω_j. We also suppose that the sliding condition holds on the interior surface of Ω_j. If we solve this problem for each Ω_j, $j = 0, 1, ..., n-1$, we can construct the components (u, v, w) which satisfy the equilibrium equation at the points of each Ω_j $j = 0, 1, , n-1$, satisfy the given displacements at the points of the circles C_j $j = 0, 1, ..., n$, satisfy the sliding condition and the given constant pressure p on the interior surface of Ω, the vector (u, v, w) being continuous on the interior and on the exterior surface of Ω. These components provide the spline-interpolation solution of the problem of reconstruction of the displacements at the elastic pressurised tube via the given exterior displacements on the circles C_j , $j = 0, 1, ..., n$. The solution fails to be continuous at the interior points of the rings $\{r < x^2 + y^2 < 1,\ h = h_j\}$

Let the given displacements on the circles C_m , be

$$(\hat{u}_m(\theta), \hat{v}_m(\theta), \hat{w}_m(\theta)),\ \ m = j, j+1,$$

let the known displacements on the circle l_j which have been obtained for the previous Ω_{j-1} be $(\tilde{u}_j(\theta), \tilde{v}_j(\theta), \tilde{w}_j(\theta))$ where

$$\hat{u}_m(\theta) + \imath\hat{v}_m(\theta) = \sum_{k=-\infty}^{+\infty} t_k^m e^{\imath k\theta},$$

$$\hat{w}_m(\theta) = \frac{g_0^m}{2} + \sum_{k=1}^{+\infty} g_k^m \cos k\theta + f_k^m \sin k\theta, \tag{5.1}$$

$$\tilde{u}_j(\theta) + \imath\tilde{v}_j(\theta) = \sum_{k=-\infty}^{+\infty} R_k^j e^{\imath k\theta},$$

$$\tilde{w}_j(\theta) = \frac{P_0^j}{2} + \sum_{k=1}^{+\infty} P_k^j \cos k\theta + Q_k^j \sin k\theta. \tag{5.2}$$

We look for the components which solve the problem for Ω_j in the form

$$\begin{aligned} u(x,y,h) &= u_0(x,y) + u_1(x,y)h + u_2(x,y)h^2, \\ v(x,y,h) &= v_0(x,y) + v_1(x,y)h + v_2(x,y)h^2, \\ w(x,y,h) &= w_0(x,y) + w_1(x,y)h + w_2(x,y)h^2, \end{aligned} \tag{5.3}$$

Let us represent the boundary values of the coefficients with the biggest number at the exterior circle as follows

$$u_2(\cos\theta, \sin\theta) + \imath v_2(\cos\theta, \sin\theta) = \sum_{k=-\infty}^{+\infty} \delta_k e^{\imath k\theta},$$

$$w_2(\cos\theta, \sin\theta) = \frac{\sigma_0}{2} + \sum_{k=1}^{+\infty} \sigma_k \cos\theta + \nu_k \sin\theta. \tag{5.4}$$

Now we obtain the boundary values of the other coefficients at the exterior circle from (5.1) and (5.4):

$$u_1(\cos\theta,\sin\theta)+\imath v_2(\cos\theta,\sin\theta)=\sum_{k=-\infty}^{+\infty}\left[\frac{t_k^j-t_k^{j+1}}{h_j-h_{j+1}}-\right.$$
$$\left.-(h_j+h_{j+1})\delta_k\right]e^{\imath k\theta},$$
$$w_1(\cos\theta,\sin\theta)=\frac{g_0^j-g_0^{j+1}}{2(h_j-h_{j+1})}- \qquad (5.5)$$
$$-\frac{\sigma_0(h_j+h_{j+1})}{2}+\sum_{k=1}^{+\infty}\left[\frac{g_k^j-g_k^{j+1}}{h_j-h_{j+1}}-(h_j+h_{j+1})\sigma_k\right]\cos\theta+$$
$$+\left[\frac{f_k^j-f_k^{j+1}}{h_j-h_{j+1}}-(h_j+h_{j+1})\nu_k\right]\sin\theta,$$

$$u_0(\cos\theta,\sin\theta)+\imath v_0(\cos\theta,\sin\theta)=\sum_{k=-\infty}^{+\infty}\left[\frac{h_jt_k^{j+1}-h_{j+1}t_k^j}{h_j-h_{j+1}}+\right.$$
$$\left.+h_jh_{j+1}\delta_k\right]e^{\imath k\theta},$$
$$w_0(\cos\theta,\sin\theta)=\frac{h_jg_0^{j+1}-h_{j+1}g_0^j}{2(h_j-h_{j+1})}-\frac{\sigma_0h_jh_{j+1}}{2}+ \qquad (5.6)$$
$$+\sum_{k=1}^{+\infty}\left[\frac{h_jg_k^{j+1}-h_{j+1}g_k^j}{h_j-h_{j+1}}+h_jh_{j+1}\sigma_k\right]\cos k\theta+$$
$$+\left[\frac{h_jf_k^{j+1}-h_{j+1}f_k^j}{h_j-h_{j+1}}+h_jh_{j+1}\nu_k\right]\sin k\theta,$$

Therefore we can find the coefficients $u_k(x,y)$, $v_k(x,y)$ and $w_k(x,y)$, $r^2\le x^2+y^2\le 1$, $k=0,1,2$, according to the method described in Chapter 3 with the help of the boundary conditions (3.6), (3.7), (5.4), (5.6) and (5.7). We restore the coefficients step-by-step beginning with the coefficients $u_2(x,y)$, $v_2(x,y)$ and $w_2(x,y)$ with the help of the representations

$$u_k(x,y)+\imath v_k(x,y)=-\frac{1}{2\mu}(-\kappa\phi_k(z)+z\overline{\phi_k'(z)}+\overline{\psi_k(z)}+\Phi_k(z,\overline{z}),$$
$$w_k(x,y)=\operatorname{Re}\rho_k(z)+\Psi_k(z,\overline{z}),$$

using the analytic in the ring $r < |z| < 1$ functions $\phi_k(z)$, $\psi_k(z)$ and $\rho_k(z)$, $k = 0, 1, 2$. These analytic functions have the following expansions

$$\phi_2(z) = \sum_{m=-\infty}^{+\infty} \gamma_m^2 z^m, \; \psi_2(z) = \sum_{m=-\infty, m\neq 0}^{+\infty} c_m^2 z^m,$$
$$\rho_2(z) = \sum_{m=-\infty}^{+\infty} D_m^2 z^m,$$
$$\phi_1(z) = \beta_1 \ln z + \sum_{m=-\infty}^{+\infty} \gamma_m^1 z^m,$$
$$\psi_1(z) = b_1 \ln z + \sum_{m=-\infty, m\neq 0}^{+\infty} c_m^1 z^m,$$
$$\rho_1(z) = B_1 \ln z + \sum_{m=-\infty}^{+\infty} D_m^1 z^m, \tag{5.7}$$
$$\phi_0(z) = \alpha_0 z \ln z + \beta_0 \ln z + \sum_{m=-\infty}^{+\infty} \gamma_m^0 z^m,$$
$$\psi_0(z) = a_0 z \ln z + b_0 \ln z + \sum_{m=-\infty, m\neq 0}^{+\infty} c_m^0 z^m,$$
$$\rho_0(z) = A_0 z \ln z + B_0 \ln z + \sum_{m=-\infty}^{+\infty} D_m^0 z^m.$$

and can be restored consequently as in Section 3.2 via the boundary conditions

$$[\kappa\phi_k(z) + z\overline{\phi_k'(z)} + \overline{\psi_k(z)}]_{z=\exp(\imath\theta)} =$$
$$-2\mu \sum_{l=-\infty}^{\infty} \delta_{l,k} e^{\imath l\theta} - \Phi_k(e^{\imath\theta}, e^{-\imath\theta}),$$
$$(-2\overline{z}\,\mathrm{Re}\,\phi_k'(z) + |z|^2\phi_k''(z) + z\psi_k'(z))_{z=r\exp(\imath\theta)} = \tag{5.8}$$
$$= [-m_k\overline{z} - (1+\frac{\lambda}{\mu})\overline{z}\,\mathrm{Re}\frac{\partial\Phi_k}{\partial z} -$$
$$-z\frac{\partial\overline{\Phi_k}}{\partial z} + \overline{z}\lambda(k+1)(\mathrm{Re}\,\rho_{k+1}(z) + \Psi_{k+1}(z,\overline{z})]_{z=r\exp(\imath\theta)},$$

and

$$\mathrm{Re}\,\rho_k(e^{\imath\theta}) = \frac{\sigma_{0,k}}{2} + \sum_{l=1}^{\infty} \sigma_{l,k}\cos l\theta + \nu_{l,k}\sin l\theta - \Psi_k(e^{\imath\theta}, e{-}\imath\theta),$$
$$\mathrm{Re}(re^{\imath\theta}\rho_k'(re^{\imath\theta})) = -\left[z\frac{\partial\Psi_k}{\partial z} + \overline{z}\frac{\partial\Psi_k}{\partial\overline{z}}\right]_{z=r\exp(\imath\theta)} + \tag{5.9}$$
$$+\frac{k+1}{2\mu}\,\mathrm{Re}\left[\overline{z}(-\kappa\phi_{k+1}(z) + z\overline{\phi_k'(z)} + \right.$$
$$\left. +\overline{\psi_{k+1}(z)} + \Psi_{k+1}(z,\overline{z}))\right]_{z=r\exp(\imath\theta)}.$$

Here we have due to (5.6) and (5.7) the following notation:

$$\delta_{l,2} = \delta_l,\ \delta_{l,1} = \frac{t_l^j - t_l^{j+1}}{h_j - h_{j+1}} - (h_j + h_{j+1})\delta_l,$$
$$\delta_{l,0} = \frac{h_j t_l^{j+1} - h_{j+1}t_l^j}{h_j - h_{j+1}} + h_j h_{j+1}\delta_l,\ m_0 = p,\ m_1 = m_2 = 0,$$
$$\sigma_{l,1} = \frac{g_l^j - g_l^{j+1}}{h_j - h_{j+1}} - (h_j + h_{j+1})\sigma_l,\ \nu_{l,1} = \frac{f_l^j - f_l^{j+1}}{h_j - h_{j+1}} - (h_j + h_{j+1})\nu_l,$$
$$\sigma_{l,0} = \frac{h_j g_l^{j+1} - h_{j+1}g_l^j}{h_j - h_{j+1}} + h_j h_{j+1}\sigma_l,\ \nu_{l,0} = \frac{h_j f_l^{j+1} - h_{j+1}f_l^j}{h_j - h_{j+1}} + h_j h_{j+1}\nu_l.$$

The expressions of some coefficients and the linear systems over the coefficients of expansions (5.8) and also the functions $\Phi_k(z,\overline{z})$ and $\Psi_k(z,\overline{z})$, $k = 1, 0$, are presented in the following subsection. These expressions are rather cumbersome but they can be successfully applied to computer calculations.

The analytic functions $\phi_k(z)$, $\psi_k(z)$, $\rho_k(z)$, $k = 0, 1, 2$ and the functions $\Phi_k(z, \overline{z})$ and $\Psi_k(z, \overline{z})$, $k = 1, 0$, are lineally dependent on the coefficients δ_m , σ_m and ν_m of the representation introduced above (5.4).

The unknown coefficients δ_m , σ_m and ν_m we find after we substitute the functions $u(x, y, h)$, $v(x, y, h)$ and $w(x, y, h)$ in the form (5.4) in the relations (5.2) and obtain the equalities

$$\tilde{u}_j(\theta) + \imath\tilde{v}_j(\theta) = u(r\cos\theta, r\sin\theta, h_j) + \imath v(r\cos\theta, r\sin\theta, h_j),$$
$$\tilde{w}_j(\theta) = w(r\cos\theta, r\sin\theta, h_j).$$

We compare the coefficients of the Fourier series in the right and in the left hand sides of the relations in the following order. After we equate the coefficient at $\exp i\theta$ of the function $u(r\cos\theta, r\sin\theta, h_j) + \imath v(r\cos\theta, r\sin\theta, h_j)$ to R_1^j and equate the free term of the function $w(r\cos\theta, r\sin\theta, h_j)$ to $P_0^j/2$ we obtain the coefficients δ_1 and σ_0 .

After we equate the coefficients at $\exp \imath k\theta$ and $\exp \imath(2-k)\theta$ of the function $u(r\cos\theta, r\sin\theta, h_j) + \imath v(r\cos\theta, r\sin\theta, h_j)$ to R_k^j and to R_{2-k}^j respectively and equate the coefficients at $\cos(k-1)\theta$ and at $\sin(k-1)\theta$ of the function $w(r\cos\theta, r\sin\theta, h_j)$ to P_{k-1}^j and to Q_{k-1}^j respectively we obtain the coefficients δ_k , δ_{2-k} , σ_{k-1} and ν_{k-1} , $k = 2, 3, \ldots$.

The corresponding systems are also presented below in the following subsection after the next one.

The number of non-zero coefficients δ_k , σ_k and ν_k is finite if the functions $\hat{u}_j(\theta) + \imath\hat{v}_j(\theta)$, $\hat{w}_j(\theta)$, $\hat{w}_j(\theta)$, $\tilde{u}_j(\theta) + \imath\tilde{v}_j(\theta)$ and $\tilde{w}_j(\theta)$ from (5.1) and (5.2) are represented with the finite Fourier series.

We must look for the solution in the form

$$u(x,y,h)=\sum_{p=0}^{l}u_p(x,y)h^p,\quad v(x,y,h)=\sum_{p=0}^{l}v_p(x,y)h^p,$$

$$w(x,y,h)=\sum_{p=0}^{l}w_p(x,y)h^p,$$

with $l\geq 3$ in order to find the spline-interpolation solution of the problem which is smooth on the exterior surface. We can provide the known values of the derivatives of all orders till $l-2$ with respect to h for the functions u, v and w at the curve C_j. Then we obtain the boundary values of the functions $u_p(x,y)$, $v_p(x,y)$ and $w_p(x,y)$, $p=0,1,...,l$, and solve the problem according to the method of interpolation solution.

5.2.2 The relations and systems over the coefficients of the analytic functions

Here we put all systems and relations connected with construction of the spline-interpolation solution for the tube. The main ideas are presented in the previous subsection and are based on the interpolation solution of the second basic problem for a hollow cylinder presented in Chapter 3. The systems have to be solved consequently.

We have $\Phi_2(z,\overline{z})=\Psi_2(z,\overline{z})\equiv 0$ and we restore the functions $\Phi_k(z,\overline{z})$ and $\Phi_k(z,\overline{z})$, $k=1,0$, below.

At the first step we restore $\phi_2(z)$ and $\psi_2(z)$ using the boundary conditions (5.9) with $k=2$.

After we compare the factors with the corresponding powers of $e^{\imath\theta}$ and have the following linear systems and relations (just as in Section 3.2) which bring the coefficients γ_m^2 and c_m^2 :

$$\begin{cases} -\kappa\gamma_2^2 + \overline{c_2^2} = -2\mu\delta_2, \\ -2\gamma_2^2 r^2 - 2\overline{c_2^2} r^{-2} = 0, \end{cases}$$

$$-\kappa\gamma_0^2 = -2\overline{\gamma_2^2} - 2\mu\delta_0,$$

$$\begin{cases} -\kappa\gamma_1^2 + \overline{\gamma_1^2} + \overline{c_{-1}^2} = -2\mu\delta_1, \\ -\gamma_1^2 r - \overline{\gamma_1^2} r - \overline{c_{-1}^2} r^{-1} = 0, \end{cases}$$

and for $m > 2$ we have

$$\begin{cases} -\kappa\gamma_m^2 + (2-m)\overline{\gamma_{2-m}^2} + \overline{c_{-m}^2} = -2\mu\delta_m, \\ -m\overline{\gamma_m^2} r^m - m(2-m)\gamma_{2-m}^2 r^{2-m} - mc_{-m}^2 r^{-m} = 0, \\ -\kappa\gamma_{2-m}^2 + m\overline{\gamma_m^2} + \overline{c_{m-2}^2} = -2\mu\delta_{2-m}, \\ -(2-m)\overline{\gamma_{2-m}^2} r^{2-m} + m(m-2)\gamma_m^2 r^m + (m-2)c_{m-2}^2 r^{m-2} = 0. \end{cases}$$

So we have the following representations of the coefficients:

$$\gamma_2^2 = \frac{2\mu\delta_2}{\kappa + r^4},$$

$$\overline{c_{-2}^2} = -\frac{2\mu\delta_2 r^4}{\kappa + r^4},$$

$$\gamma_0^2 = (\delta_0 + \frac{\overline{\delta_2}}{\kappa + r^4})\frac{2\mu}{\kappa},$$

$$\gamma_1^2 = -\frac{2(-\mu\overline{\delta_1} + r^2\mu\overline{\delta_1} - r^2\mu\delta_1 - \kappa\mu\delta_1)}{(1+\kappa)(-1+2r^2+\kappa)},$$

$$\overline{c_{-1}^2} = -\frac{2(r^2\mu\overline{\delta_1} + r^2\mu\delta_1)}{-1+2r^2+\kappa},$$

$$\gamma_m^2 = \frac{(2\mu(r^4\delta_m + r^{2m}((-2+m)(-1+r^2)\overline{\delta_{2-m}} + \kappa\delta_m)))}{(r^4\kappa + r^{4m}\kappa + r^{2m}((-2+m)m - 2(-2+m)mr^2 + (-1+m)^2 r^4 + \kappa^2))},$$

$$\overline{\gamma^2_{2-m}} = \frac{(2r^{2m}\mu((r^{2m}+\kappa)\overline{\delta_{2-m}} - m(-1+r^2)\delta_m))}{(r^4\kappa + r^{4m}\kappa + r^{2m}((-2+m)m - 2(-2+m)kr^2 + (-1+k)^2r^4 + \kappa^2))},$$

$$\overline{c^2_{-m}} = (2r^{2m}\mu((-2+m)(r^{2m}+r^2\kappa)\overline{\delta_{2-m}} - (-(-2+m)mr^2 + \\ +(-1+m)^2r^4 + r^{2m}\kappa)\delta_m))(r^4\kappa + \\ +r^{4m}\kappa + r^{2m}((-2+m)m - 2(-2+m)mr^2 + (-1+m)^2r^4 + \kappa^2))^{-1},$$

$$c^2_{-2+m} = (-2r^4\mu(\kappa\overline{\delta_{2-m}} + m\delta_m) - 2r^{2+2m}\mu((r^2 + \\ +(-2+m)m(-1+r^2))\overline{\delta_{2-m}} + m\kappa\delta_m))(r^4\kappa + r^{4m}\kappa + \\ +r^{2m}((-2+m)m - 2(-2+m)mr^2 + (-1+m)^2r^4 + \kappa^2)).$$

These coefficients depend lineally on δ_m , $\overline{\delta_{2-m}}$.

Now we restore the coefficients of $\rho_2(z)$ using the boundary conditions (5.10) with $k = 2$:

$$\begin{cases} \operatorname{Re} D^2_l + \operatorname{Re} D^2_{-l} & = \sigma^2_l, \\ \operatorname{Re} D^2_l r^l - \operatorname{Re} D^2_{-l} r^{-l} & = 0, \end{cases}$$

$$\begin{cases} \operatorname{Im} D^2_l - \operatorname{Im} D^2_{-l} & = -\nu^2_l, \\ \operatorname{Im} D^2_l r^l + \operatorname{Im} D^2_{-l} r^{-l} & = 0. \end{cases}$$

So

$$\rho_2(z) = \frac{\sigma_0}{2} + \sum_{l=1}^{\infty} \frac{(\sigma_l - \imath\nu_l)}{r^{2l}+1} z^l + \sum_{l=1}^{\infty} \frac{(\sigma_l + \imath\nu_l)r^{2l}}{r^{2l}+1} z^{-l}.$$

We restore the function $\Phi_1(z,\overline{z})$:

$$\Phi_1(z,\overline{z}) = \frac{(\lambda+\mu)}{2(\lambda+2\mu)}\left[(\lambda+3\mu)z\left(\frac{\sigma_0}{2}+\sum_{k=1}^{\infty}\frac{(\sigma_k+\imath\nu_k)}{r^{2k}+1}\overline{z}^k+\right.\right.$$
$$\left.+\sum_{k=1}^{\infty}\frac{(\sigma_k-\imath\nu_k)r^{2k}}{r^{2k}+1}\overline{z}^{-k}\right)-(\lambda+\mu)\left(\frac{\sigma_0}{2}z+\frac{(\sigma_1+\imath\nu_1)r^2}{r^2+1}\ln z+\right.$$
$$\left.\left.+\sum_{k=1}^{\infty}\frac{(\sigma_k-\imath\nu_k)}{r^{2k}+1}\frac{z^{k+1}}{k+1}+\sum_{k=2}^{\infty}\frac{(\sigma_k+\imath\nu_k)r^{2k}}{r^{2k}+1}\frac{z^{1-k}}{1-k}\right)\right].$$

We restore the function $\Psi_1(z,\overline{z})$:

$$\Psi_1(z,\overline{z}) = -\frac{1}{2\mu}\left(\overline{z}\sum_{m=-\infty}^{+\infty}\gamma_m^2 z^m + z\sum_{m=-\infty}^{+\infty}\overline{\gamma_m^2}\,\overline{z^m}\right).$$

At the second step we can restore the analytic functions $\phi_1(z)$ and $\psi_1(z)$ via the boundary condition (5.9) with $k=1$. We have the following systems over the coefficients of these functions.

$$\begin{cases} -\kappa\beta_1+\overline{b_1} = r^2\frac{(\lambda+\mu)^2}{2(\lambda+2\mu)}\frac{(\sigma_1+\imath\nu_1)}{(r^2+1)}, \\ -\beta_1+\overline{b_1} = \frac{(\lambda^2+2\lambda\mu-\mu^2)}{(\lambda+2\mu)}\frac{(\sigma_1+\imath\nu_1)r^2}{(r^2+1)} - \frac{(\lambda+\mu)(\lambda+3\mu)}{2(\lambda+2\mu)}\frac{(\sigma_1+\imath\nu_1)r^2}{(r^2+1)}, \end{cases}$$

$$\begin{cases} -\kappa\gamma_2^1+\overline{c_{-2}^1} = -2\mu\delta_{2,1} - \frac{(\lambda+\mu)(\lambda+3\mu)}{2(\lambda+2\mu)}\frac{(\sigma_1-\imath\nu_1)r^2}{(r^2+1)}+ \\ \qquad +\frac{(\lambda+\mu)^2}{4(\lambda+2\mu)}\frac{(\sigma_1-\imath\nu_1)}{(r^2+1)} - \overline{\beta_1}, \\ -2\gamma_2^1r^2-2\overline{c_{-2}^1}r^{-2} = \frac{(\lambda^2+2\lambda\mu-\mu^2)}{(\lambda+2\mu)}\frac{(\sigma_1-\imath\nu_1)r^2}{(r^2+1)}+ \\ \qquad +\frac{(\lambda+\mu)(\lambda+3\mu)}{2(\lambda+2\mu)}\frac{(\sigma_1+\imath\nu_1)r^2}{(r^2+1)}+2\overline{\beta_1}, \end{cases}$$

$$\begin{cases} -\kappa\gamma_1^1+\overline{\gamma_1^1}+\overline{c_1^1} = -2\mu\delta_{1,1} - \frac{(\lambda+\mu)(\lambda+3\mu)}{4(\lambda+2\mu)}\sigma_0 + \frac{(\lambda+\mu)^2}{4(\lambda+2\mu)}\sigma_0, \\ -\gamma_1^1r-\overline{\gamma_1^1}r-2\overline{c_1^1}r^{-1} = \frac{(\lambda^2+2\lambda\mu-\mu^2)}{2(\lambda+2\mu)}\sigma_0 r. \end{cases}$$

So

$$\beta_1 = -\frac{r^2(\lambda-\mu)(\imath\nu_1+\sigma_1)}{(1+r^2)(1+\kappa)},$$

$$\overline{b_1} = \frac{r^2((-1+\kappa)\lambda^2-2\lambda\mu-(1+5\kappa)\mu^2)(\imath\nu_1+\sigma_1)}{2(1+r^2)(1+\kappa)(\lambda+2\mu)},$$

$$\gamma_2^1 = (((1-2r^2(-1+\kappa)+\kappa+r^4(-1+3\kappa))\lambda^2+2(1+\kappa+2(-1+ \\ +r)r^2(1+r)(1+2\kappa))\lambda\mu+(1+\kappa+r^4(9+\kappa)- \\ -2r^2(7+3\kappa))\mu^2)(\imath\nu_1-\sigma_1)+8(1+r^2)(1+\kappa)\mu(\lambda+2\mu)(\frac{t_2^{j+1}-t_2^j}{h_j-h_{j+1}}- \\ -(h_j+h_{j+1})\delta_2))(4(1+r^2)(1+\kappa)(r^4+\kappa)(\lambda+2\mu))^{-1},$$

$$\overline{c_{-2}^1} = \frac{1}{4(r^4+\kappa)}(r^4((((-1+\kappa)(1+2r^2+3\kappa)\lambda^2+ \\ +2(-1+\kappa+4\kappa^2+r^2(2+4\kappa))\lambda\mu+(-1+\kappa(8+\kappa)+ \\ +2r^2(7+3\kappa))\mu^2)(\imath\nu_1-\sigma_1))((1+r^2)(1+\kappa)(\lambda+2\mu))^{-1}- \\ -8\mu(\frac{t_2^{j+1}-t_2^j}{h_j-h_{j+1}}-(h_j+h_{j+1})\delta_2))),$$

$$\overline{\gamma_0^1} = (2\mu(\lambda+\mu)((1+r^2)(\lambda+2\mu)((1+r^4)\lambda+(3+r^4)\mu)(\frac{\overline{t_0^{j+1}}-\overline{t_0^j}}{h_j-h_{j+1}}- \\ -(h_j+h_{j+1})\overline{\delta_0})+(\lambda+\mu)(((-1-2r^2+ \\ +r^4)\lambda-2(1+r^2)\mu)(\imath\nu_1-\sigma_1)+2(1+r^2)(\lambda+2\mu)(\frac{t_2^{j+1}-t_2^j}{h_j-h_{j+1}}- \\ -(h_j+h_{j+1})\delta_2))))((1+r^2)(\lambda+2\mu)(\lambda+3\mu)((1+r^4)\lambda+(3+r^4)\mu))^{-1},$$

$$\gamma_1^1 = ((\lambda+\mu)(-2(-1+r^2)\mu(\lambda+\mu)(\frac{\overline{t_1^{j+1}}-\overline{t_1^j}}{h_j-h_{j+1}}-(h_j+h_{j+1})\overline{\delta_1})+ \\ +(\mu(\lambda+\mu)+r^2(-\lambda^2-2\lambda\mu+\mu^2))\sigma_0+2\mu((1+r^2)\lambda+ \\ +(3+r^2)\mu)(\frac{t_1^{j+1}-t_1^j}{h_j-h_{j+1}}-(h_j+h_{j+1})\delta_1)))(4(\lambda+2\mu)(\mu+r^2(\lambda+\mu)))^{-1},$$

$$c_{-1}^1 = \overline{c_{-1}^1} = -\frac{1}{\mu + r^2(\lambda+\mu)}(r^2\mu((\lambda+\mu)(\frac{\overline{t_1^{j+1}} - \overline{t_1^j}}{h_j - h_{j+1}} -$$
$$-(h_j + h_{j+1})\overline{\delta_1}) + \lambda\sigma_0 + (\lambda+\mu)(\frac{t_1^{j+1} - t_1^j}{h_j - h_{j+1}} - (h_j + h_{j+1})\delta_1))).$$

We have for $m > 2$ the following system with respect to γ_m^1, γ_{2-m}^1, c_{-m}^1 and c_{m-2}^1:

$$\begin{cases}
-\kappa\gamma_m^1 + (2-m)\overline{\gamma_{2-m}^1} + \overline{c_{-m}^1} = -2\mu\delta_{m,1} - \\
-\frac{(\lambda+\mu)(\lambda+3\mu)}{2(\lambda+2\mu)}\frac{(\sigma_{k-1} - \imath\nu_{k-1})r^{2k-2}}{(r^{2k-2}+1)} + \frac{(\lambda+\mu)^2}{2k(\lambda+2\mu)}\frac{(\sigma_{k-1} - \imath\nu_{k-1})}{(r^{2k-2}+1)}, \\
-m\overline{\gamma_m^1}r^m - m(2-m)\gamma_{2-m}^1 r^{2-m} - mc_{-m}^1 r^{-m} = \\
= \frac{(\lambda^2+2\lambda\mu-\mu^2)}{(\lambda+2\mu)}\frac{(\sigma_{k-1}+\imath\nu_{k-1})r^k}{(r^{2k-2}+1)} + \frac{(\lambda+\mu)(\lambda+3\mu)(k-1)}{2(\lambda+2\mu)}\frac{(\sigma_{k-1}+\imath\nu_{k-1})r^k}{(r^{2k-2}+1)}, \\
-\kappa\gamma_{2-m}^1 + m\overline{\gamma_m^1} + \overline{c_{m-2}^1} = -2\mu\delta_{(2-m),1} - \\
-\frac{(\lambda+\mu)(\lambda+3\mu)}{2(\lambda+2\mu)}\frac{(\sigma_{k-1}+\imath\nu_{k-1})}{(r^{2k-2}+1)} + \frac{(\lambda+\mu)^2}{2(2-k)(\lambda+2\mu)}\frac{(\sigma_{k-1}+\imath\nu_{k-1})r^{2k-2}}{(r^{2k-2}+1)}, \\
-(2-m)\overline{\gamma_{2-m}^1}r^{2-m} + m(m-2)\gamma_m^1 r^m + (m-2)c_{m-2}^1 r^{m-2} = \\
= \frac{(\lambda^2+2\lambda\mu-\mu^2)}{(\lambda+2\mu)}\frac{(\sigma_{k-1}-\imath\nu_{k-1})r^k}{(r^{2k-2}+1)} - \frac{(\lambda+\mu)(\lambda+3\mu)(k-1)}{2(\lambda+2\mu)}\frac{(\sigma_{k-1}-\imath\nu_{k-1})r^k}{(r^{2k-2}+1)}.
\end{cases}$$

We do not include the solutions of the last system in this subsection because their expressions are too large. Evidently, these solutions depend lineally on σ_{m-1}, ν_{m-1}, δ_m and $\overline{\delta_{2-m}}$.

We restore the coefficients of the analytic function $\rho_1(z)$ via the boundary condition (5.10) with $k = 1$ and have the following relations.

$$\begin{cases}
B_1 - \overline{B_1} = 0, \\
B_1 + \overline{B_1} = \frac{r^2}{2\mu}(3-\kappa)(\gamma_1^2 + \overline{\gamma_1^2}) + \frac{1}{\mu}c_{-1}^2,
\end{cases}$$

so

$$B_1 = -\frac{r^2(-1+\kappa)(\overline{\delta_1}+\delta_1)}{2(-1+2r^2+\kappa)}.$$

$$D_0^1 = \frac{\sigma_0^1}{2} + \frac{1}{2\mu}(\gamma_1^2 + \overline{\gamma_1^2}) =$$

$$= \frac{\overline{\delta_1}+\delta_1}{-1+2r^2+\kappa} + \frac{g_0^j - g_0^{j+1}}{2(h_j - h_{j+1})} - \frac{\sigma_0}{2}(h_j + h_{j+1}).$$

We have for $m \geq 1$ the system

$$\begin{cases} D_m^1 - \overline{D_{-m}^1} = \sigma_m^1 - \imath\nu_m^1 + \frac{1}{\mu}(\gamma_{m+1}^2 + \overline{\gamma_{1-m}^2}), \\ m(D_m^1 r^m + \overline{D_{-m}^1 r^{-m}}) = \frac{r^2}{\mu}[(2m+3-\kappa)\gamma_{m+1}^2 + (-2m+3-\kappa)\overline{\gamma_{1-k}^2}] + \\ \quad + \frac{1}{2\mu}(c_{m-1}^2 r^m + \overline{c_{-m-1}^2 r^{-m}}). \end{cases}$$

The solutions D_m^1 and D_{-m}^1 of the last system depend lineally on σ_m, ν_m, δ_{1+m} and $\overline{\delta_{1-m}}$.

Now we find the function $\Phi_0(z,\overline{z})$:

$$\Phi_0(z,\overline{z}) = \frac{\mu}{(\lambda+\mu)}\overline{z}\Big(\sum_{m=-\infty,m\neq-1}^{+\infty} \gamma_m^2 \frac{z^{m+1}}{m+1} + \gamma_{-1}^2 \ln z\Big) -$$

$$-\frac{1}{2}z^2 \sum_{m=-\infty}^{+\infty} \overline{\gamma_m^2 z^m} - \frac{\lambda+3\mu}{4(\lambda+2\mu)} z\Big(\sum_{m=-\infty,m\neq-1}^{+\infty} \overline{c_m^2}\,\overline{\frac{z^{m+1}}{m+1}} + \overline{c_{-1}^2}\ln\overline{z}\Big) +$$

$$+\frac{\lambda+\mu}{4(\lambda+2\mu)}\Big(\sum_{m=-\infty,m\neq-1,m\neq-2}^{+\infty} c_m^2 \frac{z^{m+2}}{(m+1)(m+2)} - c_{-2}^2 \ln z +$$

$$+c_{-1}^2 z(\ln z - 1)\Big) + \frac{(\lambda+\mu)(\lambda+3\mu)}{4(\lambda+2\mu)}\Big(z\sum_{m=-\infty}^{+\infty} \overline{D_m^1 z^m} + z\overline{B_1}\ln\overline{z}\Big) -$$

$$-\frac{(\lambda+\mu)^2}{4(\lambda+2\mu)}\Big(\sum_{m=-\infty,m\neq-1}^{+\infty} D_m^1 \frac{z^{m+1}}{m+1} + D_{-1}^1 \ln z + B_1 z(\ln z - 1)\Big).$$

We find the function $\Psi_0(z,\overline{z})$:

$$\Psi_0(z,\overline{z}) = -\frac{1}{4\mu}(\overline{z}\beta_1 \ln z + \overline{z}\sum_{m=-\infty}^{+\infty} \gamma^1_m z^m + z\overline{\beta_1} + z\sum_{m=-\infty}^{+\infty} \overline{\gamma^1_m z^m}) +$$

$$+\frac{(\lambda+\mu)^2}{8\mu(\lambda+2\mu)}\left[z(\sum_{m=1}^{\infty}\frac{\sigma_m + \imath\nu_m}{r^{2m}+1}\frac{\overline{z^{m+1}}}{m+1} + \frac{\sigma_0}{2}\overline{z} + \frac{\sigma_1 - \imath\nu_1}{r^2+1}r^2\ln\overline{z} + \right.$$

$$+\sum_{m=2}^{\infty}\frac{(\sigma_m - \imath\nu_m)r^{2m}}{r^{2m}+1}\frac{\overline{z^{1-m}}}{1-m}) + \overline{z}(\sum_{m=1}^{\infty}\frac{\sigma_m - \imath\nu_m}{r^{2m}+1}\frac{z^{m+1}}{m+1} +$$

$$\left. +\frac{\sigma_0}{2}z + \frac{\sigma_1 + \imath\nu_1}{r^2+1}r^2\ln z + \sum_{m=2}^{\infty}\frac{(\sigma_m + \imath\nu_m)r^{2m}}{r^{2m}+1}\frac{z^{1-m}}{1-m})\right] -$$

$$-\frac{\lambda+2\mu}{4\mu}\left[\overline{z}(D^2_{-1}\ln z + \sum_{m=-\infty, m\neq -1}^{+\infty} D^2_m \frac{z^{m+1}}{m+1}) + \right.$$

$$\left. +z(\overline{D^2_{-1}}\ln\overline{z} + \sum_{m=-\infty, m\neq -1}^{+\infty}\overline{D^2_m}\frac{\overline{z^{m+1}}}{m+1})\right].$$

At the third step we can restore the analytic functions $\phi_0(z)$ and $\psi_0(z)$ via the boundary condition (5.9) with $k = 0$. We have the following relations and systems over the coefficients of these functions.

$$\alpha_0 = \frac{c^2_{-1} - (\lambda+\mu)B_1}{2(1+\kappa)},$$

$$a_0 = \frac{\mu}{\lambda+\mu}\overline{\gamma^2_{-1}},$$

$$\begin{cases} \kappa\overline{\beta_0} + b_0 = \frac{\lambda+\mu}{4(\lambda+2\mu)}\overline{c^2_{-2}} - \frac{(\lambda+\mu)^2}{4(\lambda+2\mu)}\overline{D^1_{-1}}, \\ -\overline{\beta_0} + b_0 = \gamma^2_2 r^4 - \overline{\gamma^2_0}\frac{\mu}{\lambda+\mu}r^2 - \overline{c^2_{-2}}\frac{\lambda+\mu}{4(\lambda+2\mu)} + \\ +c^2_0\frac{r^2}{2} + \overline{D^1_{-1}}\frac{\lambda^2+2\lambda\mu-\mu^2}{4(\lambda+2\mu)} - D^1_1\frac{\mu}{2}r^2, \end{cases}$$

$$\left\{\begin{array}{c} -\kappa\gamma_1^0 + \overline{\gamma_1^0} + \overline{c_{-1}^0} = -2\mu\delta_{1,0} - \alpha_0 - \frac{\mu}{\lambda+\mu}\gamma_1^2 + \frac{\overline{\gamma_1^2}}{2} - \\ -\frac{(\lambda+\mu)\mu}{2(\lambda+2\mu)}D_0^1 - \frac{(\lambda+\mu)^2}{4(\lambda+2\mu)}B_1, \\ -\gamma_1^0 r - \overline{\gamma_1^0}r - \overline{c_{-1}^0}r^{-1} = 2r\alpha(\ln r + 1) - pr - \frac{\mu}{2(\lambda+\mu)}\gamma_1^2 r^3 + \\ +\frac{\overline{\gamma_1^2}}{2}r^3 + (\frac{\lambda+3\mu}{4(\lambda+2\mu)}r + \frac{\lambda+\mu}{2(\lambda+2\mu)}\ln r)c_{-1}^2 + \\ +\frac{(\lambda^2+2\lambda\mu-\mu^2)}{2(\lambda+2\mu)}D_0^1 r + \frac{(\lambda^2-5\mu^2)}{4(\lambda+2\mu)}B_1 r\ln r, \end{array}\right.$$

$$\left\{\begin{array}{c} -\kappa\gamma_2^0 + \overline{c_{-2}^0} = -2\mu\delta_{2,0} - \overline{\beta_0} - \frac{\mu}{3(\lambda+\mu)}\gamma_2^2 + \frac{\overline{\gamma_0^2}}{2} - \\ -\frac{\lambda+3\mu}{4(\lambda+2\mu)}\overline{c_{-2}^2} - \frac{(\lambda+\mu)(\lambda+3\mu)}{4(\lambda+2\mu)}\overline{D_{-1}^1} + \frac{(\lambda+\mu)^2}{8(\lambda+2\mu)}D_1^1, \\ -2\gamma_2^0 r^2 - 2\overline{c_{-2}^0}r^{-2} = -\frac{\mu}{3(\lambda+\mu)}\gamma_2^2 r^4 + \frac{\mu}{2(\lambda+2\mu)}\overline{c_{-2}^2} + \\ +\frac{\lambda^2+3\lambda\mu+\mu^2}{2(\lambda+2\mu)}\overline{D_{-1}^1} + \frac{\lambda^2+2\lambda\mu-\mu^2}{4(\lambda+2\mu)}D_1^1 r^2, \end{array}\right.$$

$$\overline{\gamma_0^0} = \frac{1}{\kappa}(2\gamma_2^0 + 2\mu\overline{\delta_{0,0}} + \frac{\mu}{\lambda+\mu}\overline{\gamma_0^2} - \frac{\gamma_2^2}{2} + \frac{(\lambda+\mu)(\lambda+3\mu)}{4(\lambda+2\mu)}D_1^1),$$

$$\left\{\begin{array}{c} -\kappa\gamma_3^0 - \overline{\gamma_{-1}^0} + \overline{c_{-3}^0} = -2\mu\delta_{3,0} - \frac{\mu}{4(\lambda+\mu)}\gamma_3^2 + \frac{\overline{\gamma_{-1}^2}}{2} - \frac{\lambda+3\mu}{8(\lambda+2\mu)}\overline{c_{-3}^2} - \\ -\frac{\lambda+\mu}{24(\lambda+2\mu)}c_{-1}^2 - \frac{(\lambda+\mu)(\lambda+3\mu)}{4(\lambda+2\mu)}\overline{D_{-2}^1} + \frac{(\lambda+\mu)^2}{12(\lambda+2\mu)}D_2^1, \\ -3\gamma_3^0 r^3 + 3\overline{\gamma_{-1}^0}r^{-1} - 3\overline{c_{-3}^0}r^{-3} = -\frac{1}{2}\overline{\gamma_{-1}^2}r - \frac{\mu}{4(\lambda+\mu)}\gamma_3^2 r^5 + \\ +\frac{\lambda+5\mu}{8(\lambda+2\mu)}\overline{c_{-3}^2}r^{-1} + \frac{\lambda+\mu}{8(\lambda+2\mu)}c_1^2 r^3 + \\ +\frac{3\lambda^2+10\lambda\mu+5\mu^2}{4(\lambda+2\mu)}\overline{D_{-2}^1}r^{-1} + \frac{\lambda^2+2\lambda\mu-\mu^2}{4(\lambda+2\mu)}D_2^1 r^3, \\ 3\gamma_3^0 - \kappa\overline{\gamma_{-1}^0} + c_1^0 = -2\mu\overline{\delta_{-1,0}} + \frac{\gamma_3^2}{2} + \frac{\lambda+3\mu}{8(\lambda+2\mu)}c_1^2 - \\ -\frac{\lambda+\mu}{8(\lambda+2\mu)}\overline{c_{-3}^2} - \frac{(\lambda+\mu)(\lambda+3\mu)}{4(\lambda+2\mu)}D_2^1 - \frac{(\lambda+\mu)^2}{4(\lambda+2\mu)}\overline{D_{-2}^1}, \\ 3\gamma_3^0 r^3 + \overline{\gamma_{-1}^0}r^{-1} + c_1^0 r = \frac{3}{2}\gamma_3^2 r^5 + \frac{3\lambda+7\mu}{8(\lambda+2\mu)}c_1^2 r^3 - \\ -\frac{\lambda+\mu}{8(\lambda+2\mu)}\overline{c_{-3}^2}r^{-1} - \frac{\lambda^2+6\lambda\mu+7\mu^2}{4(\lambda+2\mu)}D_2^1 r^3 + \frac{\lambda^2+2\lambda\mu-\mu^2}{4(\lambda+2\mu)}\overline{D_{-2}^1}r^{-1}, \end{array}\right.$$

and for $m > 3$ we have

$$\begin{cases} -\kappa\gamma_m^0 + (2-m)\overline{\gamma_{2-m}^0} + \overline{c_{-m}^0} = -2\mu\delta_{m,0} - \frac{\mu}{(m+1)(\lambda+\mu)}\gamma_m^2 + \\ \quad + \frac{\overline{\gamma_{2-m}^2}}{2} - \frac{\lambda+3\mu}{4(m-1)(\lambda+2\mu)}\overline{c_{-m}^2} - \frac{\lambda+\mu}{4m(m-1)(\lambda+2\mu)}c_{m-2}^2 - \\ \quad - \frac{(\lambda+\mu)(\lambda+3\mu)}{4(\lambda+2\mu)}\overline{D_{1-m}^1} + \frac{(\lambda+\mu)^2}{4m(\lambda+2\mu)}D_{m-1}^1, \\ -m\gamma_m^0 r^m + m(m-2)\overline{\gamma_{2-m}^0}r^{2-m} - m\overline{c_{-m}^0}r^{-m} = \\ \quad = \frac{2-m}{2}\overline{\gamma_{2-m}^2}r^{4-m} - \frac{\mu}{(m+1)(\lambda+\mu)}\gamma_m^2 r^{m+2} + (\frac{\lambda+\mu}{4(1-m)(\lambda+2\mu)} + \\ \quad + \frac{\lambda+3\mu}{4(\lambda+2\mu)})\overline{c_{-m}^2}r^{2-m} + \frac{\lambda+\mu}{4(m-1)(\lambda+2\mu)}c_{m-2}^2 r^m + (\frac{\lambda^2+2\lambda\mu-\mu^2}{4(\lambda+2\mu)} + \\ \quad + \frac{(m-1)(\lambda+\mu)(\lambda+3\mu)}{4(\lambda+2\mu)})\overline{D_{1-m}^1}r^{2-m} + \frac{\lambda^2+2\lambda\mu-\mu^2}{4(\lambda+2\mu)}D_{m-1}^1 r^m, \\ m\gamma_m^0 - \kappa\overline{\gamma_{2-m}^0} + c_{m-2}^0 = -2\mu\overline{\delta_{2-m,0}} - \frac{\mu}{(3-m)(\lambda+\mu)}\overline{\gamma_{2-m}^2} + \frac{\gamma_m^2}{2} + \\ \quad + \frac{\lambda+3\mu}{4(m-1)(\lambda+2\mu)}c_{m-2}^2 - \\ \frac{\lambda+\mu}{4(m-1)(m-2)(\lambda+2\mu)}\overline{c_{-m}^2} - \frac{(\lambda+\mu)(\lambda+3\mu)}{4(\lambda+2\mu)}D_{m-1}^1 + \frac{(\lambda+\mu)^2}{4(2-m)(\lambda+2\mu)}\overline{D_{1-m}^1}, \\ m(m-2)\gamma_m^0 r^m - (2-m)\overline{\gamma_{2-m}^0}r^{2-m} + (m-2)c_{m-2}^0 r^{m-2} = \\ \quad = \frac{m}{2}\gamma_m^2 r^{m+2} - \frac{\mu}{(3-m)(\lambda+\mu)}\overline{\gamma_{2-m}^2}r^{4-m} + (\frac{\lambda+\mu}{4(m-1)(\lambda+2\mu)} + \\ \quad + \frac{\lambda+3\mu}{4(\lambda+2\mu)})c_{m-2}^2 r^m + \frac{\lambda+\mu}{4(1-m)(\lambda+2\mu)}\overline{c_{-m}^2}r^{2-m} + (\frac{\lambda^2+2\lambda\mu-\mu^2}{4(\lambda+2\mu)} - \\ \quad - \frac{(m-1)(\lambda+\mu)(\lambda+3\mu)}{4(\lambda+2\mu)})D_{m-1}^1 r^m + \frac{\lambda^2+2\lambda\mu-\mu^2}{4(\lambda+2\mu)}\overline{D_{1-m}^1}r^{2-m}. \end{cases}$$

We restore the coefficients of the analytic function $\rho_0(z)$ via the boundary condition (5.10) with $k = 0$ and have the following relations.

$$A_0 = -\frac{1}{2\mu}B_1 + \frac{(\lambda+\mu)^2}{4\mu(\lambda+2\mu)}\frac{\sigma_1 - \imath\nu_1}{r^2+1}r^2 - \frac{\lambda+2\mu}{2\mu}\overline{D_{-1}^2},$$

$$B_0 = r^2\left[\frac{2-\kappa}{2\mu}(\gamma_1^1+\overline{\gamma_1^1}) - \frac{(\lambda+\mu)^2}{2\mu(\lambda+2\mu)}\sigma_0 + \right.$$
$$\left. + \frac{(\lambda+\mu)(\lambda+3\mu)}{2\mu(\lambda+2\mu)}\frac{\sigma_1}{r^2+1} + \frac{\lambda+2\mu}{\mu}D_0^2\right] + \frac{c_{-1}^1}{\mu},$$

$$D_0^0 = \frac{\sigma_{0,0}}{2} + \frac{1}{4\mu}(\gamma_1^1+\overline{\gamma_1^1}) - \frac{(\lambda+\mu)^2}{8\mu(\lambda+2\mu)}\sigma_0 + \frac{\lambda+2\mu}{2\mu}D_0^2,$$

$$\begin{cases} D_1^0 + \overline{D_{-1}^0} = \sigma_{1,0} - \imath\nu_{1,0} + \frac{1}{2\mu}(\gamma_2^1+\overline{\gamma_0^1}) - \\ \quad - \frac{(\lambda+\mu)^2}{8\mu(\lambda+2\mu)}\frac{\sigma_1-\imath\nu_1}{r^2+1} + \frac{\lambda+2\mu}{4\mu}D_1^2, \\ D_1^0 r - \overline{D_{-1}^0}r^{-1} = -A_0 r(\ln r + 1) + \frac{7-2\kappa}{2\mu}\gamma_2^1 r^3 + \\ \quad + [3+\ln r(1-2\kappa)]\overline{\beta_1}r + \\ \quad + \frac{1-2\kappa}{2\mu}\overline{\gamma_0^1}r + \frac{\overline{c_{-2}^1}}{\mu}r^{-1} + \frac{b_1}{\mu}r\ln r + \\ \quad + [-\frac{3(\lambda+\mu)^2}{4\mu(\lambda+2\mu)}\ln r + \frac{(\lambda+\mu)(\lambda+17\mu)}{8\mu(\lambda+2\mu)}]\frac{\sigma_1-\imath\nu_1}{r^2+1}r^3 + \\ \quad + \frac{3(\lambda+2\mu)}{4\mu}D_1^2 r^3 + \frac{\lambda+2\mu}{2\mu}\overline{D_{-1}^2}r\ln r. \end{cases}$$

We have for $m > 1$

$$\begin{cases} D_m^0 + \overline{D_{-m}^0} = \sigma_{m,0} - \imath\nu_{m,0} + \frac{1}{2\mu}(\gamma_{m+1}^1 + \\ \quad + \overline{\gamma_{1-m}^1}) - \frac{(\lambda+\mu)^2}{2(1-m^2)\mu(\lambda+2\mu)}\frac{\sigma_m-\imath\nu_m}{r^{2m}+1} + \\ \quad + \frac{\lambda+2\mu}{2\mu}(\frac{D_m^2}{m+1} + \frac{\overline{D_{-m}^2}}{1-m}), \\ m(D_m^0 r^m - \overline{D_{-m}^0}r^{-m}) = \\ = \frac{3m+4-2\kappa}{2\mu}\gamma_{m+1}^1 r^{m+1} + \frac{4-3m-2\kappa}{2\mu}\overline{\gamma_{1-m}^1}r^{2-m} + \\ \quad + \frac{c_{m-1}^1}{\mu}r^m + \frac{\overline{c_{-1-m}^1}}{\mu}r^{-m} + \\ + [-\frac{3(\lambda+\mu)^2}{2(1-m^2)\mu(\lambda+2\mu)} + \frac{(\lambda+\mu)(\lambda+5\mu)}{2\mu(\lambda+2\mu)}]\frac{\sigma_m-\imath\nu_m}{r^{2m}+1}r^{m+2} + \\ + \frac{(m+2)(\lambda+2\mu)}{2(m+1)\mu}D_m^2 r^{m+2} + \frac{(2-m)(\lambda+2\mu)}{2(1-m)\mu}\overline{D_{-m}^2}r^{2-m}. \end{cases}$$

5.2.3 Restoration of the introduced coefficients

Here we present the systems which provide the values of the coefficients δ_m, σ_m and ν_m introduced in (5.4). All the coefficients of analytic functions $\phi_k(z)$, $\psi_k(z)$ and $\rho_k(z)$ found in the previous subsection and also the coefficients of the representations of the functions $\Phi_k(z,\overline{z})$ and $\Psi_k(z,\overline{z})$ are lineally dependent of δ_m, σ_m and ν_m. So we find the vector of displacements only after we obtain δ_m, σ_m and ν_m and substitute their values in the representations of the functions $\phi_k(z)$, $\psi_k(z)$, $\rho_k(z)$, $\Phi_k(z,\overline{z})$ and $\Psi_k(z,\overline{z})$.

We find the values of σ_1, ν_1, δ_2 and δ_0 after we solve the system

$$\left\{\begin{array}{c}
-\kappa\overline{\beta_0}\ln r + 2\gamma_2^0 r^2 - \kappa\overline{\gamma_0^0} + b_0\ln r + \frac{\mu}{\lambda+\mu}\overline{\gamma_0^2}r^2 - \frac{1}{2}\gamma_2^2 r^4 - \\
-\frac{\lambda+\mu}{4(\lambda+2\mu)}\overline{c_{-2}^2}\ln r + \frac{(\lambda+\mu)(\lambda+3\mu)}{4(\lambda+2\mu)}D_1^1 r^2 - \frac{(\lambda+\mu)^2}{4(\lambda+2\mu)}\overline{D_{-1}^1}\ln r + \\
+h_j[-\kappa\overline{\beta_1}\ln r + 2\gamma_2^1 r^2 - \kappa\overline{\gamma_0^1} + b_1\ln r + \frac{\lambda+\mu}{2(\lambda+2\mu)}\frac{\sigma_1 - \imath\nu_1}{r^2+1}((\lambda+ \\
+3\mu)r^2 - (\lambda+\mu)r^2\ln r)] + h_j^2[2\gamma_2^2 r^2 - \kappa\overline{\gamma_0^2}] = -2\mu\overline{R_0}, \\
-\kappa\gamma_2^0 r^2 + \overline{\beta_0} + \overline{c_{-2}^0}r^{-2} + \frac{\mu}{3(\lambda+\mu)}\gamma_2^2 r^4 - \frac{1}{2}\overline{\gamma_0^2}r^2 + \frac{\lambda+3\mu}{4(\lambda+2\mu)}\overline{c_{-2}^2} + \\
+\frac{(\lambda+\mu)(\lambda+3\mu)}{4(\lambda+2\mu)}\overline{D_{-1}^1} - \frac{(\lambda+\mu)^2}{8(\lambda+2\mu)}D_1^1 r^2 + h_j[-\kappa\gamma_2^1 r^2 + \overline{\beta_1} + \overline{c_{-2}^1}r^{-2} + \\
+\frac{(\lambda+\mu)(\lambda+5\mu)}{4(\lambda+2\mu)}\frac{\sigma_1 - \imath\nu_1}{r^2+1}r^2] + h_j^2[-\kappa\gamma_2^2 r^2 + \overline{c_{-2}^2}r^{-2}] = -2\mu R_2, \\
A_0 r\ln r + D_1^0 r + \overline{D_{-1}^0}r^{-1} - \frac{1}{2\mu}(\gamma_2^1 r^3 + \overline{\beta_1}r\ln r + \overline{\gamma_0^1}r) + \\
+\frac{(\lambda+\mu)^2}{4\mu(\lambda+2\mu)}\frac{\sigma_1 - \imath\nu_1}{r^2+1}r^3(\ln r + 1) - \frac{\lambda+2\mu}{4\mu}(D_1^2 r^3 + 2\overline{D_{-1}^2}r\ln r) + \\
+h_j[\overline{D_{-1}^1}r^{-1} + D_1^1 r - \frac{1}{2\mu}(\gamma_2^2 r^3 + \overline{\gamma_0^2}r)] + h_j^2[D_1^2 r + \overline{D_{-1}^2}r^{-1}] = \\
= P_1 - \imath Q_1.
\end{array}\right.$$

We find the values of σ_0 and δ_1 after we solve the system

$$\left\{\begin{array}{c}
-\kappa\alpha_0 r\ln r+\alpha_0 r(1+\ln r)+\overline{\gamma_1^0}r-\kappa\gamma_1^0 r+\overline{c_{-1}^0}r^{-1}+\\
+\frac{\mu}{2(\lambda+\mu)}\gamma_1^2 r^3-\frac{1}{2}\overline{\gamma_1^2}r^3+[\frac{\lambda+\mu}{4(\lambda+2\mu)}r(\ln r-1)-\frac{\lambda+3\mu}{4(\lambda+2\mu)}\ln r]c_{-1}^2+\\
+\frac{(\lambda+\mu)\mu}{2(\lambda+2\mu)}D_0^1 r+h_j[\overline{\gamma_1^1}r-\kappa\gamma_1^1 r+\overline{c_{-1}^1}r^{-1}+\frac{\mu(\lambda+\mu)}{2(\lambda+2\mu)}\sigma_0 r]+\\
+h_j^2[\overline{\gamma_1^2}r-\kappa\gamma_1^2 r+\overline{c_{-1}^2}r^{-1}]=-2\mu R_1,\\
2B_0\ln r+2D_0^0-\frac{1}{2\mu}(\gamma_1^1+\overline{\gamma_1^1})r^2+\\
+\frac{(\lambda+\mu)^2}{4\mu(\lambda+2\mu)}\sigma_0 r^2-\frac{\lambda+2\mu}{\mu}D_0^2 r^2\\
+h_j[2B_1\ln r+2D_0^1-\frac{1}{2\mu}(\gamma_1^2+\overline{\gamma_1^2})r^2]+h_j^2 2D_0^2=P_0.
\end{array}\right.$$

Recall that all the coefficients

$$\alpha_0,\ \gamma_1^0,\ \gamma_1^1,\ \gamma_1^2,\ c_{-1}^0,\ c_{-1}^1,\ c_{-1}^2,\ B_0,\ B_1,\ D_0^0,\ D_0^1,\ D_0^2$$

depend lineally on σ_0 and δ_1 .

We find the values of σ_2 , ν_2 , δ_3 and δ_{-1} after we solve the system

$$
\begin{cases}
-\overline{\gamma^0_{-1}}r^{-1} - \kappa\gamma^0_3 r^3 + \overline{c^0_{-3}}r^{-3} + \frac{\mu}{4(\lambda+\mu)}\gamma^2_3 r^5 - \\
-\frac{1}{2}\overline{\gamma^2_{-1}}r + \frac{\lambda+3\mu}{8(\lambda+2\mu)}\overline{c^2_{-3}}r^{-1} + \frac{\lambda+\mu}{24(\lambda+2\mu)}c^2_1 r^3 + \\
+\frac{(\lambda+\mu)(\lambda+3\mu)}{4(\lambda+2\mu)}\overline{D^1_{-2}}r^{-1} - \frac{(\lambda+\mu)^2}{12(\lambda+2\mu)}D^1_2 r^3 + +h_j[-\overline{\gamma^1_{-1}}r^{-1} - \\
-\kappa\gamma^1_3 r^3 + \overline{c^1_{-3}}r^{-3} + \frac{\lambda+\mu}{2(\lambda+2\mu)}\frac{\sigma_2 - \imath\nu_2}{r^4+1}((\lambda+3\mu) - \\
-\frac{\lambda+\mu}{3})r^3] + h^2_j[-\overline{\gamma^2_{-1}}r^{-1} - \kappa\gamma^2_3 r^3 + \overline{c^2_{-3}}r^{-3}] = -2\mu R_3, \\
3\gamma^0_3 r^3 - \kappa\overline{\gamma^0_{-1}}r^{-1} + c^0_1 r + \frac{\mu}{\lambda+\mu}\overline{\gamma^2_{-1}}r\ln r - \\
-\frac{1}{2}\gamma^2_3 r^5 - \frac{\lambda+3\mu}{8(\lambda+2\mu)}c^2_1 r^3 + \frac{\lambda+\mu}{8(\lambda+2\mu)}\overline{c^2_{-3}}r^{-1} + \\
+\frac{(\lambda+\mu)(\lambda+3\mu)}{4(\lambda+2\mu)}D^1_2 + \frac{(\lambda+\mu)^2}{4(\lambda+2\mu)}\overline{D^1_{-2}}r^{-1} + h_j[3\gamma^1_3 r^3 - \\
-\kappa\overline{\gamma^1_{-1}}r^{-1} + c^1_1 r + (\lambda+\mu)\frac{\sigma_2 - \imath\nu_2}{r^4+1}r^3] + \\
+h^2_j[3\gamma^2_3 r^3 - \kappa\overline{\gamma^2_{-1}}r^{-1} + c^2_1 r] = -2\mu\overline{R_{-1}}, \\
D^0_2 r^2 + \overline{D^0_{-2}}r^{-2} - \frac{1}{2\mu}(\gamma^1_3 r^4 + \overline{\gamma^1_{-1}}) - \\
-\frac{(\lambda+\mu)^2}{6\mu(\lambda+2\mu)}\frac{\sigma_2 - \imath\nu_2}{r^4+1}r^4 - \\
-\frac{\lambda+2\mu}{2\mu}(D^2_2\frac{r^4}{3} - \overline{D^2_{-2}}) + h_j[D^1_2 r^2 + \\
+\overline{D^1_{-2}}r^{-2} - \frac{1}{2\mu}(\gamma^2_3 r^4 + \overline{\gamma^2_{-1}})] + \\
+h^2_j[D^2_2 r^2 + \overline{D^2_{-2}}r^{-2}] = P_2 - \imath Q_2.
\end{cases}
$$

We find the values of σ_{m-1} , ν_{m-1} , δ_m and δ_{2-m} , $m > 3$, after we solve the system

$$\left\{\begin{array}{l}
(2-m)\overline{\gamma^0_{2-m}}r^{2-m}-\kappa\gamma^0_m r^m+\overline{c^0_{-m}}r^{-m}+\frac{\mu}{(m+1)(\lambda+\mu)}\gamma^2_m r^{2+m}- \\
-\frac{1}{2}\overline{\gamma^2_{2-m}}r^{4-m}+\frac{\lambda+3\mu}{4(m-1)(\lambda+2\mu)}\overline{c^2_{-m}}r^{2-m}+\frac{\lambda+\mu}{4m(m-1)(\lambda+2\mu)}c^2_{m-2}r^m+ \\
+\frac{(\lambda+\mu)(\lambda+3\mu)}{4(\lambda+2\mu)}\overline{D^1_{1-m}}r^{2-m}-\frac{(\lambda+\mu)^2}{4m(\lambda+2\mu)}D^1_{m-1}r^m+h_j[(2- \\
-m)\overline{\gamma^1_{2-m}}r^{2-m}-\kappa\gamma^1_m r^m+\overline{c^1_{-m}}r^{-m}+\frac{\lambda+\mu}{2(\lambda+2\mu)}\frac{\sigma_{m-1}-\imath\nu_{m-1}}{r^{2m-2}+1}((\lambda+ \\
+3\mu)-\frac{\lambda+\mu}{m})r^m]+h_j^2[(2-m)\overline{\gamma^2_{2-m}}r^{2-m}-\kappa\gamma^2_m r^m+\overline{c^2_{-m}}r^{-m}]= \\
=-2\mu R_m, \\
m\gamma^0_m r^m-\kappa\overline{\gamma^0_{2-m}}r^{2-m}+c^0_{m-2}r^{m-2}+\frac{\mu}{(3-m)(\lambda+\mu)}\overline{\gamma^2_{2-m}}r^{4-m}- \\
-\frac{1}{2}\gamma^2_m r^{2+m}+\frac{\lambda+3\mu}{4(1-m)(\lambda+2\mu)}c^2_{m-2}r^m+\frac{\lambda+\mu}{4(2-m)(1-m)(\lambda+2\mu)}\overline{c^2_{-m}}r^{2-m}+ \\
+\frac{(\lambda+\mu)(\lambda+3\mu)}{4(\lambda+2\mu)}D^1_{m-1}-\frac{(\lambda+\mu)^2}{4(2-m)(\lambda+2\mu)}\overline{D^1_{1-m}}r^{2-m}+h_j[m\gamma^1_m r^m- \\
-\kappa\overline{\gamma^1_{2-m}}r^{2-m}+c^1_{m-2}r^{m-2}+\frac{(\lambda+\mu)}{2(\lambda+2\mu)}\frac{\sigma_{m-1}-\imath\nu_{m-1}}{r^{2m-2}+1}((\lambda+3\mu)- \\
-\frac{\lambda+\mu}{2-m})r^m]+h_j^2[m\gamma^2_m r^m-\kappa\overline{\gamma^2_{2-m}}r^{2-m}+c^2_{m-2}r^{m-2}]= \\
=-2\mu\overline{R_{2-m}}, \\
D^0_{m-1}r^{m-1}+\overline{D^0_{1-m}}r^{1-m}-\frac{1}{2\mu}(\gamma^1_m r^{m+1}+\overline{\gamma^1_{2-m}}r^{3-m})- \\
-\frac{(\lambda+\mu)^2}{2m(m-2)\mu(\lambda+2\mu)}\frac{\sigma_{m-1}-\imath\nu_{m-1}}{r^{2m-2}+1}r^{m+1}- \\
-\frac{\lambda+2\mu}{2\mu}(D^2_{m-1}\frac{r^{m+1}}{m}-\overline{D^2_{1-m}}\frac{r^{3-m}}{m-2})+h_j[D^1_{m-1}r^{m-1}+ \\
+\overline{D^1_{1-m}}r^{1-m}-\frac{1}{2\mu}(\gamma^2_m r^{m+1}+\overline{\gamma^2_{2-m}}r^{3-m})]+ \\
+h_j^2[D^2_{m-1}r^{m-1}+\overline{D^2_{1-m}}r^{1-m}]=P_{m-1}-\imath Q_{m-1}.
\end{array}\right.$$

CHAPTER 6

Spline-interpolation solution for a solid of revolution

Elena A. Shirokova,
Kazan Federal University
Elena.Shirokova@ksu.ru
Pyotr N. Ivanshin,
Kazan Federal University
pivanshin@gmail.com

ABSTRACT: In this chapter we present the spline-interpolation solution of the 3D second basic elasticity problem for a solid of revolution. The methods of gluing the displacements at the ends are analogous to that for the circular cylinder.

The problem of restoration of the displacements and stresses in the solids of revolution via the given boundary displacements is very important. We change the boundary data given at the whole surface of a solid by data given at the finite number of the circles which are directrices of the body and at the ends of the body.

It is impossible to find the boundary values of the coefficients of powers of h via the given data at a finite number of circles on the surface of a solid of revolution different from a cylinder for a polynomial solution of the equilibrium equations because the boundaries of sections at different levels are also different. So we are not able to construct the interpolation solution for these elastic solids. In this chapter we present the spline-interpolation solution for solids of revolution. This method uses only two levels bounding the layer.

6.1 Spline of the first type

Let us denote by f the function generating surface S of the body of revolution M, the axis of revolution being the axis OH. We construct spline-interpolation solution of the system (1.1) for the layer

$$M_k = \{(x, y, h)|(x, y, h) \in M, h \in [h_k, h_{k+1}]\},$$

which meets the displacements given at the edges of M_k.

The solution of the system (1.1) can be constructed in the form of the spline of degree 1 on the variable h meeting the boundary conditions:

$$u(f(h_k)\cos\theta, f(h_k)\sin\theta, h_k) = -\frac{1}{2\mu}\sum_{j=0}^{P}(a_j^1\cos(j\theta) + b_j^1\sin(j\theta)),$$

$$v(f(h_k)\cos\theta, f(h_k)\sin\theta, h_k) = -\frac{1}{2\mu}\sum_{j=0}^{P}(p_j^1\cos(j\theta) + q_j^1\sin(j\theta)),$$

$$w(f(h_k)\cos\theta, f(h_k)\sin\theta, h_k) = -\frac{1}{2\mu}\sum_{j=0}^{P}(t_j^1\cos(j\theta) + g_j^1\sin(j\theta)), \quad (6.1)$$

$$u(f(h_{k+1})\cos\theta, f(h_{k+1})\sin\theta, h_{k+1}) = -\frac{1}{2\mu}\sum_{j=0}^{P}(a_j^2\cos(j\theta) + b_j^2\sin(j\theta)),$$

$$v(f(h_{k+1})\cos\theta, f(h_{k+1})\sin\theta, h_{k+1}) = -\frac{1}{2\mu}\sum_{j=0}^{P}(p_j^2\cos(j\theta) + q_j^2\sin(j\theta)),$$

$$w(f(h_{k+1})\cos\theta, f(h_{k+1})\sin\theta, h_{k+1}) = -\frac{1}{2\mu}\sum_{j=0}^{P}(t_j^2\cos(j\theta) + g_j^2\sin(j\theta)),$$

where θ is the polar angle of a point of an edge.

We construct the solution of (1.1) in the layer M_k as the linear function on h:

$$u(x,y,h) = u_0(x,y) + u_1(x,y)h,\ v(x,y,h) = v_0(x,y) + v_1(x,y)h$$
$$w(x,y,h) = w_0(x,y) + w_1(x,y)h,\ x^2+y^2 \le \max_{h\in[h_k,h_{k+1}]}|f(h)|. \quad (6.2)$$

According to the results of Chapter 1 after introducing the complex variable $z = x + \imath y$ we have the following representation of the displacement components which satisfy system (1.1):

$$u(x,y,h) + \imath v(x,y,h) = -\frac{1}{2\mu}[-\frac{\lambda+3\mu}{\lambda+\mu}\phi_0(z) + z\overline{\phi_0'(z)} + \overline{\psi_0(z)} +$$
$$+(\lambda+\mu)(-\frac{\lambda+\mu}{4(\lambda+2\mu)}\rho_1(z) + \frac{\lambda+3\mu}{4(\lambda+2\mu)}z\overline{\rho_1'(z)})] -$$
$$-\frac{1}{2\mu}[-\frac{\lambda+3\mu}{\lambda+\mu}\phi_1(z) + z\overline{\phi_1'(z)} + \overline{\psi_1(z)}]h, \quad (6.3)$$
$$w(x,y,h) = \operatorname{Re}\rho_0'(z) - \frac{z\overline{\phi_1(z)} + \overline{z}\phi_1(z)}{4\mu} + \operatorname{Re}\rho_1'(z)h,$$

here $\phi_0(z), \phi_1(z), \psi_0(z)$, $\psi_1(z)$, $\rho_0(z)$, $\rho_1(z)$ are functions analytic in the circle with the radius $\max\{f(h)|h\in[h_k,h_{k+1}]\}$. The boundary data in the form of trigonometric polynomials (6.1) provide that all these functions are polynomials. Now we find these functions using the boundary conditions (6.1).

We can reconstruct analytic functions via their real or imaginary parts restrictions to the boundary, so we use the following procedure: we replace $z\overline{\phi}(\overline{z})$ with $\overline{z}\phi(z)$ in the case of real part of the given boundary condition and by $-\overline{z}\phi(z)$ in the other case. And the boundary values of $\overline{z}$ we replace by $\frac{f(h_j)^2}{z}$.

We consider the meromorphic functions

$$F_{\pm}^{j}(z)=-\frac{1}{2\mu}[-\frac{\lambda+3\mu}{\lambda+\mu}\phi_0(z)\pm f^2(h_j)z^{-1}\phi_0'(z)\pm\psi_0(z)+$$

$$+(\lambda+\mu)(-\frac{\lambda+\mu}{4(\lambda+2\mu)}\rho_1(z)\pm\frac{\lambda+3\mu}{4(\lambda+2\mu)}f^2(h_j)z^{-1}\rho_1'(z))]-\frac{1}{2\mu}[-\frac{\lambda+3\mu}{\lambda+\mu}\phi_1(z)+$$

$$\pm f^2(h_j)z^{-1}\phi_1'(z)\pm\psi_1(z)]h_j,\ j=k,k+1.$$

The real part of $F_{+}^{j}(z)$ and the imaginary part of $F_{-}^{j}(z)$ coincide with the given in (6.1) values of u and v, respectively, at the boundaries of the disks on the levels $h=h_j$, $j=k,k+1$.

We consider also the meromorphic function

$$G^j(z)=\rho_0'(z)-\frac{f^2(h_j)z^{-1}\phi_1(z)}{2\mu}+\rho_1'(z)h_j,\ j=k,k+1.$$

The real part of this function coincides with the given in (6.1) values of w at the boundaries of the disks on the levels $h=h_j$, $j=k,k+1$.

Each of the introduced meromorphic functions has the simple pole in 0. These meromorphic functions together with the given boundary conditions (6.1) produce the following system on the unknown functions $\phi_0(z)$, $\psi_0(z)$, $\rho_0(z)$, $\phi_1(z)$, $\psi_1(z)$, $\rho_1(z)$:

$$\begin{aligned}-\frac{1}{2\mu}[-\frac{\lambda+3\mu}{\lambda+\mu}\phi_0(z)+f^2(h_k)z^{-1}\phi_0'(z)+\psi_0(z)+\\+(\lambda+\mu)(-\frac{\lambda+\mu}{4(\lambda+2\mu)}\rho_1(z)+\frac{\lambda+3\mu}{4(\lambda+2\mu)}f^2(h_k)z^{-1}\rho_1'(z))]-\\-\frac{1}{2\mu}[-\frac{\lambda+3\mu}{\lambda+\mu}\phi_1(z)+f^2(h_k)z^{-1}\phi_1'(z)+\psi_1(z)]h_k=\\=\sum_{j=0}^{N}c_j^1z^j+f(h_k)^2T_1/z-\overline{T_1}z,\end{aligned}\qquad(6.4)$$

$$-\frac{1}{2\mu}[-\frac{\lambda+3\mu}{\lambda+\mu}\phi_0(z)-f^2(h_k)z^{-1}\phi_0'(z)-$$
$$-\psi_0(z)+(\lambda+\mu)(-\frac{\lambda+\mu}{4(\lambda+2\mu)}\rho_1(z)- \tag{6.5}$$
$$-\frac{\lambda+3\mu}{4(\lambda+2\mu)}f^2(h_k)z^{-1}\rho_1'(z))]-\frac{1}{2\mu}[-\frac{\lambda+3\mu}{\lambda+\mu}\phi_1(z)-$$
$$-f^2(h_k)z^{-1}\phi_1'(z)\psi_1(z)]h_k=\sum_{j=0}^{\infty}e_j^1z^j-f(h_k)^2T_1/z-\overline{T_1}z,$$

$$\rho_0'(z)-\frac{f^2(h_k)\phi_1(z)}{2\mu z}+\rho_1'(z)h_k=\sum_{j=0}^{N}d_j^1z^j+f(h_k)^2T_3/z-\overline{T_3}z, \tag{6.6}$$

$$-\frac{1}{2\mu}[-\frac{\lambda+3\mu}{\lambda+\mu}\phi_0(z)+f^2(h_{k+1})z^{-1}\phi_0'(z)+$$
$$+\psi_0(z)+(\lambda+\mu)(-\frac{\lambda+\mu}{4(\lambda+2\mu)}\rho_1(z)+ \tag{6.7}$$
$$+\frac{\lambda+3\mu}{4(\lambda+2\mu)}f^2(h_{k+1})z^{-1}\rho_1'(z))]-\frac{1}{2\mu}[-\frac{\lambda+3\mu}{\lambda+\mu}\phi_1(z)+$$
$$+f^2(h_k)z^{-1}\phi_1'(z)+\psi_1(z)]h_{k+1}=\sum_{j=0}^{\infty}c_j^2z^j+f(h_{k+1})^2T_2/z-\overline{T_2}z,$$

$$-\frac{1}{2\mu}[-\frac{\lambda+3\mu}{\lambda+\mu}\phi_0(z)-f^2(h_{k+1})z^{-1}\phi_0'(z)-$$
$$-\psi_0(z)+(\lambda+\mu)(-\frac{\lambda+\mu}{4(\lambda+2\mu)}\rho_1(z)- \tag{6.8}$$
$$-\frac{\lambda+3\mu}{4(\lambda+2\mu)}f^2(h_{k+1})z^{-1}\rho_1'(z))]-\frac{1}{2\mu}[-\frac{\lambda+3\mu}{\lambda+\mu}\phi_1(z)-$$
$$-f^2(h_{k+1})z^{-1}\phi_1'(z)\psi_1(z)]h_{k+1}k=\sum_{j=0}^{N}e_j^2z^j-f(h_{k+1})^2T_2/z-\overline{T_2}z,$$

$$\rho_0'(z)-\frac{f^2(h_{k+1})\phi_1(z)}{2\mu z}+\rho_1'(z)h_{k+1}=\sum_{j=0}^{N}d_j^2z^j+f(h_k)^2T_4/z-\overline{T_4}z, \tag{6.9}$$

here we denote

$$c_j^{l+1} = (a_j^{l+1} - \imath b_j^{l+1})/f^j(h_{k+l}),\ e_j^{l+1} = (q_j^{l+1} + \imath p_j^{l+1})/f^j(h_{k+l}),$$

$$d_j^{l+1} = (t_j^{l+1} - \imath g_j^{l+1})/f^j(h_{k+l}),\ l = 0, 1.$$

This system is resolved consequently beginning with the function ϕ_1 and the pair ρ_0 and ρ_1 down to ϕ_0 and ψ_0, ψ_1. First we add both sides of equation (6.4) to the corresponding parts of equation (6.5). Then we add (6.7) to (6.8). Then from the former sum we subtract the latter and obtain $\phi_1(z)$. We set this function into the system consisting of equations (6.9) and (6.6) in order to get $\rho_1'(z)$ and $\rho_0'(z)$. After this we obtain ϕ_0 from either sum of the equations (6.4) and (6.5) or that of (6.7) and (6.8). Finally from equations (6.4) and (6.7) we get $\psi_0(z)$ and $\psi_1(z)$.

The summands T_m/z and $\overline{T_m}z$, $m = 1, 2, 3, 4,$ appear on the right-hand side of all equations of the system because the meromorphic functions on the left-hand side have simple poles in 0. We find the values of T_m, $m = 1, 2, 3, 4,$ using the fact that all the functions ϕ_1, ρ_0, ρ_1, ϕ_0, ψ_0 and ψ_1 are analytic.

The solution of the system (6.4), (6.5), (6.6), (6.7), (6.8), (6.9) is easily constructed for the given polynomial boundary conditions (6.1), then the functions ϕ_1, ρ_0, ρ_1, ϕ_0, ψ_0 and ψ_1 themselves are polynomials in the complex variable z, $\overline{z}$ and h. But the solution in the general form is too large to be written in this paper. So we demonstrate the method with the help of the example of the boundary conditions given as trigonometric polynomials of degree no greater than 4.

6.2 Example of the conoid

Consider the part of axial-symmetric conoid solid laying between disks of radii 1 and 2 of the height 1.

We assume that our solution meets the following boundary displacements at two levels:

$$\begin{gathered}
u(\cos\theta,\sin\theta,1)=a_0+a_1\cos\theta-A_1\sin\theta+a_2\cos 2\theta-\\
-A_2\sin 2\theta+a_3\cos 3\theta-A_3\sin 3\theta+a_4\cos 4\theta-A_4\sin 4\theta,\\
v(\cos\theta,\sin\theta,1)=B_0+b_1\sin\theta+B_1\cos\theta+b_2\sin 2\theta+\\
+B_2\cos 2\theta+b_3\sin 3\theta+B_3\cos 3\theta+b_4\sin 4\theta+B_4\cos 4\theta,\\
w(\cos\theta,\sin\theta,1)=r_0+r_1\cos\theta-R_1\sin\theta+r_2\cos 2\theta-\\
-R_2\sin 2\theta+r_3\cos 3\theta-R_3\sin 3\theta,\\
u(2\cos\theta,2\sin\theta,0)=0,\quad v(2\cos\theta,2\sin\theta,0)=0,\\
w(2\cos\theta,2\sin\theta,0)=0.
\end{gathered}\tag{6.10}$$

According to the results of the previous section we search for the solution in the form:

$$\begin{gathered}
u(x,y,h)+\imath v(x,y,h)=-\frac{1}{2\mu}[-\frac{\lambda+3\mu}{\lambda+\mu}\phi_0(z)+z\overline{\phi_0'(z)}+\\
+\overline{\psi_0(z)}+(\lambda+\mu)(-\frac{\lambda+\mu}{4(\lambda+2\mu)}\rho_1(z)+\frac{\lambda+3\mu}{4(\lambda+2\mu)}z\overline{\rho_1'(z)})]-\\
-\frac{1}{2\mu}[-\frac{\lambda+3\mu}{\lambda+\mu}\phi_1(z)+z\overline{\phi_1'(z)}+\overline{\psi_1(z)}]h,\\
w(x,y,h)=\operatorname{Re}\rho_0'(z)-\frac{z\overline{\phi_1(z)}+\overline{z}\phi_1(z)}{4\mu}+\operatorname{Re}\rho_1'(z)h,
\end{gathered}\tag{6.11}$$

Restrictions of these functions to the lower and upper cone ends together with the condition (6.10) produce the following six equations on the real and imaginary parts of the zero boundary conditions which compose the system of differential equations over the functions $\phi_0(z)$, $\phi_1(z)$, $\psi_0(z)$, $\psi_1(z)$, $\rho_0'(z)$ and $\rho_1'(z)$. Note that since we can reconstruct analytic functions via their real or imaginary parts restrictions to the boundary we use the same procedure as in the previous section: we replace $z\overline{F}(\overline{z})$ with $\overline{z}F(z)$ in the case of real part of the given boundary condition and correspondingly with

$-\overline{z}F(z)$ in the other case. Since we only work with circles as boundaries the variable $\overline{z}$ can be replaced by $\frac{R^2}{z}$, here R is the radius of the boundary circle ($R = 1, 2$).

We use these boundary conditions to write the system of equations for our unknown functions. The system which brings the analytic functions $\phi_0(z)$, $\phi_1(z)$, $\psi_0(z)$, $\psi_1(z)$, $\rho_0'(z)$ and $\rho_1'(z)$ in this case is as follows:

$$
\begin{aligned}
-1/(2\mu)(-\kappa\phi_1(z) + 1/z\phi_1'(z) + \psi_1(z)) - 1/(2\mu)(-\kappa\phi_0(z) + \\
+1/z\phi_0'(z) + \psi_0(z) - \mu(\lambda+\mu)^2/(4\mu(\lambda+2\mu))\rho_1(z) + \\
+\mu(\lambda+\mu)(\lambda+3\mu)/(4\mu(\lambda+2\mu))1/z\rho_1'(z)) = \\
= a_0 + (a_1 + \imath A_1)z + (a_2 + \imath A_2)z^2 + (a_3 + \imath A_3)z^3 + \\
+(a_4 + \imath A_4)z^4 + (\imath C_1 + c_1)z - (-\imath C_1 + c_1)/z;
\end{aligned}
\tag{6.12}
$$

$$
\begin{aligned}
-1/(2\mu)(-\kappa\phi_1(z) - 1/z\phi_1'(z) - \psi_1(z)) - 1/(2\mu)(-\kappa\phi_0(z) - \\
-1/z\phi_0'(z) - \psi_0(z) - \mu(\lambda+\mu)^2/(4\mu(\lambda+2\mu))\rho_1(z) - \\
-\mu(\lambda+\mu)(\lambda+3\mu)/(4\mu(\lambda+2\mu))1/z\rho_1'(z)) = \\
= \imath B_0 + (b_1 + \imath B_1)z + (b_2 + \imath B_2)z^2 + (b_3 + \imath B_3)z^3 + \\
+(b_4 + \imath B_4)z^4 + (\imath C_1 + c_1)z + (-\imath C_1 + c_1)/z;
\end{aligned}
\tag{6.13}
$$

$$
\begin{aligned}
\rho_1'(z) + \rho_0'(z) - (1/z\phi_1(z))/(2\mu) = \\
= r_0 + (r_1 + \imath R_1)z + (r_2 + \imath R_2)z^2 + \\
+(r_3 + \imath R_3)z^3 + (\imath D_1 + d_1)z - (-\imath D_1 + d_1)/z;
\end{aligned}
\tag{6.14}
$$

$$\begin{aligned} &-1/(2\mu)(-\kappa\phi_0(z) + 4/z\phi_0'(z) + \psi_0(z) - \\ &\quad -\mu(\lambda+\mu)^2/(4\mu(\lambda+2\mu))\rho_1(z) + \\ &+\mu(\lambda+\mu)\rho_1'(z)(\lambda+3\mu)/(z\mu(\lambda+2\mu))) = \\ &\quad = ({}_1C_2 + c_2)z - (-{}_1C_2 + c_2)4/z; \end{aligned} \tag{6.15}$$

$$\begin{aligned} &-1/(2\mu)(-\kappa\phi_0(z) - 4/z\phi_0'(z) - \psi_0(z) - \\ &\quad -\mu(\lambda+\mu)^2/(4\mu(\lambda+2\mu))\rho_1(z) - \\ &-\mu(\lambda+\mu)\rho_1'(z)(\lambda+3\mu)/(z\mu(\lambda+2\mu))) = \\ &\quad = ({}_1C_2 + c_2)z + (-{}_1C_2 + c_2)4/z; \end{aligned} \tag{6.16}$$

$$\rho_0'(z) - (4/z\phi_1(z))/(2\mu) = ({}_1D_2 + d_2)z - (-{}_1D_2 + d_2)4/z \tag{6.17}$$

Note that the functions ϕ_0, ϕ_1, ψ_0, ψ_1, ρ_0', ρ_1' on the left-hand side of the equations are analytic. Since we replace $\overline{z}$ by R^2/z with $R = 1, 2$, we have meromorphic functions at the left-hand sides of equations (6.12), (6.13), (6.14), (6.15), (6.16) and (6.17) with simple poles in 0. So we must describe the behavior of these functions on the right-hand side of the equations without losing boundary conditions. In order to do this we introduce the additional constants d_1, D_1, d_2, D_2, c_1, C_1, c_2, C_2.

Let us consider the constants as follows:

$$a_0 = 0.01,\ B_0 = -0.02,$$

$$A_1 = 0.1,\ a_1 = 0,\ a_2 = 0.2,$$

$$A_2 = -0.1,\ a_3 = 0.1,\ A_3 = 0,\ a_4 = -0.01,$$

$$A_4 = 0,\ B_1 = -0.2,\ b_1 = 0,\ b_2 = -0.02,$$

$$B_2 = 0,\ b_3 = 0,\ B_3 = 0.1,\ b_4 = 0.02,$$

$$B_4 = 0,\ r_0 = 0.1,\ r_1 = 0.01,\ R_1 = 0,$$

$$r_2 = -0.02,\ R_2 = 0,\ r_3 = -0.1,\ R_3 = 0,\ \lambda = 15/13(10)^{11},\ \mu = 10/13(10)^{11}.$$

Then according to the formulas of the system solution (6.11) we have

$$(u+\imath v)(z,\overline{z},h) = -0.12037((0.00310+$$

$$+0.81425\imath)+\overline{z})((0.95074-1.58348\imath)+\overline{z})+0.03009(\overline{z})((0.00310+$$

$$+0.81425\imath)+\overline{z})((0.95074-1.58348\imath)+\overline{z})z+$$

$$+h(-0.015((-5.58808-2.29566\imath)+\overline{z})((-0.28626-1.98247\imath)+$$

$$+\overline{z})((-0.01438+0.53146\imath)+\overline{z})((2.55539+$$

$$+0.41334\imath)+\overline{z})-0.01111((-0.27943+$$

$$+0.66233\imath)+\overline{z})((0.50475+0.24914\imath)+$$

$$+\overline{z})((7.27467-8.41147\imath)+\overline{z})z+(0.09-$$

$$-0.05\imath)z^2+(0.05+0.05\imath)z^3+0.005z^4);$$

$$w(z,\overline{z},h) = ((0.1-0.05556\imath)-(0.025+0.0139\imath)\overline{z}^2-(0.0139-0.0139\imath)\overline{z}^3-$$

$$-0.00139\overline{z}^4)z+((0.1+0.05556\imath)-(0.025-0.0139\imath)z^2-(0.0139+0.0139\imath)z^3-$$

$$-0.00139z^4)\overline{z}+\overline{z}^2(\overline{z}0.00556+((0.0555556-0.0555556\imath)))+$$

$$+((0.0555556+0.0555556\imath)+0.00556z)z^2+h(-0.054167(z^3+\overline{z}^3)-$$

$$-(0.0517-0.0417\imath)\overline{z}^2-(0.0517+0.0417\imath)z^2-(0.07-0.042\imath)z-$$

$$-(0.07+0.042\imath)\overline{z}+0.1)$$

The deformation of the side surface of the conoid is shown in figure 6.1.

Also in figure 6.2 we can see how rather large deformation of the upper level becomes less pronounced as we pass from top to bottom.

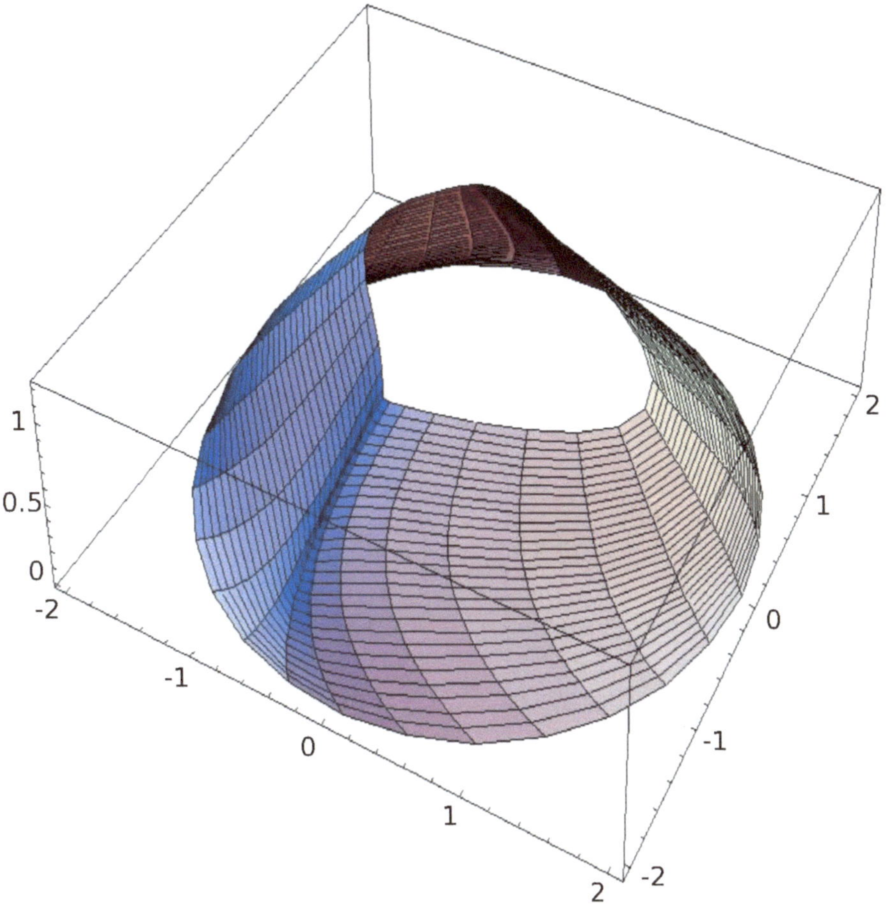

Figure 6.1: Deformations of the conoid side surface.

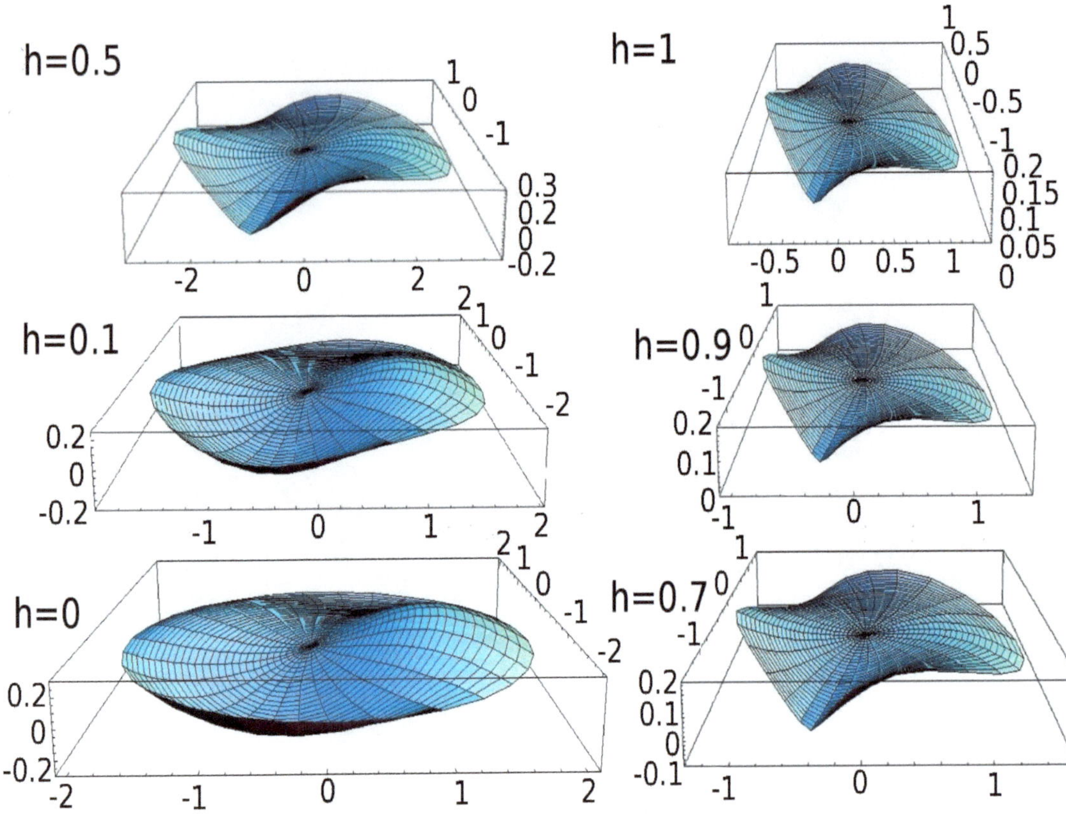

Figure 6.2: Deformations of the conoid sections by the planes parallel to XOY

6.3 Spline of the second type

The linear spline-interpolation solution gives the solution of system (1.1) which satisfies data given at a finite number directrices of a body of revolution. But this solution ignores the data at the points of the ends of this body, moreover this solution is discontinuous at the points of any disk which lays on the plane where one of the given directrices lays.

6.3.1 Gluing over one end

Here we construct the solution of the system (1.1) which meets the given data at the points of the edge of one of the ends of the axial-symmetric layer M_k and meets the values equal to the given function at the other end of the layer.

Let us formulate the problem. Given the boundary displacements at the points of the boundary circle C_{k+1} on the level $h = h_{k+1}$ and also the displacements (U_k, V_k, W_k) at the points of the disc D_k on the level $h = h_k$, it should be possible to restore the displacements in the axial-symmetric layer M_k which satisfy equilibrium equations (1.1) and take the given boundary displacements.

We suppose, as in the case of the cylinder, that the given functions $U_k + \imath V_k$ and W_k are three-harmonic in the plane XOY and also the given three-harmonic function $U_k + \imath V_k$ has the following representation:

$$U_A + \imath V_A = F_0(z) + \overline{G_0(z)} + |z|^2[F_1(z) + \overline{G_1(z)}] + |z|^4[A_0^2 + \overline{G_0(z)}],$$

$$W_A = \frac{1}{2}[H_0(z) + \overline{H_0(z)}] + \frac{|z|^2}{2}[H_1(z) + \overline{H_1(z)}] + \\ + \frac{|z|^4}{2}[H_2(z) + \overline{H_2(z)}],$$

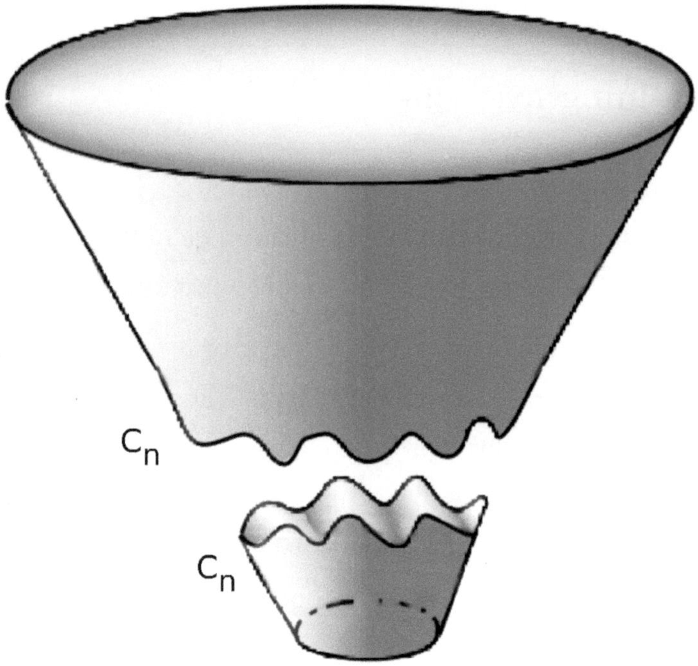

Figure 6.3: Gluing the displacements at the fragments over the common end

where

$$F_j(z) = \sum_{p=0}^{n} A_p^j z^p, \; G_j(z) = \sum_{p=1}^{n} B_p^j z^p$$

$$H_j(z) = \sum_{p=0}^{n} D_p^j z^p,$$

A_0^2 is a complex constant.

Note that the coefficient of $|z|^4$ in the representation of the function $U_k + \imath V_k$ does not contain the analytic summand. The class of such functions is rather wide, it contains all biharmonic functions.

Assume that the radius of the circle C_{k+1} equals 1 and $h_{k+1} = 1$. This allows us to represent the given data on the second (upper) level boundary circle in terms of polar angle of a point of the circle as follows:

$$\begin{aligned} u(x,y,h)|_{h=1,x^2+y^2=1} &= u(\cos\theta, \sin\theta, 1) = \\ &= \sum_{k=0}^{M} (\alpha_k^1 \cos k\theta - \dot{\alpha}_k^1 \sin k\theta), \\ v(x,y,h)|_{h=1,x^2+y^2=1} &= w(\cos\theta, \sin\theta, 1) = \\ &= \sum_{k=0}^{M} (\dot{\beta}_k^1 \cos k\theta + \beta_k^1 \sin k\theta), \\ w(x,y,h)|_{h=1,x^2+y^2=1} &= w(\cos\theta, \sin\theta, 1) = \\ &= \sum_{k=0}^{M} (\gamma_k^1 \cos k\theta - \dot{\gamma}_k^1 \sin k\theta). \end{aligned} \tag{6.18}$$

We assume that $h_k = 0$ and search for the solution in the form

$$\begin{aligned} u(x,y,h) &= u_0(x,y) + u_1(x,y)h + u_2(x,y)h^2 + u_3(x,y)h^3, \\ v(x,y,h) &= v_0(x,y) + v_1(x,y)h + v_2(x,y)h^2 + v_3(x,y)h^3, \\ w(x,y,h) &= w_0(x,y) + w_1(x,y)h + w_2(x,y)h^2 + w_3(x,y)h^3. \end{aligned}$$

After we substitute these displacement components in the equilibrium equation we get representations (1.6) and (1.7) for $u_k(x,y)+\ \imath v_k(x,y)$ and $w_k(x,y)$ which contain the functions $\Phi_k(z,\overline{z})$ and $\Psi_k(z,\overline{z})$ from (1.8) and (1.9), $k=0,1,2$.

So we have due to the results of Chapter 1 the following representation for the first pair of coordinates:

$$(u+\imath v)(z,\overline{z},h)=-\frac{\overline{\psi_0}(\overline{z})}{2\mu}+\frac{3\lambda\overline{z}f_2(z)}{4(\lambda+2\mu)^2}+\frac{\lambda^2\overline{z}f_2(z)}{4\mu(\lambda+2\mu)^2}+\frac{\mu\overline{z}f_2(z)}{2(\lambda+2\mu)^2}-$$

$$-\frac{\lambda\overline{z}f_2(z)}{(\lambda+\mu)(\lambda+2\mu)}-\frac{\lambda^2\overline{z}f_2(z)}{4\mu(\lambda+\mu)(\lambda+2\mu)}-\frac{5\mu\overline{z}f_2(z)}{4(\lambda+\mu)(\lambda+2\mu)}-\frac{g_2(z)}{8(\lambda+2\mu)}-$$

$$-\frac{\lambda g_2(z)}{8\mu(\lambda+2\mu)}+\frac{\kappa\phi_0(z)}{2\mu}+\frac{\lambda r_1(z)}{4(\lambda+2\mu)}+\frac{\lambda^2 r_1(z)}{8\mu(\lambda+2\mu)}+\frac{\mu r_1(z)}{8(\lambda+2\mu)}+\frac{3\lambda^2\overline{z}R_3(z)}{8(\lambda+2\mu)^2}+$$

$$+\frac{9\lambda\mu\overline{z}R_3(z)}{8(\lambda+2\mu)^2}+\frac{3\mu^2\overline{z}R_3(z)}{4(\lambda+2\mu)^2}+\frac{3\lambda z^2\overline{f_2}'(\overline{z})}{8(\lambda+2\mu)^2}+\frac{\lambda^2 z^2\overline{f_2}'(\overline{z})}{8\mu(\lambda+2\mu)^2}+\frac{\mu z^2\overline{f_2}'(\overline{z})}{4(\lambda+2\mu)^2}+$$

$$+\frac{3z^2\overline{f_2}'(\overline{z})}{8(\lambda+2\mu)}+\frac{\lambda z^2\overline{f_2}'(\overline{z})}{8\mu(\lambda+2\mu)}+\frac{3z\overline{g_2}'(\overline{z})}{8(\lambda+2\mu)}+\frac{\lambda z\overline{g_2}'(\overline{z})}{8\mu(\lambda+2\mu)}-\frac{z\overline{\phi_0}'(\overline{z})}{2\mu}-\frac{\lambda z\overline{r_1}'(\overline{z})}{2(\lambda+2\mu)}-$$

$$-\frac{\lambda^2 z\overline{r_1}'(\overline{z})}{8\mu(\lambda+2\mu)}-\frac{3\mu z\overline{r_1}'(\overline{z})}{8(\lambda+2\mu)}+\frac{9\lambda^2 z^2\overline{R_3}'(\overline{z})}{8(\lambda+2\mu)^2}+\frac{3\lambda^3 z^2\overline{R_3}'(\overline{z})}{16\mu(\lambda+2\mu)^2}+\frac{33\lambda\mu z^2\overline{R_3}'(\overline{z})}{16(\lambda+2\mu)^2}+$$

$$+\frac{9\mu^2 z^2\overline{R_3}'(\overline{z})}{8(\lambda+2\mu)^2}+h\Big(-\frac{\overline{\psi_1}(\overline{z})}{2\mu}+\frac{3\overline{z}f_3(z)}{4(\lambda+2\mu)}+\frac{3\lambda\overline{z}f_3(z)}{4\mu(\lambda+2\mu)}-\frac{3\lambda\overline{z}f_3(z)}{(\lambda+\mu)(\lambda+2\mu)}-$$

$$-\frac{3\lambda^2\overline{z}f_3(z)}{4\mu(\lambda+\mu)(\lambda+2\mu)}-\frac{15\mu\overline{z}f_3(z)}{4(\lambda+\mu)(\lambda+2\mu)}-\frac{3g_3(z)}{8(\lambda+2\mu)}-\frac{3\lambda g_3(z)}{8\mu(\lambda+2\mu)}+\frac{\kappa\phi_1(z)}{2\mu}+$$

$$+\frac{\lambda r_2(z)}{2(\lambda+2\mu)}+\frac{\lambda^2 r_2(z)}{4\mu(\lambda+2\mu)}+\frac{\mu r_2(z)}{4(\lambda+2\mu)}+\frac{3z^2\overline{f_3}'(\overline{z})}{2(\lambda+2\mu)}+\frac{3\lambda z^2\overline{f_3}'(\overline{z})}{4\mu(\lambda+2\mu)}+\frac{9z\overline{g_3}'(\overline{z})}{8(\lambda+2\mu)}+$$

$$+\frac{3\lambda z\overline{g_3}'(\overline{z})}{8\mu(\lambda+2\mu)}-\frac{z\overline{\phi_1}'(\overline{z})}{2\mu}-\frac{\lambda z\overline{r_2}'(\overline{z})}{\lambda+2\mu}-\frac{\lambda^2 z\overline{r_2}'(\overline{z})}{4\mu(\lambda+2\mu)}-\frac{3\mu z\overline{r_2}'(\overline{z})}{4(\lambda+2\mu)}\Big)+$$

$$+h^2\Big(\frac{\kappa f_2'(z)}{2\mu}+\frac{3\lambda R_3'(z)}{4(\lambda+2\mu)}+\frac{3\lambda^2 R_3'(z)}{8\mu(\lambda+2\mu)}+\frac{3\mu R_3'(z)}{8(\lambda+2\mu)}-\frac{z\overline{f_2}''(\overline{z})}{2\mu}-\frac{\overline{g_2}''(\overline{z})}{2\mu}-$$

$$-\frac{3\lambda z\overline{R_3}''(\overline{z})}{2(\lambda+2\mu)}-\frac{3\lambda^2 z\overline{R_3}''(\overline{z})}{8\mu(\lambda+2\mu)}-\frac{9\mu z\overline{R_3}''(\overline{z})}{8(\lambda+2\mu)}\Big)+h^3\Big(\frac{\kappa f_3'(z)}{2\mu}-\frac{z\overline{f_3}''(\overline{z})}{2\mu}-\frac{\overline{g_3}''(\overline{z})}{2\mu}\Big);$$

The third coordinate then has the form:

$$w(z,\overline{z},h)=\frac{3\lambda z^2\overline{f_3}(\overline{z})}{16\mu^2}+\frac{3z^2\overline{f_3}(\overline{z})}{8\mu}-$$

$$-\frac{3\lambda z^2\overline{f_3}(\overline{z})}{8(\lambda+\mu)(\lambda+2\mu)}-\frac{3\lambda^3 z^2\overline{f_3}(\overline{z})}{16\mu^2(\lambda+\mu)(\lambda+2\mu)}-$$

$$-\frac{9\lambda^2 z^2\overline{f_3}(\overline{z})}{16\mu(\lambda+\mu)(\lambda+2\mu)}-\frac{3\lambda z\overline{g_3}(\overline{z})}{8(\lambda+\mu)(\lambda+2\mu)}-\frac{3\lambda^2 z\overline{g_3}(\overline{z})}{16\mu(\lambda+\mu)(\lambda+2\mu)}-$$

$$-\frac{3\mu z\overline{g_3}(\overline{z})}{16(\lambda+\mu)(\lambda+2\mu)}-\frac{z\overline{\phi_1}(\overline{z})}{4(\lambda+\mu)}-\frac{\lambda z\overline{\phi_1}(\overline{z})}{4\mu(\lambda+\mu)}-\frac{1}{2}z\overline{r_2}(\overline{z})-\frac{\lambda z\overline{r_2}(\overline{z})}{4\mu}+$$

$$+\frac{3\lambda^2 z\overline{r_2}(\overline{z})}{8(\lambda+\mu)(\lambda+2\mu)}+\frac{\lambda^3 z\overline{r_2}(\overline{z})}{8\mu(\lambda+\mu)(\lambda+2\mu)}+\frac{3\lambda\mu z\overline{r_2}(\overline{z})}{8(\lambda+\mu)(\lambda+2\mu)}+$$

$$+\frac{\mu^2 z\overline{r_2}(\overline{z})}{8(\lambda+\mu)(\lambda+2\mu)}+\frac{\overline{\rho_0}(\overline{z})}{2}+\frac{3\lambda\overline{z}^2 f_3(z)}{16\mu^2}+\frac{3\overline{z}^2 f_3(z)}{8\mu}-\frac{3\lambda\overline{z}^2 f_3(z)}{8(\lambda+\mu)(\lambda+2\mu)}-$$

$$-\frac{3\lambda^3\overline{z}^2 f_3(z)}{16\mu^2(\lambda+\mu)(\lambda+2\mu)}-\frac{9\lambda^2\overline{z}^2 f_3(z)}{16\mu(\lambda+\mu)(\lambda+2\mu)}-\frac{3\lambda\overline{z}g_3(z)}{8(\lambda+\mu)(\lambda+2\mu)}-$$

$$-\frac{3\lambda^2\overline{z}g_3(z)}{16\mu(\lambda+\mu)(\lambda+2\mu)}-\frac{3\mu\overline{z}g_3(z)}{16(\lambda+\mu)(\lambda+2\mu)}-\frac{\overline{z}\phi_1(z)}{4(\lambda+\mu)}-\frac{\lambda\overline{z}\phi_1(z)}{4\mu(\lambda+\mu)}-$$

$$-\frac{1}{2}\overline{z}r_2(z)-\frac{\lambda\overline{z}r_2(z)}{4\mu}+\frac{3\lambda^2\overline{z}r_2(z)}{8(\lambda+\mu)(\lambda+2\mu)}+\frac{\lambda^3\overline{z}r_2(z)}{8\mu(\lambda+\mu)(\lambda+2\mu)}+$$

$$+\frac{3\lambda\mu\overline{z}r_2(z)}{8(\lambda+\mu)(\lambda+2\mu)}+\frac{\mu^2\overline{z}r_2(z)}{8(\lambda+\mu)(\lambda+2\mu)}+\frac{\rho_0(z)}{2}+h(-\frac{z\overline{f_2}'(\overline{z})}{2(\lambda+\mu)}-\frac{\lambda z\overline{f_2}'(\overline{z})}{2\mu(\lambda+\mu)}+$$

$$+\frac{\overline{r_1}'(\overline{z})}{2}-\frac{3}{2}z\overline{R_3}'(\overline{z})-\frac{3lz\overline{R_3}'(\overline{z})}{4\mu}+\frac{3\lambda z\overline{R_3}'(\overline{z})}{4(\lambda+2\mu)}+\frac{3\lambda^2 z\overline{R_3}'(\overline{z})}{8\mu(\lambda+2\mu)}+\frac{3\mu z\overline{R_3}'(\overline{z})}{8(\lambda+2\mu)}-$$

$$-\frac{\overline{z}f_2'(z)}{2(\lambda+\mu)}-\frac{\lambda\overline{z}f_2'(z)}{2\mu(\lambda+\mu)}+\frac{r_1'(z)}{2}-\frac{3}{2}\overline{z}R_3'(z)-\frac{3\lambda\overline{z}R_3'(z)}{4\mu}+\frac{3\lambda\overline{z}R_3'(z)}{4(\lambda+2\mu)}+$$

$$+\frac{3\lambda^2\overline{z}R_3'(z)}{8\mu(\lambda+2\mu)}+\frac{3\mu\overline{z}R_3'(z)}{8(\lambda+2\mu)})+h^2(-\frac{3z\overline{f_3}'(\overline{z})}{4\mu}+\frac{\overline{r_2}'(\overline{z})}{2}-\frac{3\overline{z}f_3'(z)}{4\mu}+\frac{r_2'(z)}{2})+$$

$$+h^3(\frac{\overline{R_3}''(\overline{z})}{2}+\frac{R_3''(z)}{2}),$$

here we suppose that the introduced analytic in the plane XOY functions have the following representations

$$\phi_0(z)=\sum_{k=0}^{m} z^k\phi_{0,k},\ \phi_1(z)=\sum_{k=0}^{m} z^k\phi_{1,k},\ \psi_0(z)=\sum_{k=0}^{m} z^k\psi_{0,k},$$
$$\psi_1(z)=\sum_{k=0}^{m} z^k\psi_{1,k},\ r_1(z)=\sum_{k=0}^{m} z^k r_{1,k},\ \rho_0(z)=\sum_{k=2}^{m} z^k\rho_{0,k}, \tag{6.19}$$
$$f_2(z)=\sum_{k=1}^{m} z^k f_{2,k},\ f_3(z)=\sum_{k=1}^{m} z^k f_{3,k},\ g_2(z)=\sum_{k=2}^{m} z^k g_{2,k},$$
$$g_3(z)=\sum_{k=2}^{m} z^k g_{3,k},\ r_2(z)=\sum_{k=1}^{m} z^k r_{2,k},\ R_3(z)=\sum_{k=2}^{m} z^k R_{3,k}.$$

Recall that the radius of the circle C_{k+1} equals 1 and $h_{k+1}=1$. Then we get the equations on the real and imaginary parts of $(u+\imath v)|_{h=1,x^2+y^2=1}$.

For example, the boundary condition on the first component $u|_{h=1,x^2+y^2=1}$ of displacement vector gives

$$-\frac{\psi_0(z)}{2\mu}+\frac{3\lambda f_2(z)}{4z(\lambda+2\mu)^2}+\frac{\lambda^2 f_2(z)}{4z\mu(\lambda+2\mu)^2}+\frac{\mu f_2(z)}{2z(\lambda+2\mu)^2}-$$
$$-\frac{\lambda f_2(z)}{z(\lambda+\mu)(\lambda+2\mu)}-\frac{\lambda^2 f_2(z)}{4z\mu(\lambda+\mu)(\lambda+2\mu)}-\frac{5\mu f_2(z)}{4z(\lambda+\mu)(\lambda+2\mu)}-\frac{g_2(z)}{8(\lambda+2\mu)}-$$
$$-\frac{\lambda g_2(z)}{8\mu(\lambda+2\mu)}+\frac{\kappa\phi_0(z)}{2\mu}+\frac{\lambda r_1(z)}{4(\lambda+2\mu)}+\frac{\lambda^2 r_1(z)}{8\mu(\lambda+2\mu)}+\frac{\mu r_1(z)}{8(\lambda+2\mu)}+\frac{3\lambda^2 R_3(z)}{8z(\lambda+2\mu)^2}+$$
$$+\frac{9\lambda\mu R_3(z)}{8z(\lambda+2\mu)^2}+\frac{3\mu^2 R_3(z)}{4z(\lambda+2\mu)^2}+\frac{3\lambda f_2'(z)}{8z^2(\lambda+2\mu)^2}+\frac{\lambda^2 f_2'(z)}{8z^2\mu(\lambda+2\mu)^2}+\frac{\mu f_2'(z)}{4z^2(\lambda+2\mu)^2}+$$
$$+\frac{3f_2'(z)}{8z^2(\lambda+2\mu)}+\frac{\lambda f_2'(z)}{8z^2\mu(\lambda+2\mu)}+\frac{3g_2'(z)}{8z(\lambda+2\mu)}+$$
$$+\frac{\lambda g_2'(z)}{8z\mu(\lambda+2\mu)}-\frac{\phi_0'(z)}{2z\mu}-\frac{\lambda r_1'(z)}{2z(\lambda+2\mu)}-$$
$$-\frac{\lambda^2 r_1'(z)}{8z\mu(\lambda+2\mu)}-\frac{3\mu r_1'(z)}{8z(\lambda+2\mu)}+\frac{9\lambda^2 R_3'(z)}{8z^2(\lambda+2\mu)^2}+$$
$$+\frac{3\lambda^3 R_3'(z)}{16z^2\mu(\lambda+2\mu)^2}+\frac{33\lambda\mu R_3'(z)}{16z^2(\lambda+2\mu)^2}+$$

$$+\frac{9\mu^2 R_3'(z)}{8z^2(\lambda+2\mu)^2}+(-\frac{\psi_1(z)}{2\mu}+\frac{3f_3(z)}{4z(\lambda+2\mu)}+\frac{3\lambda f_3(z)}{4z\mu(\lambda+2\mu)}-\frac{3\lambda f_3(z)}{z(\lambda+\mu)(\lambda+2\mu)}-$$

$$-\frac{3\lambda^2 f_3(z)}{4z\mu(\lambda+\mu)(\lambda+2\mu)}-\frac{15\mu f_3(z)}{4z(\lambda+\mu)(\lambda+2\mu)}-\frac{3g_3(z)}{8(\lambda+2\mu)}-\frac{3\lambda g_3(z)}{8\mu(\lambda+2\mu)}+$$

$$+\frac{\kappa\phi_1(z)}{2\mu}+\frac{\lambda r_2(z)}{2(\lambda+2\mu)}+\frac{\lambda^2 r_2(z)}{4\mu(\lambda+2\mu)}+\frac{\mu r_2(z)}{4(\lambda+2\mu)}+$$

$$+\frac{3f_3'(z)}{2z^2(\lambda+2\mu)}+\frac{3\lambda f_3'(z)}{4z^2\mu(\lambda+2\mu)}+\frac{9g_3'(z)}{8z(\lambda+2\mu)}+$$

$$+\frac{3\lambda g_3'(z)}{8z\mu(\lambda+2\mu)}-\frac{\phi_1'(z)}{2z\mu}-\frac{\lambda r_2'(z)}{z(\lambda+2\mu)}-\frac{\lambda^2 r_2'(z)}{4z\mu(\lambda+2\mu)}-\frac{3\mu r_2'(z)}{4z(\lambda+2\mu)})+$$

$$+(\frac{\kappa f_2'(z)}{2\mu}+\frac{3\lambda R_3'(z)}{4(\lambda+2\mu)}+\frac{3\lambda^2 R_3'(z)}{8\mu(\lambda+2\mu)}+\frac{3\mu R_3'(z)}{8(\lambda+2\mu)}-\frac{f_2''(z)}{2z\mu}-\frac{g_2''(z)}{2\mu}-$$

$$-\frac{3\lambda R_3''(z)}{2z(\lambda+2\mu)}-\frac{3\lambda^2 R_3''(z)}{8z\mu(\lambda+2\mu)}-\frac{9\mu R_3''(z)}{8z(\lambda+2\mu)})+(\frac{\kappa f_3'(z)}{2\mu}-\frac{f_3''(z)}{2z\mu}-\frac{g_3''(z)}{2\mu})=$$

$$=\sum_{k=0}^{M}(\alpha_k^1+\imath\dot{\alpha}_k^1)z^k+T_1/z+\overline{T_1}z+D_1/z^2+\overline{D_1}z^2.$$

Similarly we consider the rest of the components.

Here we must set $M \leq m-4$ in order to provide vanishing of remainders of the series (6.19).

Now assume that the radius of the disk D_k equals Q. Then we get the following equations for $(u+\imath v)|_{h=0,x^2+y^2\leq Q^2}$:

$$-\frac{\overline{\psi_0(\overline{z})}}{2\mu}+\frac{3\lambda\overline{z}f_2(z)}{4(\lambda+2\mu)^2}+\frac{\lambda^2\overline{z}f_2(z)}{4\mu(\lambda+2\mu)^2}+\frac{\mu\overline{z}f_2(z)}{2(\lambda+2\mu)^2}-$$

$$-\frac{\lambda\overline{z}f_2(z)}{(\lambda+\mu)(\lambda+2\mu)}-\frac{\lambda^2\overline{z}f_2(z)}{4\mu(\lambda+\mu)(\lambda+2\mu)}-\frac{5\mu\overline{z}f_2(z)}{4(\lambda+\mu)(\lambda+2\mu)}-\frac{g_2(z)}{8(\lambda+2\mu)}-$$

$$-\frac{\lambda g_2(z)}{8\mu(\lambda+2\mu)}+\frac{\kappa\phi_0(z)}{2\mu}+\frac{\lambda r_1(z)}{4(\lambda+2\mu)}+\frac{\lambda^2 r_1(z)}{8\mu(\lambda+2\mu)}+\frac{\mu r_1(z)}{8(\lambda+2\mu)}+\frac{3\lambda^2\overline{z}R_3(z)}{8(\lambda+2\mu)^2}+$$

$$+\frac{9\lambda\mu\overline{z}R_3(z)}{8(\lambda+2\mu)^2}+\frac{3\mu^2\overline{z}R_3(z)}{4(\lambda+2\mu)^2}+\frac{3\lambda z^2\overline{f_2}'(\overline{z})}{8(\lambda+2\mu)^2}+\frac{\lambda^2 z^2\overline{f_2}'(\overline{z})}{8\mu(\lambda+2\mu)^2}+\frac{\mu z^2\overline{f_2}'(\overline{z})}{4(\lambda+2\mu)^2}+$$

$$+\frac{3z^2\overline{f_2}'(\overline{z})}{8(\lambda+2\mu)}+\frac{\lambda z^2\overline{f_2}'(\overline{z})}{8\mu(\lambda+2\mu)}+\frac{3z\overline{g_2}'(\overline{z})}{8(\lambda+2\mu)}+\frac{\lambda z\overline{g_2}'(\overline{z})}{8\mu(\lambda+2\mu)}-\frac{z\overline{\phi_0}'(\overline{z})}{2\mu}-\frac{\lambda z\overline{r_1}'(\overline{z})}{2(\lambda+2\mu)}-$$

$$-\frac{\lambda^2 z\overline{r_1}'(\overline{z})}{8\mu(\lambda+2\mu)}-\frac{3\mu z\overline{r_1}'(\overline{z})}{8(\lambda+2\mu)}+\frac{9\lambda^2 z^2\overline{R_3}'(\overline{z})}{8(\lambda+2\mu)^2}+\frac{3\lambda^3 z^2\overline{R_3}'(\overline{z})}{16\mu(\lambda+2\mu)^2}+\frac{33\lambda\mu z^2\overline{R_3}'(\overline{z})}{16(\lambda+2\mu)^2}+$$

$$+\frac{9\mu^2 z^2\overline{R_3}'(\overline{z})}{8(\lambda+2\mu)^2}=F_0(z)+\overline{G_0(z)}+|z|^2[F_1(z)+\overline{G_1(z)}]+|z|^4[A_0^2+\overline{G_0(z)}]$$

Similarly we reconstruct the relation on the third component.

We use the relations on the lower end and on the upper boundary circle of the layer in order to obtain the coefficients of the polynomials ϕ_0, ϕ_1, ψ_0, ψ_1, $f_2(z)$, $f_3(z)$, $g_2(z)$, $g_3(z)$, $r_2(z)$, $R_3(z)$, $r_1(z)$ and $\rho_0(z)$.

On the lower end we deal with polyharmonic functions so we have also coefficients of the products of the type $z^p\overline{z}^q$. We equate the coefficients of the same powers of $z^q\overline{z}^p$ and get the linear system which consists of five sets of equations (three on the upper circle and two on the lower end) with respect to the coefficients $\phi_{0,k}$, $\phi_{1,k}$, $\psi_{0,k}$, $\psi_{1,k}$, $\rho_{0,k}$, $r_{1,k}$, $f_{l,k}$, $g_{l,k}$, $r_{l,k}$, $l=2,3$, $k\in\{0,1,\ldots,m\}$. Two sets of equations on the lower end split into 8 independent subsets as coefficients of different $z^p\overline{z}^q$. Recall that there are three equations of the coefficients of the same powers of $e^{\imath\theta}$ which appear when we compare the values of $u(\cos\theta,\sin\theta,1)+\imath v(\cos\theta,\sin\theta,1)$ and $w(\cos\theta,\sin\theta,1)$ with the given values of the displacements at the points of the circle C_{k+1}. Thus the number of the sets of the equations of the system is 11 and the number of the sets of the free coefficients is 12, so the solution of the system depends on one of the sets of the coefficients, namely, on the coefficients $g_{3,k}$, $k=1,2,...,m-1$. Hence we have some additional degrees of freedom in our solution. These additional variables can be used, for example, for smoothing of the spline-interpolation solution at the edge C_k.

The solution should be taken in the form of the polynomial of the odd power more than 3 in h if the order of harmonicity of the given displacements at the end D_A is more than 3.

6.3.2 Point-wise gluing of the ends

Suppose that we approximate the given boundary conditions at the ends $h = h_k$ and $h = h_{k+1}$ by polynomials with fixed values at the given set of points at these ends. So we glue the displacements defined for the adjacent layers at the given finite set of points at the corresponding sections. We can assume that the spline meets null displacements at the edges of the layer, otherwise we add the spline of the first type with the corresponding data at the edges.

Let us construct the spline of the second type for the section $[h_k, h_{k+1}]$. We must construct the solution of (1.1) vanishing on the boundary circles, i.e. the left-hand side in equations (6.1) is 0. Let us add to these conditions the given displacements values at the finite set of $(2m+1)$ points of the disk $x^2 + y^2 < f(h_k)^2$ on the level $h = h_k$ and the disc $x^2 + y^2 < f(h_{k+1})^2$ on the level $h = h_{k+1}$. For $h = h_0, h = h_N$ these points are the points of the given displacements at the ends of S.

The additional properties imply introduction of the new parameters. Thus we construct the solution of (1.1) in the form

$$u(x,y,h) + \imath v(x,y,h) = u_0(x,y) + \imath v_0(x,y) + [u_1(x,y) + \imath v_1(x,y)]h +$$

$$+[-\frac{\lambda+3\mu}{\lambda+\mu}\sum_{j=0}^{m} A_j z^j + z\sum_{j=1}^{m}\overline{A_j}j\overline{z}^{j-1} + \sum_{j=1}^{m}\overline{B_j}\overline{z}^j]h^2, \tag{6.20}$$

$$w(x,y,h) = w_0(x,y) + w_1(x,y)h + \frac{1}{2}[\sum_{j=0}^{m} C_j z^j + \sum_{j=0}^{m}\overline{C_j}\overline{z}^j]h^2,$$

here $A_k, k = 0,1,...,m$, B_k, C_k, $k = 1,\dots,m$, are arbitrary complex constants and C_0 is a real constant, so we have $6m+3$ arbitrary real constants.

Now we get the following representation of the solutions by putting (6.20) into (1.1):

$$u(x,y,h)+\imath v(x,y,h)=\frac{\mu}{\lambda+\mu}\overline{z}\sum_{j=1}^{m}A_j\frac{z^{j+1}}{j+1}-$$

$$-\frac{z^2}{2}\sum_{j=1}^{m}\overline{A_j}\overline{z}^j-\frac{\lambda+3\mu}{4(\lambda+2\mu)}z\sum_{j=1}^{m}\overline{B_j}\frac{\overline{z}^{j+1}}{j+1}+$$

$$+\frac{\lambda+\mu}{4(\lambda+2\mu)}\sum_{j=1}^{m}B_j\frac{z^{j+2}}{(j+1)(j+2)}+$$

$$+\frac{(\lambda+3\mu)^2}{4(\lambda+\mu)(\lambda+2\mu)}A_0|z|^2-\frac{\lambda+3\mu}{8(\lambda+2\mu)}\overline{A_0}z^2-$$

$$-\frac{(\lambda+\mu)(\lambda+3\mu)}{8\mu(\lambda+2\mu)}z\overline{H_1'(z)}+\frac{(\lambda+\mu)^2}{8\mu(\lambda+2\mu)}H_1(z)+$$

$$+\frac{\lambda+3\mu}{2\mu(\lambda+\mu)}\Phi_0(z)-\frac{1}{2\mu}z\overline{\Phi_0'(z)}-\frac{1}{2\mu}\overline{\Psi_0(z)}+$$

$$+[-\frac{(\lambda+\mu)(\lambda+3\mu)}{4\mu(\lambda+2\mu)}z\sum_{j=1}^{m}\overline{C_j}\overline{z}^j+$$

$$+\frac{(\lambda+\mu)^2}{4\mu(\lambda+2\mu)}\sum_{j=1}^{m}C_j\frac{z^{j+1}}{j+1}+\frac{\lambda+3\mu}{2\mu(\lambda+\mu)}\Phi_1(z)-$$

$$-\frac{1}{2\mu}z\overline{\Phi_1'(z)}-\frac{1}{2\mu}\overline{\Psi_1(z)}]h+[-\frac{\lambda+3\mu}{\lambda+\mu}\sum_{j=0}^{m}A_jz^j+ \tag{6.21}$$

$$+z\sum_{j=1}^{m}\overline{A_j}j\overline{z}^{j-1}+\sum_{j=1}^{m}\overline{B_j}\overline{z}^j]h^2,$$

$$w(x,y,h)=-\frac{\lambda^2+6\lambda\mu+7\mu^2}{8\mu(\lambda+2\mu)}[z\sum_{j=1}^{m}\overline{C_j}\frac{\overline{z}^{j+1}}{j+1}+$$

$$+\overline{z}\sum_{j=1}^{m}C_j\frac{z^{j+1}}{j+1}]-\frac{1}{4\mu}(\overline{z}\Phi_1(z)+z\overline{\Phi_1(z)})-$$

$$-\frac{(\lambda+2\mu)}{4\mu}(C_0+\overline{C_0})|z|^2+\frac{1}{2}(H_0'(z)+\overline{H_0'(z)})+[\overline{z}\sum_{j=1}^{m}A_jz^j+z\sum_{j=1}^{m}\overline{A_j}\overline{z}^j+$$

$$+\frac{1}{2}(H_1'(z)+\overline{H_1'(z)})]h+\frac{1}{2}[\sum_{j=0}^{m}C_jz^j+\sum_{j=0}^{m}\overline{C_j}\overline{z}^j]h^2,$$

$\Phi_0(z)$, $\Phi_1(z)$, $\Psi_0(z)$, $\Psi_1(z)$, $H_0(z)$, $H_1(z)$ being analytic functions.

Now we apply zero boundary conditions (6.1) to the solution (6.21) and get the following boundary relations for the functions $\Phi_0(z)$, $\Phi_1(z)$, $\Psi_0(z)$, $\Psi_1(z)$, $H_0(z)$, $H_1(z)$:

$$\frac{(\lambda+\mu)(\lambda+3\mu)}{8\mu(\lambda+2\mu)}f(h_k)e^{\imath\theta}\overline{H_1'(f(h_k)e^{\imath\theta})}-\frac{(\lambda+\mu)^2}{8\mu(\lambda+2\mu)}H_1(f(h_k)e^{\imath\theta})-$$

$$-\frac{\lambda+3\mu}{2\mu(\lambda+\mu)}\Phi_0(f(h_k)e^{\imath\theta})+\frac{1}{2\mu}f(h_k)e^{\imath\theta}\overline{\Phi_0'(f(h_k)e^{\imath\theta})}+\frac{1}{2\mu}\overline{\Psi_0(f(h_k)e^{\imath\theta})}-$$

$$-\left[\frac{\lambda+3\mu}{2\mu(\lambda+\mu)}\Phi_1(f(h_k)e^{\imath\theta})-\frac{1}{2\mu}f(h_k)e^{\imath\theta}\overline{\Phi_1'(f(h_k)e^{\imath\theta})}-\frac{1}{2\mu}\overline{\Psi_1(f(h_k)e^{\imath\theta})}\right]h_k=$$

$$=\left[\frac{\mu}{\lambda+\mu}\overline{z}\sum_{j=1}^{m}A_j\frac{z^{j+1}}{j+1}-\frac{z^2}{2}\sum_{j=1}^{m}\overline{A_j}\overline{z}^j-\frac{\lambda+3\mu}{4(\lambda+2\mu)}z\sum_{j=1}^{m}\overline{B_j}\frac{\overline{z}^{j+1}}{j+1}+\right.$$

$$+\frac{\lambda+\mu}{4(\lambda+2\mu)}\sum_{j=1}^{m}B_j\frac{z^{j+2}}{(j+1)(j+2)}+\frac{(\lambda+3\mu)^2}{4(\lambda+\mu)(\lambda+2\mu)}A_0|z|^2-\frac{\lambda+3\mu}{8(\lambda+2\mu)}\overline{A_0}z^2+$$

$$+[-\frac{(\lambda+\mu)(\lambda+3\mu)}{4\mu(\lambda+2\mu)}z\sum_{j=1}^{m}\overline{C_j}\overline{z}^j+\frac{(\lambda+\mu)^2}{4\mu(\lambda+2\mu)}\sum_{j=1}^{m}C_j\frac{z^{j+1}}{j+1}]h_k+$$

$$\left.+[-\frac{\lambda+3\mu}{\lambda+\mu}\sum_{j=0}^{m}A_jz^j+z\sum_{j=1}^{m}\overline{A_j}j\overline{z}^{j-1}+\sum_{j=1}^{m}\overline{B_j}\overline{z}^j]h_k^2\right]_{z=f(h_k)e^{\imath\theta}},$$

$$\frac{1}{4\mu}(f(h_k)e^{-\imath\theta}\Phi_1(f(h_k)e^{\imath\theta})+f(h_k)e^{\imath\theta}\overline{\Phi_1(f(h_k)e^{\imath\theta})})-$$

$$-\frac{1}{2}(H_0'(f(h_k)e^{\imath\theta})+\overline{H_0'(f(h_k)e^{\imath\theta})})-\frac{1}{2}(H_1'(f(h_k)e^{\imath\theta})+\overline{H_1'(f(h_k)e^{\imath\theta})})h_k=$$

$$=\left[-\frac{\lambda^2+6\lambda\mu+7\mu^2}{8\mu(\lambda+2\mu)}[z\sum_{j=1}^{m}\overline{C_j}\frac{\overline{z}^{j+1}}{j+1}+\overline{z}\sum_{j=1}^{m}C_j\frac{z^{j+1}}{j+1}]-\right.$$

$$-\frac{(\lambda+2\mu)}{4\mu}(C_0+\overline{C_0})|z|^2+[\overline{z}\sum_{j=1}^{m}A_jz^j+z\sum_{j=1}^{m}\overline{A_j}\overline{z}^j]h_k+$$

$$\left.+\frac{1}{2}[\sum_{j=0}^{m}C_jz^j+\sum_{j=0}^{m}\overline{C_j}\overline{z}^j]h_k^2\right]_{z=f(h_k)e^{\imath\theta}}.$$

$$\frac{(\lambda+\mu)(\lambda+3\mu)}{8\mu(\lambda+2\mu)}f(h_{k+1})e^{\imath\theta}\overline{H_1'(f(h_{k+1})e^{\imath\theta})}-\frac{(\lambda+\mu)^2}{8\mu(\lambda+2\mu)}H_1(f(h_{k+1})e^{\imath\theta})-$$

$$-\frac{\lambda+3\mu}{2\mu(\lambda+\mu)}\Phi_0(f(h_{k+1})e^{\imath\theta})+\frac{1}{2\mu}f(h_{k+1})e^{\imath\theta}\overline{\Phi_0'(f(h_{k+1})e^{\imath\theta})}+$$

$$+\frac{1}{2\mu}\overline{\Psi_0(f(h_{k+1})e^{\imath\theta})}-$$

$$-\left[\frac{\lambda+3\mu}{2\mu(\lambda+\mu)}\Phi_1(f(h_{k+1})e^{\imath\theta})-\frac{1}{2\mu}f(h_{k+1})e^{\imath\theta}\overline{\Phi_1'(f(h_{k+1})e^{\imath\theta})}-\right.$$

$$\left.-\frac{1}{2\mu}\overline{\Psi_1(f(h_{k+1})e^{\imath\theta})}\right]h_{k+1}=$$

$$=\left[\frac{\mu}{\lambda+\mu}\overline{z}\sum_{j=1}^{m}A_j\frac{z^{j+1}}{j+1}-\frac{z^2}{2}\sum_{j=1}^{m}\overline{A_j}\overline{z}^j-\frac{\lambda+3\mu}{4(\lambda+2\mu)}z\sum_{j=1}^{m}\overline{B_j}\frac{\overline{z}^{j+1}}{j+1}+\right.$$

$$+\frac{\lambda+\mu}{4(\lambda+2\mu)}\sum_{j=1}^{m}B_j\frac{z^{j+2}}{(j+1)(j+2)}+$$

$$+\frac{(\lambda+3\mu)^2}{4(\lambda+\mu)(\lambda+2\mu)}A_0|z|^2-\frac{\lambda+3\mu}{8(\lambda+2\mu)}\overline{A_0}z^2+$$

$$+[-\frac{(\lambda+\mu)(\lambda+3\mu)}{4\mu(\lambda+2\mu)}z\sum_{j=1}^{m}\overline{C_j}\overline{z}^j+\frac{(\lambda+\mu)^2}{4\mu(\lambda+2\mu)}\sum_{j=1}^{m}C_j\frac{z^{j+1}}{j+1}]h_{k+1}+$$

$$\left.+[-\frac{\lambda+3\mu}{\lambda+\mu}\sum_{j=0}^{m}A_jz^j+z\sum_{j=1}^{m}\overline{A_j}j\overline{z}^{j-1}+\sum_{j=1}^{m}\overline{B_j}\overline{z}^j]h_{k+1}^2\right]_{z=f(h_{k+1})e^{\imath\theta}},$$

$$\frac{1}{4\mu}(f(h_{k+1})e^{-\imath\theta}\Phi_1(f(h_{k+1})e^{\imath\theta})+f(h_{k+1})e^{\imath\theta}\overline{\Phi_1(f(h_{k+1})e^{\imath\theta})})-$$

$$-\frac{1}{2}(H_0'(f(h_{k+1})e^{\imath\theta})+\overline{H_0'(f(h_{k+1})e^{\imath\theta})})-$$

$$-\frac{1}{2}(H_1'(f(h_{k+1})e^{\imath\theta})+\overline{H_1'(f(h_{k+1})e^{\imath\theta})})h_{k+1}=$$

$$=\left[-\frac{\lambda^2+6\lambda\mu+7\mu^2}{8\mu(\lambda+2\mu)}[z\sum_{j=1}^{m}\overline{C_j}\frac{\overline{z}^{j+1}}{j+1}+\overline{z}\sum_{j=1}^{m}C_j\frac{z^{j+1}}{j+1}]-\right.$$

$$-\frac{(\lambda+2\mu)}{4\mu}(C_0+\overline{C_0})|z|^2+[\overline{z}\sum_{j=1}^{m}A_jz^j+z\sum_{j=1}^{m}\overline{A_j}\overline{z}^j]h_{k+1}+$$

$$\left.+\frac{1}{2}[\sum_{j=0}^{m}C_jz^j+\sum_{j=0}^{m}\overline{C_j}\overline{z}^j]h_{k+1}^2\right]_{z=f(h_{k+1})e^{\imath\theta}}.$$

The last boundary relations provide the system analogous to (6.4)–(6.9) with respect to the functions $\Phi_0(z)$, $\Phi_1(z)$, $\Psi_0(z)$, $\Psi_1(z)$, $H_0(z)$, $H_1(z)$. These analytic functions contain linearly the unknown constants A_j, B_j, C_j from (6.20). The unknown constants can be found via the given displacements values at the given set of $2m+1$ points of the ends of the layer using the linear system of $6m+3$ equations with $6m+3$ variables.

6.4 Smooth spline-interpolation

Suppose that we in addition to gluing splines over the boundary circles and the points at the sections have to glue the derivatives with respect to t of the splines at adjacent fragments.

So we have to solve the following problem for every fragment M_k. Given the vectors of displacements at the edges C_k, C_{k+1}:

$$(\tilde{u}_k(\theta), \tilde{v}_k(\theta), \tilde{w}_k(\theta)), \qquad (6.22)$$

$$(\tilde{u}_{k+1}(\theta), \tilde{v}_{k+1}(\theta), \tilde{w}_{k+1}(\theta))\theta \in [0, 2\pi], \ j = 1, 2, ..., l,$$

the displacements (U_j, V_j, W_j), $j = 1, ..., 2m+1$ at the points of the ends and the boundary values of n first derivatives of the displacements with respect to t at the curve C_k:

$$(\widehat{u}_p(\theta), \widehat{v}_p(\theta), \widehat{w}_p(\theta)), \ p = 1, ..., n, \qquad (6.23)$$

it should be possible to find the displacements which satisfy system (1.1) in M_k.

We search for the solution in the form

$$u(x, y, h) + \imath v(x, y, h) = \sum_{l=0}^{n+1} (u_l(x, y) + \imath v_l(x, y))h^l +$$

$$+[-\frac{\lambda + 3\mu}{\lambda + \mu} \sum_{j=0}^{m} A_j z^j + z \sum_{j=1}^{m} \overline{A_j} j \overline{z}^{j-1} + \sum_{j=1}^{m} \overline{B_j} \overline{z}^j] h^{n+2}, \qquad (6.24)$$

$$w(x, y, h) = \sum_{l=0}^{n+1} w_l(x, y)h^l + \frac{1}{2}[\sum_{j=0}^{m} C_j z^j + \sum_{j=0}^{m} \overline{C_j} \overline{z}^j] h^{n+2}.$$

We substitute (6.24) in boundary relations (6.22) and (6.23) and obtain for every component the following system where $r_k = f(h_k)$, $r_{k+1} =$

$f(h_{k+1})$:

$$\sum_{l=0}^{n+1} u_l(r_k\cos\theta, r_k\sin\theta)h_k^l = \tilde{u}_k(\theta) - \mathrm{Re}\left[-\frac{\lambda+3\mu}{\lambda+\mu}\sum_{j=0}^{m} A_j z^j + \right.$$

$$\left. +z\sum_{j=1}^{m}\overline{A_j}j\overline{z}^{j-1} + \sum_{j=1}^{m}\overline{B_j}\overline{z}^j\right]_{z=r_k e^{\imath\theta}} h_k^{n+2},$$

$$\sum_{l=1}^{n+1} l u_l(r_k\cos\theta, r_k\sin\theta)h_k^{l-1} = \widehat{u}_1(\theta) - \mathrm{Re}\left[-\frac{\lambda+3\mu}{\lambda+\mu}\sum_{j=0}^{m} A_j z^j + \right.$$

$$\left. +z\sum_{j=1}^{m}\overline{A_j}j\overline{z}^{j-1} + \sum_{j=1}^{m}\overline{B_j}\overline{z}^j\right]_{z=r_k e^{\imath\theta}} (n+2)h_k^{n+1},$$

$$\sum_{l=2}^{n+1} l(l-1) u_l(r_k\cos\theta, r_k\sin\theta)h_k^{l-2} = \widehat{u}_2(\theta) - \mathrm{Re}\left[-\frac{\lambda+3\mu}{\lambda+\mu}\sum_{j=0}^{m} A_j z^j + \right.$$

$$\left. +z\sum_{j=1}^{m}\overline{A_j}j\overline{z}^{j-1} + \sum_{j=1}^{m}\overline{B_j}\overline{z}^j\right]_{z=r_k e^{\imath\theta}} (n+2)(n+1)h_k^{n},$$

....

$$\sum_{l=n}^{n+1} l...(l-n+1) u_l(r_k\cos\theta, r_k\sin\theta)h_k^{l-n} = \widehat{u}_n(\theta) - \mathrm{Re}\left[-\frac{\lambda+3\mu}{\lambda+\mu}\sum_{j=0}^{m} A_j z^j + \right.$$

$$\left. +z\sum_{j=1}^{m}\overline{A_j}j\overline{z}^{j-1} + \sum_{j=1}^{m}\overline{B_j}\overline{z}^j\right]_{z=r_k e^{\imath\theta}} \frac{(n+2)!}{2}h_k^{2},$$

$$\sum_{l=0}^{n+1} u_l(r_{k+1}\cos\theta, r_{k+1}\sin\theta)h_{k+1}^l = \tilde{u}_{k+1}(\theta) - \mathrm{Re}\left[-\frac{\lambda+3\mu}{\lambda+\mu}\sum_{j=0}^{m} A_j z^j + \right.$$

$$\left. +z\sum_{j=1}^{m}\overline{A_j}j\overline{z}^{j-1} + \sum_{j=1}^{m}\overline{B_j}\overline{z}^j\right]_{z=r_{k+1} e^{\imath\theta}} h_{k+1}^{n+2},$$

Note that the boundary values $u_l(r_k\cos\theta, r_k\sin\theta)$, $k=0,...,n$, can be lineally expressed in terms of the values $u_{n+1}(r_k\cos\theta, r_k\sin\theta)$ via the first $n+1$ equations of the last system.

Due to the results of Chapter 1 we represent displacement components with the help of the analytic functions $\phi_l(z)$, $\psi_l(z)$ and $\rho_l(z)$, $l = 0, ..., n+1$ as in (1.6), (1.7).

The systems with boundary relations concerning the coefficients of all components give analogously to the last system the boundary conditions for the analytic functions $\phi_l(z)$, $\phi_l(z)$, $\rho_l(z)$, $l = 0, ..., n+1$, and these analytic functions can be restored as the polynomials in the complex variable z if the given data are Fourier polynomials. Note that the coefficients of powers of z in the functions $\phi_l(z)$, $\phi_l(z)$ are linearly dependent on the introduced constants A_k , B_k , C_k .

We find these constants when we satisfy the given displacements at the given $(2m+1)$ points at the ends of M_k .

CHAPTER 7

Spline-interpolation solution for asymmetric cones or conoids

Pyotr N. Ivanshin,
Kazan Federal University
pivanshin@gmail.com

ABSTRACT: In this chapter we construct the spline interpolation solution of the second basic problem of elasticity theory for asymmetric solids.

The results of Chapter 6 applicable for the elastic body bounded by an axial symmetric cone fail to be applied for the body bounded by an asymmetric cone or conoid. We consider the asymmetric cones which have circles in the sections by the planes orthogonal to the axis OH. We change some assumptions and formulas in order to construct the spline-interpolation solutions for this type of elastic solids. The main idea — to divide the cone by the planes orthogonal to OH-axis and to change the data on the conic surface by the data on the edges of the sections — remains the same. We construct the splines which satisfy equations (1.1) and the given data for

Elena A. Shirokova and Pyotr N. Ivanshin (Eds)

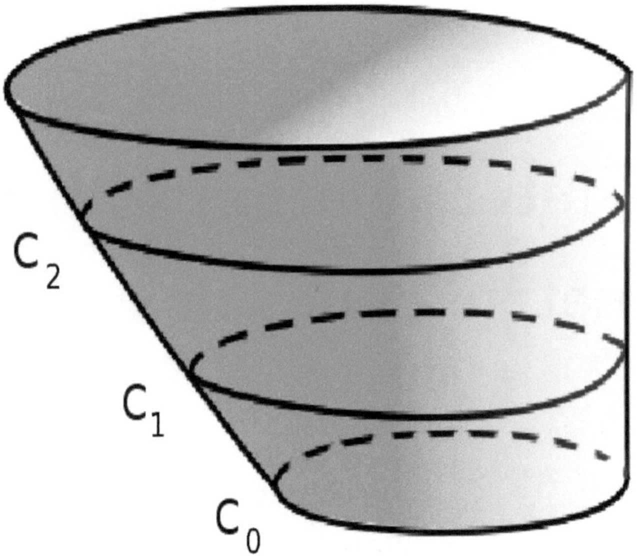

Figure 7.1: Division of the conoid

every layer of the cone. We consider two types of splines: the first one is linear with respect to h displacement vector components which meet the given data at the edges of the layer. The spline of the second type meets the given displacements at the inner points of one or the both ends, it can meet null displacements at the edges of the layer.

The asymmetry of the solid implies necessity of more complex considerations. For example, when we solve the boundary value problems in the case of asymmetric bodies we have to reconstruct the meromorphic functions with poles at two different points while in the case of axial-symmetric solids we restrict ourselves to one point. The immediate consequence of these two-pole constructions is lengthening of the corresponding solution formulas.

7.1 Spline of the first type

Consider the asymmetric cone or conoid M which directrix is the circle in the plane orthogonal to the axis OH. We construct spline-interpolation solution of the system (1.1) for the layer

$$M_k = \{(x, y, h)|(x, y, h) \in M, h \in [h_k, h_{k+1}]\},$$

which meets the displacements given at the edges of M_k:

$$u(f(h_k)\cos\theta, f(h_k)\sin\theta, h_k) = -\frac{1}{2\mu}\sum_{j=0}^{P}(a_{1,j}\cos(j\theta) - A_{1,j}\sin(j\theta)),$$

$$v(f(h_k)\cos\theta, f(h_k)\sin\theta, h_k) = -\frac{1}{2\mu}\sum_{j=0}^{P}(B_{1,j}\cos(j\theta) + b_{1,j}\sin(j\theta)),$$

$$w(f(h_k)\cos\theta, f(h_k)\sin\theta, h_k) = -\frac{1}{2\mu}\sum_{j=0}^{P}(d_{1,j}\cos(j\theta) - D_{1,j}\sin(j\theta)), \quad (7.1)$$

$$u(f(h_{k+1})\cos\theta, f(h_{k+1})\sin\theta, h_{k+1}) = -\frac{1}{2\mu}\sum_{j=0}^{P}(a_{2,j}\cos(j\theta) - A_{2,j}\sin(j\theta)),$$

$$v(f(h_{k+1})\cos\theta, f(h_{k+1})\sin\theta, h_{k+1}) = -\frac{1}{2\mu}\sum_{j=0}^{P}(B_{2,j}\cos(j\theta) + b_{2,j}\sin(j\theta)),$$

$$w(f(h_{k+1})\cos\theta, f(h_{k+1})\sin\theta, h_{k+1}) = -\frac{1}{2\mu}\sum_{j=0}^{P}(d_{2,j}\cos(j\theta) - D_{2,j}\sin(j\theta)),$$

where θ is the polar angle of a point of the corresponding circle.

We search for the solution of the system (1.1) on the interval (h_k, h_{k+1}) in the form

$$u(x, y, h) = u_0(x, y) + u_1(x, y)h, \ v(x, y, h) = v_0(x, y) + v_1(x, y)h$$
$$w(x, y, h) = w_0(x, y) + w_1(x, y)h, \quad (7.2)$$

just as in the previous chapter. So the solution of (1.1) again has the form

$$u(x,y,h)+\imath v(x,y,h)=-\frac{1}{2\mu}[-\frac{\lambda+3\mu}{\lambda+\mu}\phi_0(z)+z\overline{\phi_0'(z)}+\overline{\psi_0(z)}+$$
$$+(\lambda+\mu)(-\frac{\lambda+\mu}{4(\lambda+2\mu)}\rho_1(z)+\frac{\lambda+3\mu}{4(\lambda+2\mu)}z\overline{\rho_1'(z)})]-$$
$$-\frac{1}{2\mu}[-\frac{\lambda+3\mu}{\lambda+\mu}\phi_1(z)+z\overline{\phi_1'(z)}+\overline{\psi_1(z)}]h, \qquad (7.3)$$
$$w(x,y,h)=\operatorname{Re}\rho_0'(z)-\frac{z\overline{\phi_1(z)}+\overline{z}\phi_1(z)}{4\mu}+\operatorname{Re}\rho_1'(z)h,$$

where the functions $\phi_0(z), \phi_1(z), \psi_0(z)$, $\psi_1(z)$, $\rho_0(z)$, $\rho_1(z)$ in (7.3) are analytic in the projection of the layer M_k onto the plane XOY .

Assume that the boundary circle on the level $h=h_k$ has the radius 1 and the center 0 and that on the level $h=h_{k+1}$ has the radius R and the center s .

Let us present the convenient representation of the displacement components on the level $h=h_{k+1}$:

$$u(x,y,h_{k+1})+\imath v(x,y,h_{k+1})=-\frac{1}{2\mu}[-\frac{\lambda+3\mu}{\lambda+\mu}\tilde{\phi}_0(\zeta)+\zeta\overline{\tilde{\phi}_0'(\zeta)}+$$
$$+\overline{\tilde{\psi}_0(\zeta)}+(\lambda+\mu)(-\frac{\lambda+\mu}{4(\lambda+2\mu)}\tilde{\rho}_1(\zeta)+\frac{\lambda+3\mu}{4(\lambda+2\mu)}\zeta\overline{\tilde{\rho}_1'(\zeta)})]-$$
$$-\frac{1}{2\mu}[-\frac{\lambda+3\mu}{\lambda+\mu}\tilde{\phi}_1(\zeta)+\zeta\overline{\tilde{\phi}_1'(\zeta)}+\overline{\tilde{\psi}_1(\zeta)}]h_{k+1}, \qquad (7.4)$$
$$w(x,y,h_{k+1})=\operatorname{Re}\tilde{\rho}_0'(\zeta)-\frac{\zeta\overline{\tilde{\phi}_1(\zeta)}+\overline{\zeta}\tilde{\phi}_1(\zeta)}{4\mu}+\operatorname{Re}\tilde{\rho}_1'(\zeta)h_{k+1}.$$

Here

$$\zeta=z-s, |\zeta|\le R,\ \tilde{\phi}_j(z-s)=\phi_j(z),\ \tilde{\psi}_j(z-s)=\psi_j(z),$$
$$\tilde{\rho}_j(z-s)=\rho_j(z), j=0,1.$$

Let us now put the relations (7.2) into the boundary conditions (7.1) in order to get the system of equations which allows us to determine the functions ϕ_0, ϕ_1, ψ_0, ψ_1, ρ_0, ρ_1.

The first three equations of this system are better represented in the variable $\zeta = z - s$ since the solution at that level is represented in form (7.4). Here we again use the standard relation $\overline{\zeta} = R^2/\zeta$, at the corresponding boundary circle. So at the left-hand side of these three equations we obtain meromorphic functions with simple poles at the point $\zeta = 0$.

The last three equations of the system initially do not differ from that in the set of equations from the previous chapter (6.4), (6.5), (6.6). Recall that reconstruction method implies change of $\overline{z}$ by $1/z$. So we again get meromorphic functions with simple pole at the point $z = 0$ at the left-hand side of the relations defined by boundary conditions.

We must choose either z or ζ as variable on both level disks in order to solve the full system. Without loss of generality we choose ζ. So the last set of meromorphic functions will have simple pole at the point $-s/R$. Now the system on the functions ϕ_0, ϕ_1, ψ_0, ψ_1, ρ_0, ρ_1 has the form:

$$-\frac{1}{2\mu}(-k\tilde{\phi}_0(\zeta) + (R/\zeta + s)\frac{\tilde{\phi}_0'(\zeta)}{R} + \tilde{\psi}_0(\zeta) - \frac{(\lambda+\mu)^2\tilde{\rho}_1(\zeta)}{4(\lambda+2\mu)} + \\ + \frac{(\lambda+\mu)(\lambda+3\mu)\tilde{\rho}_1'(\zeta)(R/\zeta+s)}{4R(\lambda+2\mu)} + h_k(-k\tilde{\phi}_1(\zeta) + \\ +(R/\zeta+s)\frac{\tilde{\phi}_1'(\zeta)}{R} + \tilde{\psi}_1(\zeta))) = \sum_{j=1}^{n}(a_{1j} + \imath A_{1j})\zeta^j + \frac{\overline{C_1}}{\zeta} - C_1\zeta, \tag{7.5}$$

$$-\frac{1}{2\mu}(-k\tilde{\phi}_0(\zeta) - (R/\zeta + s)\frac{\tilde{\phi}_0'(\zeta)}{R} - \tilde{\psi}_0(\zeta) - \frac{(\lambda+\mu)^2\tilde{\rho}_1(\zeta)}{4(\lambda+2\mu)} - \\ - \frac{(\lambda+\mu)(\lambda+3\mu)\tilde{\rho}_1'(\zeta)(R/\zeta+s)}{4R(\lambda+2\mu)} + h_k(-k\tilde{\phi}_1(\zeta) - \\ -(R/\zeta+s)\frac{\tilde{\phi}_1'(\zeta)}{R} - \tilde{\psi}_1(\zeta))) = \sum_{j=1}^{n}(b_{1j} + \imath B_{1j})\zeta^j - \frac{\overline{C_1}}{\zeta} - C_1\zeta, \tag{7.6}$$

$$-\frac{\tilde{\rho}_0'(\zeta)}{R}-\frac{(R/\zeta+s)\tilde{\phi}_1(\zeta)}{2\mu}+h_k\frac{\tilde{\rho}_1'(\zeta)}{R}=\sum_{j=1}^{n}(d_{1j}+\imath D_{1j})\zeta^j+\frac{\overline{C_3}}{\zeta}-C_3\zeta, \quad (7.7)$$

$$\begin{aligned}&-\frac{1}{2\mu}(-\kappa\tilde{\phi}_0(\zeta)+1/(R\zeta+s)\frac{\tilde{\phi}_0'(\zeta)}{R}+\tilde{\psi}_0(\zeta)-\frac{(\lambda+\mu)^2\tilde{\rho}_1(\zeta)}{4(\lambda+2\mu)}+\\&+\frac{(\lambda+\mu)(\lambda+3\mu)\tilde{\rho}_1'(\zeta)}{4R(R\zeta+s)(\lambda+2\mu)}+h_{k+1}(-\kappa\tilde{\phi}_1(\zeta)+1/(R\zeta+s)\frac{\tilde{\phi}_1'(\zeta)}{R}+\\&+\tilde{\psi}_1(\zeta)))=\sum_{j=1}^{n}(a_{2j}+\imath A_{2j})\zeta^j+\frac{\overline{C_2}}{R\zeta+s}-C_2(R\zeta+s),\end{aligned} \quad (7.8)$$

$$\begin{aligned}&-\frac{1}{2\mu}(-\kappa\tilde{\phi}_0(\zeta)-1/(R\zeta+s)\frac{\tilde{\phi}_0'(\zeta)}{R}-\tilde{\psi}_0(\zeta)-\frac{(\lambda+\mu)^2\tilde{\rho}_1(\zeta)}{4(\lambda+2\mu)}-\\&-\frac{(\lambda+\mu)(\lambda+3\mu)\tilde{\rho}_1'(\zeta)}{4(R\zeta+s)R(\lambda+2\mu)}+h_{k+1}(-\kappa\tilde{\phi}_1(\zeta)-1/(R\zeta+s)\frac{\tilde{\phi}_1'(\zeta)}{R}-\\&-\tilde{\psi}_1(\zeta)))=\sum_{j=1}^{n}(b_{2j}+\imath B_{2j})\zeta^j-\frac{\overline{C_2}}{R\zeta+s}-C_1(R\zeta+s),\end{aligned} \quad (7.9)$$

$$\begin{aligned}&-\frac{\tilde{\rho}_0'(\zeta)}{R}-\frac{\tilde{\phi}_1(\zeta)}{2\mu(R\zeta+s)}+h_{k+1}\frac{\tilde{\rho}_1'(\zeta)}{R}=\\&=\sum_{j=1}^{n}(d_{2j}+\imath D_{2j})\zeta^j+\frac{\overline{C_4}}{R\zeta+s}-C_4(R\zeta+s).\end{aligned} \quad (7.10)$$

The method of system (7.5)–(7.10) resolution is similar to that of the system (6.4)—(6.9): it is the same process of excluding and finding the unknown analytic functions $\tilde{\phi}_1$, $\tilde{\phi}_0$, $\tilde{\psi}_1$, $\tilde{\psi}_0$, $\tilde{\rho}_1$, $\tilde{\rho}_0$ and unknown variables C_1 , C_2 , C_3 , C_4 . Namely, first we sum equations (7.5) and (7.6), then we sum equations (7.8) and (7.9). After this we subtract the first of these sums from the second. This allows us to obtain $\tilde{\phi}_1(\zeta)$. Then we set $\phi_1(z)$ in the system consisting of equations (7.7) and (7.10) and solve it with respect to $\tilde{\rho}_1'(\zeta)$ and $\tilde{\rho}_0'(\zeta)$. At the same step again similarly to the solution of system (6.4)—(6.9) we use analyticity of $\tilde{\rho}_1'(\zeta)$ and $\tilde{\rho}_0'(\zeta)$ to find C_3 and C_4 . The only difference from the solution of the similar system on C_3 and C_4 of the previous chapter is that we must consider two poles in 0 and in $-s/R$, i.e. for computational purposes we must equate to 0 the coefficients of $1/\zeta$ and $1/(R\zeta+s)$. After this we compute $\tilde{\rho}_1(\zeta)$ and put it together with $\tilde{\phi}_1(\zeta)$ into the sum of (7.5) and (7.6) in order to get $\tilde{\phi}_0(\zeta)$. Finally by setting all of the already computed functions in the system consisting of equations (7.5) and (7.8) we find the functions $\tilde{\psi}_0(\zeta)$, $\tilde{\psi}_1(\zeta)$ and the constants C_1 and C_2.

Boundary conditions (7.1) provide the polynomial solution which can be found rather easily. Again the solution of this system is too large to be presented here.

Notice that the solutions of systems (6.4)—(6.9) and (7.5)—(7.10) lead to three biharmonic functions u, v, w, being the coordinates of the displacement vector.

7.2 Spline of the second type

Now we describe the construction of the spline of the second type which satisfies equations (1.1), which has null displacements at the edges of the layer. The additional parameters in the components representations allow us to provide point-wise gluing at a finite number of the inner points of the sections of the adjacent layers.

7.2.1 Point-wise gluing

Suppose that we search for the solution of system (1.1) which meets null data condition at the points of the edges at the levels $h=h_k$ and $h=h_{k+1}$ and also has the given displacements at $2m+1$ inner points of the sections at the same levels. The boundary conditions have the form

$$\begin{aligned}
u(a_k + r_k\cos\theta, r_k\sin\theta, h_k) &= 0,\\
v(a_k + r_k\cos\theta, r_k\sin\theta, h_k) &= 0,\\
w(a_k + r_k\cos\theta, r_k\sin\theta, h_k) &= 0,\\
u(a_{k+1} + r_{k+1}\cos\theta, r_{k+1}\sin\theta, h_{k+1}) &= 0,\\
v(a_{k+1} + r_{k+1}\cos\theta, r_{k+1}\sin\theta, h_{k+1}) &= 0,\\
w(a_{k+1} + r_{k+1}\cos\theta, r_{k+1}\sin\theta, h_{k+1}) &= 0,
\end{aligned} \tag{7.11}$$

where θ is the polar angle of a point of an edge.

We use the same representation of the displacement components as in Section 6.3.2:

$$u(x,y,h) + \imath v(x,y,h) = \frac{\mu}{\lambda+\mu}\overline{z}\sum_{j=1}^{m} A_j\frac{z^{j+1}}{j+1} -$$

$$-\frac{z^2}{2}\sum_{j=1}^{m}\overline{A_j}\overline{z}^j - \frac{\lambda+3\mu}{4(\lambda+2\mu)}z\sum_{j=1}^{m}\overline{B_j}\frac{\overline{z}^{j+1}}{j+1}+$$

$$+\frac{\lambda+\mu}{4(\lambda+2\mu)}\sum_{j=1}^{m} B_j\frac{z^{j+2}}{(j+1)(j+2)}+$$

$$+\frac{(\lambda+3\mu)^2}{4(\lambda+\mu)(\lambda+2\mu)}A_0|z|^2 - \frac{\lambda+3\mu}{8(\lambda+2\mu)}\overline{A_0}z^2 -$$

$$-\frac{(\lambda+\mu)(\lambda+3\mu)}{8\mu(\lambda+2\mu)}z\overline{H_1'(z)} + \frac{(\lambda+\mu)^2}{8\mu(\lambda+2\mu)}H_1(z)+$$

$$+\frac{\lambda+3\mu}{2\mu(\lambda+\mu)}\Phi_0(z) - \frac{1}{2\mu}z\overline{\Phi_0'(z)} - \frac{1}{2\mu}\overline{\Psi_0(z)}+$$

$$+[-\frac{(\lambda+\mu)(\lambda+3\mu)}{4\mu(\lambda+2\mu)}z\sum_{j=1}^{m}\overline{C_j}\overline{z}^j +$$

$$+\frac{(\lambda+\mu)^2}{4\mu(\lambda+2\mu)}\sum_{j=1}^{m} C_j\frac{z^{j+1}}{j+1} + \frac{\lambda+3\mu}{2\mu(\lambda+\mu)}\Phi_1(z) -$$

$$-\frac{1}{2\mu}z\overline{\Phi_1'(z)}-\frac{1}{2\mu}\overline{\Psi_1(z)}]h+[-\frac{\lambda+3\mu}{\lambda+\mu}\sum_{j=0}^{m}A_jz^j+ \tag{7.12}$$

$$+z\sum_{j=1}^{m}\overline{A_j}j\overline{z}^{j-1}+\sum_{j=1}^{m}\overline{B_j}\overline{z}^j]h^2,$$

$$w(x,y,h)=-\frac{\lambda^2+6\lambda\mu+7\mu^2}{8\mu(\lambda+2\mu)}[z\sum_{j=1}^{m}\overline{C_j}\frac{\overline{z}^{j+1}}{j+1}+$$

$$+\overline{z}\sum_{j=1}^{m}C_j\frac{z^{j+1}}{j+1}]-\frac{1}{4\mu}(\overline{z}\Phi_1(z)+z\overline{\Phi_1(z)})-$$

$$-\frac{(\lambda+2\mu)}{4\mu}(C_0+\overline{C_0})|z|^2+\frac{1}{2}(H_0'(z)+\overline{H_0'(z)})+[\overline{z}\sum_{j=1}^{m}A_jz^j+z\sum_{j=1}^{m}\overline{A_j}\overline{z}^j+$$

$$+\frac{1}{2}(H_1'(z)+\overline{H_1'(z)})]h+\frac{1}{2}[\sum_{j=0}^{m}C_jz^j+\sum_{j=0}^{m}\overline{C_j}\overline{z}^j]h^2,$$

$\Phi_0(z)$, $\Phi_1(z)$, $\Psi_0(z)$, $\Psi_1(z)$, $H_0(z)$, $H_1(z)$ being analytic functions, C_0 being a real constant, A_k , $k=0,...,m$, B_k and C_k , $k=1,...,m$, being complex constants.

Now we apply zero boundary conditions (7.11) to the solution (7.12) and get the system analogous to (7.5)—(7.10) with respect to the functions $\Phi_0(z)$, $\Phi_1(z)$, $\Psi_0(z)$, $\Psi_1(z)$, $H_0(z)$, $H_1(z)$. These analytic functions contain linearly the unknown constants A_j , B_j , C_j from (6.20). The unknown constants can be found via the given displacements values at the given set of $2m+1$ points of the ends of the layer using the linear system of $6m+3$ equations with $6m+3$ variables.

Note that the different centers of the boundary disks provide two different poles of the corresponding meromorphic functions.

7.2.2 Solution of the second basic problem for conoid with one shifted base point

According to the general formula we search for the solution in the following form:

$$
\begin{aligned}
(u+\imath v)(z,\overline{z},h) = {} & (\lambda+3\mu)^2/(4(\lambda+\mu)(\lambda+2\mu))(a_0+\imath A_0)|z|^2 - \\
& -(\lambda+3\mu)/(8(\lambda+2\mu))(a_0-\imath A_0)z^2 - \\
& -(\lambda+\mu)(\lambda+3\mu)/(8\mu(\lambda+2\mu))\overline{\rho_1'}(\overline{z})z + \\
& +(\lambda+\mu)^2/(8\mu(\lambda+2\mu))\rho_1(z) + (\lambda+3\mu)/(2\mu(\lambda+\mu))\phi_0(z) - \\
& -1/(2\mu)\overline{\phi_0'}(\overline{z})z - 1/(2\mu)\overline{\psi_0}(\overline{z}) + ((\lambda+3\mu)/(2\mu(\lambda+\mu))\phi_1(z) - \\
& -1/(2\mu)\overline{\phi_1'}(\overline{z})z - 1/(2\mu)\overline{\psi_1}(\overline{z}))h - (\lambda+3\mu)/(\lambda+\mu)(a_0+\imath A_0)h^2,
\end{aligned}
$$

$$
\begin{aligned}
w(z,\overline{z},h) = {} & -1/(4\mu)(\overline{z}\phi_1(z) + z\overline{\phi_1}(\overline{z})) - (\lambda+2\mu)/(2\mu)C_0(z\overline{z}) + \\
& +(\rho_0'(z)+\overline{\rho_0'}(\overline{z}))/2 + C_0h^2 + (\rho_1'(z)+\overline{\rho_1'}(\overline{z}))/2h.
\end{aligned}
$$

Restrictions of these functions to the lower and upper cone ends produce the six equations on the real and imaginary parts of the zero boundary conditions which compose the system of differential equations over the functions $\phi_0(z)$, $\phi_1(z)$, $\psi_0(z)$, $\psi_1(z)$, $\rho_0(z)$ and $\rho_1(z)$. Note that since we can reconstruct only analytic functions by their real or imaginary parts restrictions to the boundary we must use the following procedure: we replace $z\overline{f}(\overline{z})$ with $\overline{z}f(z)$ in the case of real part of the given boundary condition and by $-\overline{z}f(z)$ in the other case. Since we only work with circles as boundaries the variable $\overline{z}$ can be replaced by $\frac{R^2}{z-z_0}$, here R is the radius of the boundary circle, and z_0 is its center.

$$\begin{aligned}
&(\lambda+3\mu)^2/(4(\lambda+\mu)(\lambda+2\mu))(a_0+\imath A_0)4- \\
&-(\lambda+3\mu)/(8(\lambda+2\mu))(a_0-\imath A_0)z^2- \\
&-(\lambda+\mu)(\lambda+3\mu)/(8\mu(\lambda+2\mu))\rho_1'(z)4/z+ \\
&+(\lambda+\mu)^2/(8\mu(\lambda+2\mu))\rho_1(z)+ \\
&+(\lambda+3\mu)/(2\mu(\lambda+\mu))\phi_0(z)-1/(2\mu)\phi_0'(z)4/z-1/(2\mu)\psi_0(z)+ \\
&+((\lambda+3\mu)/(2\mu(\lambda+\mu))\phi_1(z)-1/(2\mu)\phi_1'(z)4/z-1/(2\mu)\psi_1(z))- \\
&-(\lambda+3\mu)/(\lambda+\mu)(a_0+\imath A_0)-(c_1-\imath C_1)4/z+(c_1+\imath C_1)z=0;
\end{aligned} \tag{7.13}$$

$$\begin{aligned}
&(\lambda+3\mu)^2/(4(\lambda+\mu)(\lambda+2\mu))(a_0+\imath A_0)4- \\
&-(\lambda+3\mu)/(8(\lambda+2\mu))(a_0-\imath A_0)z^2+ \\
&+(\lambda+\mu)(\lambda+3\mu)/(8\mu(\lambda+2\mu))\rho_1'(z)4/z+ \\
&+(\lambda+\mu)^2/(8\mu(\lambda+2\mu))\rho_1(z)+ \\
&+(\lambda+3\mu)/(2\mu(\lambda+\mu))\phi_0(z)+1/(2\mu)\phi_0'(z)4/z+1/(2\mu)\psi_0(z)+ \\
&+((\lambda+3\mu)/(2\mu(\lambda+\mu))\phi_1(z)+1/(2\mu)\phi_1'(z)4/z+1/(2\mu)\psi_1(z))- \\
&-(\lambda+3\mu)/(\lambda+\mu)(a_0+\imath A_0)+(c_1-\imath C_1)4/z+(c_1+\imath C_1)z=0;
\end{aligned} \tag{7.14}$$

$$\begin{aligned}
&(\lambda+3\mu)^2/(4(\lambda+\mu)(\lambda+2\mu))2(a_0)(z)- \\
&-(\lambda+3\mu)/(8(\lambda+2\mu))(a_0-\imath A_0)z^2- \\
&-(\lambda+\mu)(\lambda+3\mu)/(8\mu(\lambda+2\mu))\rho_1'(z)(1+1/(z-1))+ \\
&+(\lambda+\mu)^2/(8\mu(\lambda+2\mu))\rho_1(z)+(\lambda+3\mu)/(2\mu(\lambda+\mu))\phi_0(z)- \\
&-1/(2\mu)\phi_0'(z)(1+1/(z-1))-1/(2\mu)\psi_0(z)- \\
&-(c_3-\imath C_3)/(z-1)+(c_3+\imath C_3)(z-1)=0;
\end{aligned} \tag{7.15}$$

$$(\lambda+3\mu)^2/(4(\lambda+\mu)(\lambda+2\mu))2(\imath A_0)(z)-$$
$$-(\lambda+3\mu)/(8(\lambda+2\mu))(a_0-\imath A_0)z^2+ \tag{7.16}$$
$$+(\lambda+\mu)(\lambda+3\mu)/(8\mu(\lambda+2\mu))\rho_1'(z)(1+1/(z-1))+$$
$$+(\lambda+\mu)^2/(8\mu(\lambda+2\mu))\rho_1(z)+(\lambda+3\mu)/(2\mu(\lambda+\mu))\phi_0(z)+$$
$$+1/(2\mu)\phi_0'(z)(1+1/(z-1))+1/(2\mu)\psi_0(z)+$$
$$+(c_3-\imath C_3)/(z-1)+(c_3+\imath C_3)(z-1)=0;$$

$$-1/(2\mu)(4/z\phi_1(z))-(\lambda+2\mu)/(2\mu)C_0 4+(\rho_0'(z))+C_0+ \tag{7.17}$$
$$+\rho_1'(z)-(c_2-\imath C_2)4/z+(c_2+\imath C_2)z=0;$$

$$-1/(2\mu)((1+1/(z-1))\phi_1(z))-(\lambda+2\mu)/(2\mu)C_0 2(z)+ \tag{7.18}$$
$$+(\rho_0'(z))-(c_4-\imath C_4)/(z-1)+(c_4+\imath C_4)(z-1)=0.$$

In fact the system (7.13), (7.14), (7.15), (7.16), (7.17) and (7.18) can be easily solved in the following way. First we sum (7.13) and (7.14) excluding $\phi_0'(z)$ $\phi_1'(z)$ and $\rho_1'(z)$. Then we sum (7.16) with (7.17) with similar consequences. At last we subtract the relation ((7.16) +(7.17)) from the relation ((7.13) + (7.14)). This allows us immediately to get $\phi_1(z)$. This function we set in the system of (7.15) and (7.18) and solve the last set of equations with respect to $\rho_0'(z)$ and $\rho_1'(z)$ simultaneously getting $c_2+\imath C_3$ and $c_4+\imath C_4$. Then we find $\phi_0(z)$ from either ((7.13) + (7.14)) or ((7.16) +(7.17)). Finally we obtain $\psi_0(z)$ and $\psi_1(z)$ together with $c_1+\imath C_1$ and $c_2+\imath C_2$.

Thus

$$(u+\imath v)(z,\overline{z},h)=-\frac{1}{(\lambda+\mu)(\lambda^2-5\mu^2)}(C_0(\lambda+\mu)^2(\lambda+2\mu)(-\overline{z}+(-1+\overline{z})z+$$
$$+h(-4+\overline{z}+z))+a_0(\lambda+3\mu)(h^2(\lambda^2-5\mu^2)-h(\lambda^2+\mu^2(3-2\overline{z}-2z))+$$
$$+2\mu^2(-\overline{z}+(-1+\overline{z})z))+\imath A_0(\lambda^2-5\mu^2)(h^2(\lambda+3\mu)+\mu(\overline{z}+z-\overline{z}z)-$$

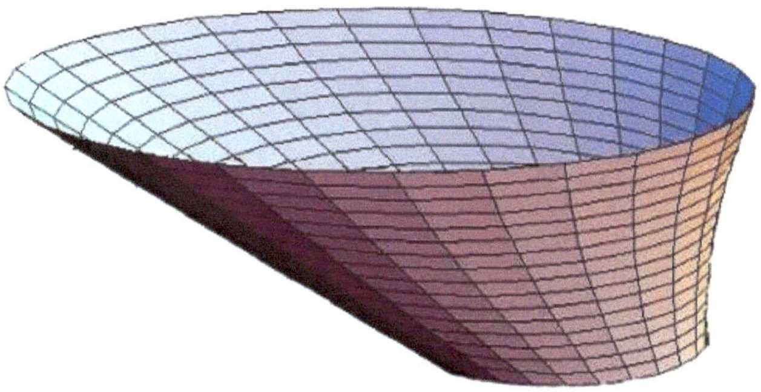

Figure 7.2: Side surface of the conoid with one shifted point at the base

$$-h(\lambda + \mu(-1 + \overline{z} + z)))),$$

$$w(z, \overline{z}, h) = \frac{1}{\lambda^2 - 5\mu^2}(a_0\mu(\lambda + 3\mu)(-\overline{z} + (-1 + \overline{z})z + h(-4 + \overline{z} + z)) +$$

$$+C_0(h^2(\lambda^2 - 5\mu^2) + (\lambda + 2\mu)(\lambda + 3\mu)(\overline{z}(-1 + z) - z) + h(\lambda^2(-5 + \overline{z} + z) +$$

$$+5\lambda\mu(-4 + \overline{z} + z) + \mu^2(-19 + 6\overline{z} + 6z)))).$$

Now in order to find constants a_0, A_0 and C_0 we apply the boundary conditions at the point on the upper section of the cone:

$$a_0 = -\frac{5}{72}(9x_0 + 5z_0),$$

$$A_0 = -\frac{5y_0}{8},$$

$$C_0 = \frac{1}{28}(5x_0 + 4z_0).$$

The side surface of the conoid is given on figure 7.2.

On figure 7.3 the dotted lines represent the shift of the solid lines in the solution of the problem.

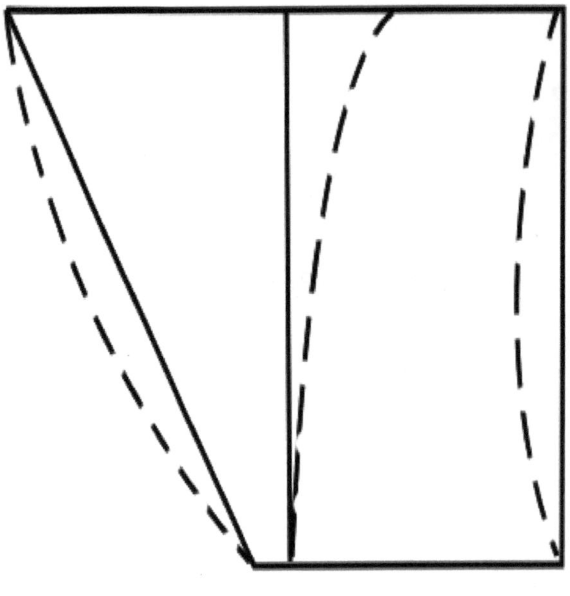

Figure 7.3: Section of the deformed conoid by the plane $x = 0$.

7.2.3 Cone with shifted base point

Consider the system of equations similar to one discussed in the previous section:

$$(\lambda+3\mu)^2/(4(\lambda+\mu)(\lambda+2\mu))(a_0+\imath A_0)-(\lambda+3\mu)/(8(\lambda+2\mu))(a_0-\imath A_0)z^2-$$
$$-(\lambda+\mu)(\lambda+3\mu)/(8\mu(\lambda+2\mu))(h_1-\imath H_1)z+(\lambda+\mu)^2/(8\mu(\lambda+2\mu))(h_1+\imath H_1)z+$$
$$+(\lambda+3\mu)/(2\mu(\lambda+\mu))\phi_0(z)-1/(2\mu z)\phi_0'(z)-1/(2\mu)\psi_0(z)+$$
$$+((\lambda+3\mu)/(2\mu(\lambda+\mu))(\phi_1+\imath\Phi_1)-1/(2\mu)(\psi_1+\imath\Psi_1))-$$
$$-(\lambda+3\mu)/(\lambda+\mu)(a_0+\imath A_0)-$$
$$-(c_1-\imath C_1)/z+(c_1+\imath C_1)z=0,$$
$$(\lambda+3\mu)^2/(4(\lambda+\mu)(\lambda+2\mu))(a_0+\imath A_0)-(\lambda+3\mu)/(8(\lambda+2\mu))(a_0-\imath A_0)z^2-$$
$$-(\lambda+\mu)(\lambda+3\mu)/(8\mu(\lambda+2\mu))(h_1-\imath H_1)z+(\lambda+\mu)^2/(8\mu(\lambda+2\mu))(h_1+\imath H_1)z+$$
$$+(\lambda+3\mu)/(2\mu(\lambda+\mu))\phi_0(z)+1/(2\mu z)\phi_0'(z)+1/(2\mu)\psi_0(z)+$$
$$+((\lambda+3\mu)/(2\mu(\lambda+\mu))(\phi_1+\imath\Phi_1)+1/(2\mu)(\psi_1+\imath\Psi_1))-$$
$$-(\lambda+3\mu)/(\lambda+\mu)(a_0+\imath A_0)+$$
$$+(c_1-\imath C_1)/z+(c_1+\imath C_1)z=0,$$
$$-1/(2\mu z)(\phi_1+\imath\Phi_1)-(\lambda+2\mu)/(2\mu)C_0+h_0'(z)+C_0+(h_1+\imath H_1)-\overline{c_2}/z+c_2z=0.$$

So

$$(u+\imath v)(z,\overline{z},h)=\frac{1}{2\mu(\lambda+\mu)}(2a_0\mu(\lambda-h^2\lambda+\mu(2-3h^2+\overline{z}z))-$$
$$-\imath(\imath(-1+h)(\lambda(\phi_1+\imath(\Phi_1+\imath\psi_1+\Psi_1))+\mu(3\phi_1+\imath(3\Phi_1+\imath\psi_1+\Psi_1)))+$$
$$+2A_0\mu((-1+h^2)\lambda+\mu(-2+3h^2-\overline{z}z)))),$$

and

$$w(z,\overline{z},h)=\frac{2(-1+h)h_1\mu+C_0(\lambda-\lambda\overline{z}z+2\mu(h^2-\overline{z}z))}{2\mu}.$$

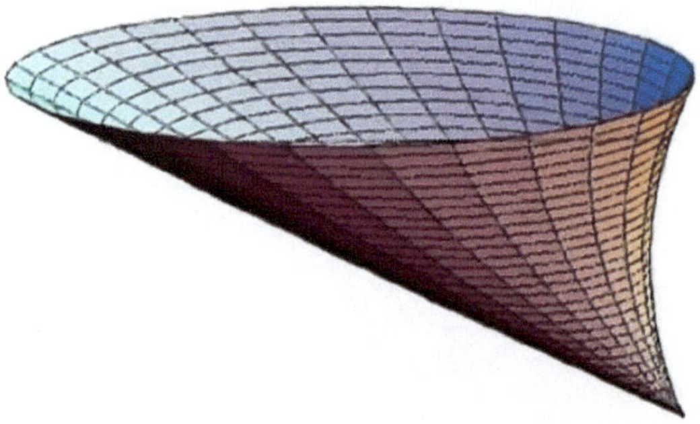

Figure 7.4: Side surface of the cone with shifted base point

In order to find constants $\phi_1 + \imath\Phi_1$, $\psi_1 + \imath\Psi_1$, $a_0 + \imath A_0$, C_0 and h_1 one must use the other boundary conditions, i.e. displacement values at the fixed points. Consider first the case of $h = 0$ and $z = 1$ — the tip of the cone. Assume that $\lambda = 15/13 * 10^{11}$, $\mu = 10/13 * 10^{11}$. Then

$$(180000000000000 a_0 + 180000000000000 \imath A_0 -$$

$$-13(9\phi_1 + 9\imath\Phi_1 - 5\psi_1 + 5\imath\Psi_1))/100000000000000 = 0,$$

$$-C_0 - h_1 = 0.$$

Now let $h = 1$, $z = 0$. Then

$$\frac{1}{560}(-224a_0 - 224\imath A_0 - 125(h_1 + \imath H_1)) = M_x + \imath M_y,$$

$$\frac{7C_0}{4} = M_z.$$

The side surface of the cone is given on figure 7.4.

On figure 7.5 the dotted lines represent the shift of the solid lines in the solution of the problem.

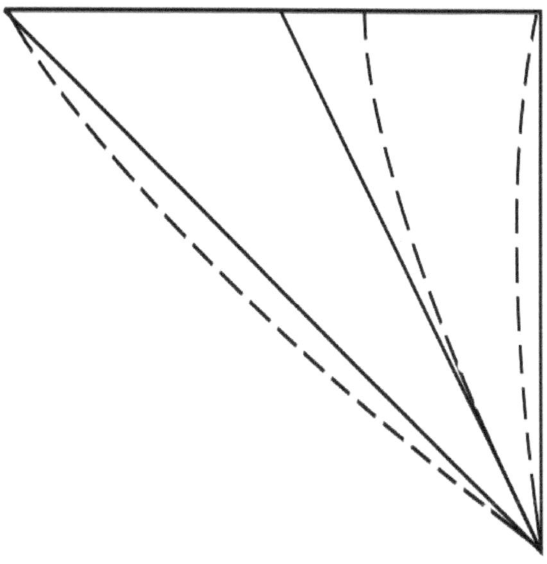

Figure 7.5: Section of the deformed cone by the plane $x = 0$

7.2.4 Smooth spline with point-wise gluing

We construct p-differentiable on S spline for the layer $h \in [0, h_0]$ with the help of the spline of the second type. In order to provide additional conditions we must rise the degree of the spline. Thus the solution of the system (1.1) in the form:

$$u(x,y,h) + \imath v(x,y,h) = \sum_{j=0}^{p+1}(u_j(x,y) + \imath v_j(x,y))h^j + [-\frac{\lambda+3\mu}{\lambda+\mu}\sum_{k=0}^{n} A_k z^k +$$

$$+z\sum_{k=1}^{n}\overline{A_k}k\overline{z}^{k-1} + \sum_{k=1}^{n}\overline{B_k}\overline{z}^k]h^{p+2} \quad (7.19)$$

$$w(x,y,h) = \sum_{j=0}^{p+1} w_j(x,y)h^j + \frac{1}{2}[\sum_{k=0}^{n} C_k z^k + \sum_{j=0}^{n}\overline{C_k}\overline{z}^k]h^{p+2},$$

can meet not only the gluing conditions of the previous section but also the conditions

$$u(x,y,h_0)|_{(x-a_0)^2+y^2=r_0{}^2} = \sum_{j=0}^{m}(\alpha_{0,j}\cos j\theta + \dot{\alpha}_{0,j}\sin j\theta),$$

$$v(x,y,h_0)|_{(x-a_0)^2+y^2=r_0{}^2} = \sum_{j=0}^{m}(\beta_{0,j}\cos j\theta + \dot{\beta}_{0,j}\sin j\theta),$$

$$w(x,y,h_0)|_{(x-a_0)^2+y^2=r_0{}^2} = \sum_{j=0}^{m}(\gamma_{0,j}\cos j\theta + \dot{\gamma}_{0,j}\sin j\theta);$$

$$u(x,y,0)|_{x^2+y^2=1} = \sum_{j=0}^{m}(a_{0,j}\cos j\theta + A_{0,j}\sin j\theta),$$

$$v(x,y,0)|_{x^2+y^2=1} = \sum_{j=0}^{m}(b_{0,j}\cos j\theta + B_{0,j}\sin j\theta),$$

$$w(x,y,0)|_{x^2+y^2=1} = \sum_{j=0}^{m}(c_{0,j}\cos j\theta + C_{0,j}\sin j\theta);$$

$$\frac{\partial^k}{\partial h^k} u(x,y,h)|_{x^2+y^2=1,h=0} = \sum_{j=0}^{m} (a_{k,j}\cos j\theta + A_{k,j}\sin j\theta),$$

$$\frac{\partial^k}{\partial h^k} v(x,y,h)|_{x^2+y^2=1,h=0} = \sum_{j=0}^{m} (b_{k,j}\cos j\theta + B_{k,j}\sin j\theta),$$

$$\frac{\partial^k}{\partial h^k} w(x,y,h)|_{x^2+y^2=1,h=0} = \sum_{j=0}^{m} (c_{k,j}\cos j\theta + C_{k,j}\sin j\theta),\ k = 1,\ldots,n.$$

We achieve the result by solution of the corresponding system of boundary equations similar to one for the cylinder. We again obtain $3(n+2)$ equations on the set of $3(n+2)$ unknown functions: $\phi_k(z)$, $\psi_k(z)$, $\rho_k(z)$, $k = 0,\ldots,n+1$.

7.3 The spline for the layer with a singular point

Assume that the layer for which $h \in [h_k, h_{k+1}]$ has one end degenerated into the point, say for $h = h_{k+1}$. Let this point have the coordinates (x_0, y_0, h_{k+1}). Let the other end of this layer be the disc $|z| < r_k$.

In order to construct the spline of the first type we search for the solution of (1.1) in the following form

$$u(x,y,h) + \imath v(x,y,h) = (-1/(2\mu))(-\kappa\phi_0(z) + z\overline{\phi_0'(z)} + \overline{\psi_0(z)} +$$

$$+\mu(\lambda+\mu)z((\rho_1+\overline{\rho_1})/(4(\lambda+2\mu)) - (\rho_1-\overline{\rho_1})/(4\mu))) + h(-1/(2\mu))(-\kappa\phi_1+\overline{\psi_1}),$$

$$w(x,y,h) = (\rho_0(z) + \overline{\rho_0(z)})/2 - (z\overline{\phi_1} + \overline{z}\phi_1)/(4\mu) + h(\rho_1+\overline{\rho_1})/2,$$

here $\phi_0(z)$, $\psi_0(z)$, $\rho_0(z)$ are unknown analytic in the disc $|z| < r_k$ functions, ϕ_1, ψ_1, ρ_1 are unknown constants.

Let the boundary conditions be given as follows:

$$u(r_k \cos\theta, r_k \sin\theta, h_k) = \sum_{j=0}^{P}(A_j \cos j\theta - a_j \sin j\theta);$$

$$v(r_k \cos\theta, r_k \sin\theta, h_k) = \sum_{j=0}^{P}(B_j \cos j\theta + b_j \sin j\theta);$$

$$w(r_k \cos\theta, r_k \sin\theta, h_k) = \sum_{j=0}^{P}(C_j \cos j\theta - c_j \sin j\theta),$$

and $u(x_0, y_0, h_{k+1}) = u_0$, $v(x_0, y_0, h_{k+1}) = v_0$, $w(x_0, y_0, h_{k+1}) = w_0$.

Then in order to determine the functions ϕ_0, ψ_0, ρ_0 we get the following system of equations:

$$(-1/(2\mu))(-\kappa\phi_0(z) + r_k^2/z\phi_0'(z) + \psi_0(z) + \mu(\lambda+\mu)z((\rho_1+\overline{\rho_1})/(4(\lambda+2\mu))-$$

$$-(\rho_1-\overline{\rho_1})/(4\mu))) + h_k(-1/(2\mu))(-\kappa\phi_1+\psi_1) - (A_0+\imath a_0) - (A_1+\imath a_1)z/r_k-$$

$$-\sum_{j=0}^{P}(A_j + ia_j)z^j/r_k^j - K/z + \overline{K}z/r_k^2 = 0;$$

$$(-1/(2\mu))(-\kappa\phi_0(z) - r_k^2/z\phi_0'(z) - \psi_0(z) + \mu(\lambda+\mu)z((\rho_1+\overline{\rho_1})/(4(\lambda+2\mu))-$$

$$-(\rho_1-\overline{\rho_1})/(4\mu))) + h_k(-1/(2\mu))(-\kappa\phi_1-\psi_1) - (B_0+\imath b_0) - (B_1+\imath b_1)z/r_k-$$

$$-\sum_{j=0}^{P}(B_j + \imath b_j)z^j/r_k^j + K/z + \overline{K}z/r_k^2 = 0;$$

$$(\rho_0(z)) - (r_k^2/z\phi_1)/(2\mu) + h_k(\rho_1) - (C_0+\imath c_0) - (C_1+\imath c_1)z/r_k-$$

$$-\sum_{j=0}^{P}(C_j + \imath c_j)z^j/r_k^j - L/z + \overline{L}z/r_k^2 = 0.$$

Now we can present the spline of the first type:

$$w(x,y,h) = \frac{1}{2}(2C_0 + \overline{\rho_1}h + h\rho_1 - h_k(\overline{\rho_1}+\rho_1) - \imath c_1 r_k^{-1}\overline{z} + C_1 r_k^{-1}\overline{z} + ic_1 r_k^{-1}z+$$

$$+C_1 r_k^{-1} z + \sum_{j=2}^{P}(-\imath c_j r_k^{-j}\overline{z}^j + C_j r_k^{-j}\overline{z}^j + (\imath c_j + C_j) r_k^{-j} z^j))$$

$$u(x,y,h) + \imath v(x,y,h) = \frac{1}{2\mu}(h_k\overline{\psi_1} - \overline{\psi_1}h + \imath a_0\mu + A_0\mu + \imath b_0\mu +$$

$$+(-\imath a_0 + A_0 + \imath b_0 - B_0)\mu +$$

$$+B_0\mu - \frac{h_k(\lambda+3\mu)\phi_1}{\lambda+\mu} + \frac{h(\lambda+3\mu)\phi_1}{\lambda+\mu} + \frac{(-\imath a_1 + A_1 + \imath b_1 - B_1)\mu\overline{z}}{r_k} +$$

$$+\sum_{j=2}^{P}\frac{(-\imath a_j + A_j - \imath b_j + B_j) j\mu(\lambda+\mu) r_k^{2-j}\overline{z}^{-2+j}}{\lambda+3\mu} +$$

$$+\sum_{j=2}^{P}(-\imath a_j + A_j + \imath b_j - B_j)\mu r_k^{-j}\overline{z}^j + \frac{(\lambda+\mu)(\rho_1(\lambda+\mu) - \overline{\rho_1}(\lambda+3\mu))z}{4(\lambda+2\mu)} +$$

$$+\frac{(\lambda+3\mu)(2\imath(a_1+b_1)\mu + 2A_1(\lambda+2\mu) + 2B_1(\lambda+2\mu) + \overline{\rho_1}(\lambda+\mu)r_k)z}{4(\lambda+2\mu)r_k} +$$

$$+\frac{1}{4}(\lambda+\mu)(\frac{-2(A_1+B_1)\lambda + 2\imath(a_1 + 2\imath A_1 + b_1 + 2\imath B_1)\mu - \rho_1(\lambda+\mu)r_k}{(\lambda+2\mu)r_k} +$$

$$+\sum_{j=2}^{P}\frac{4\imath(a_j + \imath A_j + b_j + \imath B_j) j\mu r_k^{-j}\overline{z}^{-1+j}}{\lambda+3\mu})z + \sum_{j=2}^{P}(\imath a_j + A_j + \imath b_j + B_j)\mu r_k^{-j} z^j).$$

Here the constants ρ_1, ψ_1, ϕ_1 are uniquely defined by the boundary conditions at the singular point of the layer.

$$\frac{1}{2}(2C_0 + (h_{k+1} - h_k)(\overline{\rho_1} + \rho_1)) = w_0,$$

$$\frac{1}{2\mu}(-h_{k+1}\overline{\psi_1} + \frac{h_{k+1}(\lambda+3\mu)\phi_1}{\lambda+\mu} +$$

$$+\frac{1}{(\lambda+\mu)(\lambda+2\mu)(\lambda+3\mu)}(h_k(\lambda+2\mu)(\lambda+3\mu)(\overline{\psi_1}(\lambda+\mu) - (\lambda+3\mu)\phi_1) +$$

$$+\mu(\lambda+\mu)((\lambda+2\mu)(2(A_0 + \imath b_0)(\lambda+3\mu))))) = u_0 + \imath v_0.$$

The spline of the second type for the layer with singular point can be constructed similarly: the coefficients of the highest degrees of h in the representations of $u(x,y,h)+\imath v(x,y,h)$ and $w(x,y,h)$ should be the constants to be found.

CHAPTER 8

Approximation estimates and properties of the interpolation and spline-interpolation solutions

Elena A. Shirokova,
Kazan Federal University
Elena.Shirokova@ksu.ru
Pyotr N. Ivanshin,
Kazan Federal University
pivanshin@gmail.com

ABSTRACT: The interpolation and spline-interpolation solutions of the 3D basic problem of elasticity are the approximate solutions which satisfy the given boundary displacements only at a subset of the boundary surface of an elastic body. So these solutions are compared with the exact solution in this chapter.

According to the well-known estimate for polynomial interpolation of a smooth function defined on a interval the estimate of the difference between the interpolation and the exact solution of the problem for a cylinder at any generatrix can be compared with $1/((n+1)!)$ where n is the power of the interpolation polynomial in the third variable h . Section 8.1 contains the estimate of the difference between the interpolation and the exact solutions in the whole cylinder in the space $L_2(M)$. The similar estimate is also true for the smooth spline-interpolation solution of a higher power with respect to h considered in Section 8.3.

When we construct the spline-interpolation solution we glue the displacements at the edges of adjacent layers. Gluing also the derivatives of the orders $1, ..., m$ in the direction of the normal to the boundary of the common section allows to approximate the displacements at the inner points of the common section. Such gluing increases the power of the spline. The question of spline-interpolation solution convergence while increasing of the power of the spline to the given displacements at the ends of the layer is considered in Section 8.2

8.1 Comparison of the interpolation solution with the exact solution

Let us compare the solution of the second basic problem of elasticity for the cylinder M with the displacements $(U_0, V_0, W_0))$ given on the boundary surface S , and the interpolation solution, presented in this paper, with the displacements, given at a finite number of boundary curves on the levels $h = h_k$ and at the ends $h = a$ and $h = b$ and coinciding with $(U_0, V_0, W_0))$ at the corresponding points. The interpolation solution meets the elasticity equations inside M , meets the given displacements at the given boundary curves and at the finite number of points of the ends (if we apply point-wise gluing). It approximates the displacements at the other points of S.

We denote by u, v, w the differences between the displacements values of the exact solution of the second basic problem and that of the interpolation solution in M and also assume that all the partial derivatives of the displacement coordinates are integrable in M. We assume that the given functions (U_0, V_0, W_0) are continuous together with their partial derivatives. So then we consider not U_0, V_0 or W_0 themselves but differences $\tilde{U}_0$, $\tilde{V}_0$ or $\tilde{W}_0$ between them and their approximations on S by polynomials and polyharmonic functions.

Let us estimate $\delta(\sigma)$ — the difference between the coordinate values of the second basic problem solution stress tensor and that of the interpolation one S.

At the ends $h = a$ and $h = b$ of the body M the functions U_0, V_0, W_0 are approximated by polynomials with the accuracy

$$\|\frac{\partial^{n+1} f}{\partial x^{n+1}}\|_0 \omega_n(x)/(n+1)! + \|\frac{\partial^{n+1} f}{\partial y^{n+1}}\|_0 \omega_n(y)/(n+1)! +$$

$$+\|\frac{\partial^{2n+2} f}{\partial x^{n+1} \partial y^{n+1}}\|_0 \omega_n(x)\omega_n(y)/((n+1)!)^2$$

[5], here $\omega_n(x)$ is the bounded function depending on the step of the splitting of the corresponding disk and f equals either of the three functions U_0, V_0 or W_0.

Because of all the estimates given above the value of the integral

$$\int\int_S (\delta(\sigma)_n, r) dS,$$

can be made arbitrary small. Now Green theorem enables us to write the following relation:

$$\Delta = \int\int_S (\delta(\sigma)_n, r) dS = \int\int\int_M div(uX_n + vY_n + wZ_n) dV =$$

$$= \iiint\limits_{M} \lambda\theta^2 + 2\mu((\partial_x u)^2 + (\partial_y v)^2 + (\partial_z w)^2 + (\partial_x v + \partial_y u)^2 +$$

$$+(\partial_x w + \partial_z u)^2 + (\partial_z v + \partial_y w)^2)dV,$$

here X_n, Y_n, Z_n are differences between components of stress vector acting on the plane given by normal vector n for the considered solutions and $\theta = \partial_x u + \partial_y v + \partial_z w$. Our assumptions on U_0, V_0, W_0 imply that the components X_n, Y_n, Z_n may have discontinuities only of the first kind, that is why we can apply the Green theorem and Δ can be made arbitrary small. Thus for every element of the stress tensor we get the estimates

$$\|\lambda(\partial_x u + \partial_y v + \partial_z w) + 2\mu(\partial_x u)\|_{L^2(M)} = \|\lambda\theta + 2\mu(\partial_x u)\|_{L^2(M)} = \|\sigma_{11}\|_{L^2(M)} \leq$$

$$\leq (\sqrt{\lambda} + \sqrt{2\mu})\sqrt{\Delta},$$

$$\|\lambda\theta + 2\mu(\partial_y v)\|_{L^2(M)} = \|\sigma_{22}\|_{L^2(M)} \leq (\sqrt{\lambda} + \sqrt{2\mu})\sqrt{\Delta},$$

$$\|\lambda\theta + 2\mu(\partial_z w)\|_{L^2(M)} = \|\sigma_{33}\|_{L^2(M)} \leq (\sqrt{\lambda} + \sqrt{2\mu})\sqrt{\Delta},$$

$$\|\sigma_{ij}\|_{L^2(M)} \leq \sqrt{\mu}\sqrt{\Delta},\ i \neq j,\ i,j = 1,2,3.$$

Hence we get $\|\sigma_{jk}\|_{L^2(M)} \to 0$ as $\Delta \to 0$.

8.2 $C^0(S)$-spline and L_2 approximation

Let M be a solid of revolution. Let us construct the spline-interpolation solution of (1.1). We will represent this solution as the spline of degree n with respect to the variable h. We assume that the solution vanishes at the boundary circles on the planes $h = h_k$, $h = h_{k+1}$ of the layer. We need to minimise the difference at the section of M by the plane $h = h_k$ between the current spline and the previous spline in the layer $[h_{k-1}, h_k]$. Assume that the radius of the circle at the level $h = h_k$ is 1 and that at the level

$h = h_{k+1}$ is r. The displacement components have the form

$$u = \sum_{k=0}^{n} u_k(x,y)h^k, v = \sum_{k=0}^{n} v_k(x,y)h^k, w = \sum_{k=0}^{n} w_k(x,y)h^k. \tag{8.1}$$

Here according to the results of Chapter 1

$$-2\mu(u_k(x,y) + \imath v_k(x,y)) = -\kappa\phi_k(z) + z\overline{\phi_k'(z)} + \overline{\psi_k(z)} + \Phi_k(z,\overline{z}),$$

$$w_k(x,y) = \operatorname{Re}\rho_k(z) + \Psi_k(z,\overline{z}),$$

where the additional functions $\Phi_k(z,\overline{z})$ and $\Psi_k(z,\overline{z})$ should be obtained with the help of the recurrent relations

$$\Phi_k(z,\overline{z}) = \frac{(k+1)(k+2)}{8(\lambda+2\mu)}\left[\frac{2(\lambda^2+4\lambda\mu+5\mu^2)}{(\lambda+\mu)}\overline{z}\overline{\int\phi_{k+2}(z)dz}-\right.$$
$$-(\lambda+3\mu)z^2\overline{\phi_{k+2}(z)} - (\lambda+3\mu)z\overline{\int\psi_{k+2}(z)dz}+$$
$$+(\lambda+\mu)\int dz\int\psi_{k+2}(z)dz - (\lambda+3\mu)\int dz\int\Phi_{k+2}(z,\overline{z})d\overline{z}+$$
$$\left.+(\lambda+\mu)\int dz\overline{\int\Phi_{k+2}(z,\overline{z})d\overline{z}}\right]+$$
$$+\frac{(\lambda+\mu)(k+1)}{4(\lambda+2\mu)}\left[4\mu\int\Psi_{k+1}(z,\overline{z})dz+\right.$$
$$\left.+(\lambda+3\mu)z\overline{\rho_{k+1}(z)} - (\lambda+\mu)\int\rho_{k+1}(z)dz\right],$$

$$\Psi_k(z,\overline{z}) = -\frac{(\lambda+\mu)(k+1)}{4\mu^2}\left[-\frac{2\mu}{(\lambda+\mu)}\operatorname{Re}(\overline{z}\phi_{k+1}(z))+\right.$$
$$\left.+\operatorname{Re}(\int\Phi_{k+1}(z,\overline{z})d\overline{z})\right]-$$
$$-\frac{(k+1)(k+2)(\lambda+2\mu)}{4\mu}\left[\operatorname{Re}(\overline{z}\int\rho_{k+2}(z)dz) + \int dz\int\Psi_{k+2}(z,\overline{z})d\overline{z}\right].$$

We use these recurrent formulas starting at $k = n$ and ending at $k = 0$. Each application of them produces the functions $u_k(x,y)$, $v_k(x,y)$ and

$w_k(x,y)$ whose order of harmonicity is greater by 1 than that of the previously constructed u_{k+2}, v_{k+2} and w_{k+2} respectively. Also at each step we have 3 new analytic functions $\phi_k(z)$, $\psi_k(z)$ and $\rho_k(z)$. Hence the displacement components (u,v,w) are $([\frac{n}{2}]+2)$-harmonic functions.

We assume without loss of generality that we have to approximate three zero functions on the disc $h=h_k$ by the spline (u,v,w) described by (8.1). These functions restricted to the plane $h=h_k$ can be represented as follows:

$$u|_{h=h_k} = \sum_{k=0}^{[\frac{n}{2}]+2} |z|^{2k}(U_k(z)+\overline{U_k(z)}),$$

$$v|_{h=h_k} = \sum_{k=0}^{[\frac{n}{2}]+2} |z|^{2k}(V_k(z)+\overline{V_k(z)}), \tag{8.2}$$

$$w|_{h=h_k} = \sum_{k=0}^{[\frac{n}{2}]+2} |z|^{2k}(W_k(z)+\overline{W_k(z)}),$$

here U_k, V_k, W_k are analytic functions in the disc $|z|\le\max\{1,r\}$ [15]. So since analytic and anti-analytic functions take maximal values at the boundary of their domains, in order to estimate the approximation we consider their values on the circles. It is well known that L_2 norm is most convenient one for such observations because it can be represented via the coefficients of series expansions.

At the boundary circle on the plane $h=h_{k+1}$ the displacement functions have the following form:

$$u(re^{\imath\theta})|_{h=h_{k+1}} = \sum_{k=0}^{[\frac{n}{2}]+2} \alpha_k(U_k(re^{\imath\theta})+\overline{U_k(re^{\imath\theta})}),$$

$$v(re^{\imath\theta})|_{h=h_{k+1}} = \sum_{k=0}^{[\frac{n}{2}]+2} \beta_k(V_k(re^{\imath\theta})+\overline{V_k(re^{\imath\theta})}),$$

$$w(re^{\imath\theta})|_{h=h_{k+1}} = \sum_{k=0}^{[\frac{n}{2}]+2} \gamma_k(W_k(re^{\imath\theta}) + \overline{W_k(re^{\imath\theta})}),$$

here $\alpha_k \sim A_k Q_{n,k}$, $\beta_k \sim B_k Q_{n,k}$, $\gamma_k \sim C_k Q_{n,k}$, $(n \to \infty)$,

$$Q_{n,k} = r^{2k}(h_k - h_{k+1})^{n-k}((\lambda+\mu)^k + \frac{(\lambda+2\mu)^k(\lambda+\mu)^k}{\mu^k})\frac{n!}{(n-k)!},$$

the terms A_k, B_k, C_k being bounded in these expressions. It suffices to consider only the coefficient of Ψ_{n-k} in order to estimate the behavior of α_k, β_k and γ_k. The behavior of the other summands is similar to $\frac{L}{n}Q_{n,k}$.

Now let us show that by increasing the degree of h of our spline (8.1) we get better approximation to $(0,0,0)$ at the section D of M on the level $h = h_k$ in the space $L^2(D)$.

First choose an arbitrary component of displacement vector, say u. Assume that $U_k(e^{i\theta}) = \sum_{j=-\infty}^{\infty} U_k^j e^{ij\theta}$ for $U_k(z)$ from (8.2). Consider the system of boundary conditions on the levels $h = h_k$ and $h = h_{k+1}$:

$$\sum_{k=0}^{[\frac{n}{2}]+2} (U_k^j + \overline{U_k^{-j}}) = 0,$$

$$\sum_{k=0}^{[\frac{n}{2}]+2} \alpha_k(U_k^j + \overline{U_k^{-j}}) = m_j, j \in \mathbb{Z},$$

$m_j = 0$ for large numbers of j in the case of polynomial boundary conditions.

Note that since the coefficient of each power of h in (8.1) provides us with the set of analytic functions we also get the variables necessary for the minimisation of the function $|u|$ on the common disc $h = h_k$.

We minimise the functional

$$F(u) = \sum_{j=-\infty}^{\infty} \sum_{k=0}^{[\frac{n}{2}]+2} (U_k^j)^2$$

natural in the space $L^2(\mathbb{S}^1)$, $\mathbb{S}^1$ being the unite circle. We consider Lagrange function

$$\sum_{j=-\infty}^{\infty}\sum_{k=0}^{[\frac{n}{2}]+2}(U_k^j)^2-\sum_{j=-\infty}^{\infty}\lambda_j\sum_{k=0}^{[\frac{n}{2}]+2}(U_k^j+\overline{U_k^{-j}})-\sum_{j=-\infty}^{\infty}\mu_j(\sum_{k=0}^{[\frac{n}{2}]+2}\alpha_k(U_k^j+\overline{U_k^{-j}})-m_j).$$

The extremal point $\tilde{u}$ of the functional $F(u)$ has the following coordinates:

$$\tilde{U}_k^{-j}=\overline{\tilde{U}_k^j}=\lambda_j+\mu_j\alpha_k,$$

$$\lambda_j=-\frac{\mu_j\sum_{k=0}^{[\frac{n}{2}]+2}\alpha_k}{n+1},$$

$$\mu_j=\frac{m_j}{2\sum_{k=0}^{[\frac{n}{2}]+2}\alpha_k(\frac{\sum_{l=0}^{[\frac{n}{2}]+2}(\alpha_k-\alpha_l)}{n+1})}.$$

So

$$F(\tilde{u})=\sum_{j=-\infty}^{\infty}\sum_{k=0}^{[\frac{n}{2}]+2}\frac{m_j^2(\sum_{l=0}^{[\frac{n}{2}]+2}(\alpha_k-\alpha_l))^2}{4(\sum_{q=0}^{[\frac{n}{2}]+2}\alpha_q(\sum_{l=0}^{[\frac{n}{2}]+2}(\alpha_q-\alpha_l)))^2}\sim$$

$$\sim A\frac{(b-a)^n r^n n^2}{n!}\to 0,(n\to\infty)$$

since the difference $\alpha_k-\alpha_j\sim\frac{n!}{(n-k)!}\varepsilon(\lambda,\mu,r,h_k-h_{k+1})^{n-k}-\frac{n!}{(n-j)!}\varepsilon(\lambda,\mu,r,h_k-h_{k+1})^{n-j}$, ($n\to\infty$), here $\varepsilon(\lambda,\mu,r,h_k-h_{k+1})$ is the bounded value.

Thus

$$\sum_{k=0}^{[\frac{n}{2}]+2}\|\tilde{U}_k\|_{L^2}^2\sim C\frac{(h_{k+1}-h_k)^n r^n n^2}{n!}\to 0,(n\to\infty).$$

Hence

$$\|\tilde{u}\|_{L^2(D)}^2\le\pi\max\{1,r^{2n}\}(\sum_{k=0}^{[\frac{n}{2}]+2}\|\tilde{U}_k\|_{L^2})^2\le$$

$$\le\pi\max\{1,r^{2n}\}(\sum_{k=0}^{[\frac{n}{2}]+2}\|\tilde{U}_k\|_{L^2}^2+n(n-1)\max\{\|\tilde{U}_k\|_{L^2}^2|k=0,\dots,[\frac{n}{2}]+2\})\le$$

$$\leq \pi \max\{1, r^{2n}\}(n(n-1)+1) \sum_{k=0}^{[\frac{n}{2}]+2} \|\tilde{U}_k\|_{L^2}^2 \to 0, (n \to \infty).$$

For v and w the similar procedure gives the same result.

Therefore we approximate the values of spline on the section $h = h_k$, on the ends $h = a$ and $h = b$ with the help of the spline of the second type by choosing the degree n and corresponding analytic functions.

8.3 On $C^p(S)$-spline with L_2 approximation

The sum of the splines of the first and the second type is differentiable any time in any layer and continuous everywhere on S. Suppose that we want to have the sum of the splines of two types which is p times differentiable over h on S.

Assume that the radius of the section of S by the plane $h = h_k$ equals r. We construct the $C^p(S)$-spline of the second type as the solution of the system (1.1) in the layer $[h_k, h_{k+1}]$ which has p derivatives over h equal to that of the spline in the layer $[h_{k-1}, h_k]$ at their common boundary. Thus we add to the initial boundary conditions (6.1) the conditions of the following type:

$$u_{h^j}^{(j)}(r\cos\theta, r\sin\theta, h_k) = q_{1j}(\theta),$$

$$v_{h^j}^{(j)}(r\cos\theta, r\sin\theta, h_k) = q_{2j}(\theta), \tag{8.3}$$

$$w_{h^j}^{(j)}(r\cos\theta, r\sin\theta, h_k) = q_{3j}(\theta),\ \ \theta \in [0, 2\pi], j = 1, \ldots, p,$$

here the right-hand sides of relations contain values of the derivatives of the displacements found for the previous layer on the common edge $h = h_k$, $x^2 + y^2 = r^2$ (for $j = 1$ we must take into account also the value of the derivative of the first type spline).

We search for the solution in the following form:

$$u(x,y,h) = \sum_{j=0}^{p+n} u_j(x,y)h^j;$$

$$v(x,y,h) = \sum_{j=0}^{p+n} v_j(x,y)h^j; \tag{8.4}$$

$$w(x,y,h) = \sum_{j=0}^{p+n} w_j(x,y)h^j, \ (x,y) \in \mathbf{R}^2.$$

The coefficients $u_j(x,y)$, $v_j(x,y)$ and $w_j(x,y)$ in (8.4) can be found by the formulas of Chapter 1 (with $(n+p)$ instead of n) as the solution of system (1.1) constructed by the method of [10]. We obtain the solution of system (1.1) in every layer with the given displacements at the boundary circles on the levels $h = h_k$, which is p-differentiable on S but discontinuous on the levels $h = h_k$ inside M if $n = 1$. The additional powers of h help us approximate the given displacements at the section of M by the plane $h = h_k$ as it was done in the previous paragraph. The difference is that now we use $p+2$ (instead of 2) conditions while minimising the corresponding functional. The given functions U_0, V_0, W_0 can be approximated by the spline of degree n with respect to h on the surface S with any accuracy. Note that for the step of the splitting

$$d = \max\{|h_k - h_{k+1}||k = 0, \ldots, N-1\}, h_0 = a < h_1 < \ldots < h_N = b$$

the estimate of $\tilde{U}_0$, $\tilde{V}_0$ or $\tilde{W}_0$ — the difference between the values of U_0, V_0, W_0 and spline-interpolated ones equals

$$C \max\{||\tilde{U}^{(n+1)}_{0,h^{n+1}}||_0, ||\tilde{V}^{(n+1)}_{0,h^{n+1}}||_0, ||\tilde{W}^{(n+1)}_{0,h^{n+1}}||_0\}d$$

due to [9].

CHAPTER 9

Dynamic problems

Elena A. Shirokova,
Kazan Federal University
Elena.Shirokova@ksu.ru
Pyotr N. Ivanshin,
Kazan Federal University
pivanshin@gmail.com

ABSTRACT: In this chapter we present the most natural generalisation of the interpolation and spline-interpolation solutions of the static problems to dynamic ones. At first we present the interpolation solution of the 2D dynamic problem. And for the 3D case we simply add the time variable to the solutions of static problems and construct the solution in the form of polynomial in two variables, namely h and t in each layer of solid M.

9.1 Interpolation solution of the 2D problem

Let D be a finite domain in XOY plane with smooth boundary $C = \{(x(s), y(s))\ ,\ s \in [0, S]\}$, $x(0) = x(S)$, $y(0) = y(S)$. Let $\vec{a}(x, y, t) = (u(x, y, t), v(x, y, t))$, $(x, y) \in D \cup C$, $t \in [t_0, T]$ be the vector of displacements in XOY plane. The second basic plane dynamic problem of elasticity [7] can be formulated as follows: given the functions $f_1(s, t), f_2(s, t)$, $t \in [t_0, T]$, $s \in [0, S]$, $g_1(x, y), g_2(x, y)$, $h_1(x, y), h_2(x, y),$ $(x, y) \in D$, there should be found the components of the vector of displacements $u(x, y, t)$ and $v(x, y, t)$ at $D \times [t_0, T]$ which satisfy the dynamic equations

$$\begin{aligned}(\lambda + \mu)\, \tfrac{\partial \theta}{\partial x} + \mu \Delta u = \rho \tfrac{\partial^2 u}{\partial t^2},\\ (\lambda + \mu)\, \tfrac{\partial \theta}{\partial y} + \mu \Delta v = \rho \tfrac{\partial^2 v}{\partial t^2},\end{aligned} \tag{9.1}$$

where

$$\theta = \frac{\partial u}{\partial x} + \frac{\partial v}{\partial y},$$

λ and μ being Lamé constants, and boundary and initial conditions:

$$\begin{aligned}(u(x, y, t), v(x, y, t))|_{(x,y) \in \partial D} = (f_1(s, t), f_2(s, t)),\\ (u(x, y, t_0), v(x, y, t_0)) = (g_1(x, y), g_2(x, y)),\\ (\partial u/\partial t, \partial v/\partial t)|_{t=t_0} = (h_1(x, y), h_2(x, y)).\end{aligned}$$

Here we change the known at every moment of $[t_0, T]$ boundary displacements by the boundary displacements known at the finite number of moments of the time segment $[t_0, T]$. The boundary conditions at the other moments are replaced by the interpolation polynomials in the time variable t which are obtained simultaneously with constructing the differentiable solution of the equations (9.1). This solution of the boundary value problem is named the interpolation solution and is presented in [13].

The interpolation method of solution of the $2D$ dynamic problem is the analog of the method of interpolation solution of the second basic problem of elasticity for cylinders presented in Chapters 1,2,3,4. The coordinates of the desired vector of displacements are supposed to be polynomials in t. The coefficients of the polynomials are the solutions of plane boundary value problems for the elliptic differential equations. These boundary value problems are solved successively.

The method is programmable and applicable for problems of dynamic.

9.1.1 Interpolation of the boundary data

Let us construct the solution of the following boundary problem: given the boundary conditions

$$(\breve{u}(x,y,t_j),\breve{v}(x,y,t_j))\,|_{(x,y)\in C}=(\tilde{u}_j(s),\tilde{v}_j(s))\,,j=0,1,...,n, \tag{9.2}$$

at the moments $t_j\in[t_0,T]$, $j=0,1,...,n$, there should be found the differentiable functions $\breve{u}(x,y,t)$, $\breve{v}(x,y,t)$ which satisfy in D the differential equations (9.1).

We search for the solution of this boundary value problem in the form

$$\breve{u}(x,y,t)=\sum_{k=0}^{n}u_k(x,y)t^k,\quad \breve{v}(x,y,t)=\sum_{k=0}^{n}v_k(x,y)t^k. \tag{9.3}$$

Obviously we obtain from (9.2) the boundary values of the unknown coefficients:

$$u_k(x,y)|_{(x,y)\in C}=\hat{u}_k(s),\quad v_k(x,y)|_{(x,y)\in C}=\hat{v}_k(s),k=0,1,...,n,$$

because the systems

$$\sum_{k=0}^{n} \hat{u}_k(s)t_j^k = \tilde{u}_j(s), j = 0, 1, ..., n$$

and

$$\sum_{k=0}^{n} \hat{v}_k(s)t_j^k = \tilde{v}_j(s), j = 0, 1, ..., n$$

with Vandermonde determinant are uniquely resolved.

Now we substitute the components of the vector of displacements in the form (9.2) to the equations (9.1). After we compare the coefficients of the same powers of t at the right and the left -hand side of each equation we have the following equations over the unknown coefficients:

$$\begin{cases} \frac{\partial}{\partial x}\left[(\lambda+2\mu)\left(\frac{\partial u_k}{\partial x}+\frac{\partial v_k}{\partial y}\right)\right]-\frac{\partial}{\partial y}\left[\mu\left(\frac{\partial v_k}{\partial x}-\frac{\partial u_k}{\partial y}\right)\right]= \\ \qquad = \rho(k+1)(k+2)u_{k+2}, \\ \frac{\partial}{\partial y}\left[(\lambda+2\mu)\left(\frac{\partial u_k}{\partial x}+\frac{\partial v_k}{\partial y}\right)\right]+\frac{\partial}{\partial x}\left[\mu\left(\frac{\partial v_k}{\partial x}-\frac{\partial u_k}{\partial y}\right)\right]= \\ \qquad = \rho(k+1)(k+2)v_{k+2}, \end{cases}$$

where $u_{n+1}(x,y) \equiv u_{n+2}(x,y) \equiv v_{n+1}(x,y) \equiv v_{n+2}(x,y) \equiv 0$, $(x,y) \in D$. After we multiply the second equation of the last system by $\imath/2$, the first by $1/2$ and sum them together we have the equivalent relation

$$\frac{\partial}{\partial \bar{z}}\left[(\lambda+2\mu)\left(\frac{\partial u_k}{\partial x}+\frac{\partial v_k}{\partial y}\right)+\right.$$
$$\left.+\imath\mu\left(\frac{\partial v_k}{\partial x}-\frac{\partial u_k}{\partial y}\right)\right]=\frac{\rho(k+1)(k+2)}{2}\left[u_{k+2}+\imath v_{k+2}\right]. \tag{9.4}$$

Here $z = x + \imath y$ and the derivatives with respect to z and $\bar{z}$ are computed as in Chapter 1:

$$\frac{\partial}{\partial z} = \frac{1}{2}\left(\frac{\partial}{\partial x} - \imath\frac{\partial}{\partial y}\right),$$
$$\frac{\partial}{\partial \overline{z}} = \frac{1}{2}\left(\frac{\partial}{\partial x} + \imath\frac{\partial}{\partial y}\right).$$

We solve the equations (9.4) successively beginning with $k = n$. We have for $k = n, n-1$

$$(\lambda + 2\mu)\left(\frac{\partial u_k}{\partial x} + \frac{\partial v_k}{\partial y}\right) +$$
$$+\imath\mu\left(\frac{\partial v_k}{\partial x} - \frac{\partial u_k}{\partial y}\right) = f_k(z),$$

where $f_k(z)$ is an analytic in D function. After we separate the real and imaginary part we have

$$\frac{\partial}{\partial z}(u_k + \imath v_k) = \frac{\operatorname{Re} f_k(z)}{2(\lambda+2\mu)} + \frac{\imath \operatorname{Im} f_k(z)}{2\mu} =$$
$$= \frac{f_k + \overline{f_k}}{4(\lambda+2\mu)} + \frac{f_k - \overline{f_k}}{4\mu}.$$

Therefore

$$u_k + \imath v_k = \frac{\lambda+3\mu}{4\mu(\lambda+2\mu)}\int f_k(z)dz - \frac{\lambda+\mu}{4\mu(\lambda+2\mu)} z\overline{f_k(z)} + \overline{g_k(z)},$$

where $g_k(z)$ is an analytic in D function. Now after we introduce the following analytic in D functions

$$\frac{\lambda+\mu}{2(\lambda+2\mu)}\int f_k(z)dz \equiv \phi_k(z), \quad -2\mu g_k(z) \equiv \psi_k(z),$$
$$\frac{\lambda+3\mu}{\lambda+\mu} \equiv \kappa,$$

we obtain the following representation of the complex combination of the desired coefficients:

$$-2\mu(u_k + \imath v_k) = -\kappa\phi_k(z) + z\overline{\phi_k'(z)} + \overline{\psi_k(z)},$$

— just the same as the representation of the complex displacements in the second basic plane problem of the theory of elasticity [7].

The boundary condition

$$\left[-\kappa\phi_k(z) + z\overline{\phi_k'(z)} + \overline{\psi_k(z)}\right]\Big|_{z\in C} = \\ = -2\mu\left(\hat{u}_k(s) + \imath\hat{v}_k(s)\right), s \in [0, S],$$

is also the same as that in the second basic plane problem of the theory of elasticity. So the functions $\phi_k(z)$ and $\psi_k(z)$, $k = n$, $n-1$, can be uniquely restored [7].

The method of solution of the equation (9.4) for $k \leq n-2$ is the same. We only obtain the additional summand at the right side of complex combination of the desired coefficients:

$$-2\mu(u_k + \imath v_k) = -\kappa\phi_k(z) + z\overline{\phi_k'(z)} + \overline{\psi_k(z)} + \Phi_k(z, \overline{z}),$$

where

$$\begin{aligned} \Phi_k(z,\overline{z}) &= -2\mu(k+1)(k+2)S[u_{k+2} + \imath v_{k+2}], \\ S[u + \imath v] &= k_1 \int \mathrm{Re}\left[\int (u + \imath v) d\overline{z}\right] dz + \\ &+ \imath k_2 \int \mathrm{Im}\left[\int (u + \imath v) d\overline{z}\right] dz, \\ k_1 &= \frac{\rho}{4(\lambda + 2\mu)}, k_2 = \frac{\rho}{4\mu}. \end{aligned} \tag{9.5}$$

Note that the function $\Phi_k(z,\overline{z})$ is known because the complex coefficient $u_{k+2} + \imath v_{k+2}$ is already obtained. The boundary condition for the analytic in D functions $\phi_k(z)$ and $\psi_k(z)$, $k < n-1$, is also the same as that in the second basic plane problem of the theory of elasticity:

$$\left[-\kappa\phi_k(z) + z\overline{\phi_k'(z)} + \overline{\psi_k(z)}\right]\Big|_{z\in C} =$$
$$= -2\mu\left(\hat{u}_k(s) + \imath\hat{v}_k(s)\right) - \Phi_k(z(s), \overline{z(s)}), s \in [0, l].$$

After we restore all the coefficients $u_k(x,y)$, $v_k(x,y)$, $k = 0, 1, ..., n,$ we obtain the components of the vector of displacements $(\breve{u}(x,y,t), \breve{v}(x,y,t))$ from (9.3). These components are differentiable, they satisfy the boundary conditions (9.2), but we have not used the initial conditions, so this interpolation solution is adequate only for the moments far from $t = t_0$.

The interpolation solution $(\breve{u}(x,y,t), \breve{v}(x,y,t))$ when

$$(x, y) = (x(s^*), y(s^*)) \in C$$

is the polynomial in the time variable t with the given values at $t = t_k$, $k = 0, ..., n$. So the estimate of an error of real boundary displacement $(U(t), V(t))$ interpolation

$$(r_1(t), r_2(t)) = (U(t) - \breve{u}(x(s^*), y(s^*), t), V(t) - \breve{v}(x(s^*), y(s^*), t))$$

is for $(n+1)$-times differentiable real boundary displacements just the same as for the interpolation polynomials:

$$|r_1(t)| \leq \tfrac{1}{(n+1)!} \max_{t_0 \leq t \leq T} |U^{(n+1)}(t)| \textstyle\prod_{k=0}^{n} |t - t_k|,$$
$$|r_2(t)| \leq \tfrac{1}{(n+1)!} \max_{t_0 \leq t \leq T} |V^{(n+1)}(t)| \textstyle\prod_{k=0}^{n} |t - t_k|.$$

Note that the boundary displacements can be given not at the same moments for all boundary points. One can choose different moments for different boundary points: $t_j = t^j(s)$, $j = 0, ..., n$, $s \in [0, S]$. For each value s^*, $s^* \in [0, S]$, the points $t^j(s^*)$, $j = 0, ..., n$, must be the points of good polynomial interpolation of the boundary displacements for the corresponding boundary point over $[t_0, T]$. The method of constructing the interpolation solution for this case is the same.

We also can use this method for the case when the boundary displacements are given not at all boundary points but at a finite number of boundary points.

9.1.2 Analysis of the interpolation solution

We proof that all the functions $u_k(x, y)$ and $v_k(x, y)$, $k = 0, 1, ..., n$, constructed above are polyharmonic in D, and also that the order of harmonicity adds 1 as we pass from $u_k(x, y)$ and $v_k(x, y)$ to $u_{k-2}(x, y)$ and $v_{k-2}(x, y)$.

We note first of all that the functions $u_n(x, y)$, $v_n(x, y)$, $u_{n-1}(x, y)$ and $v_{n-1}(x, y)$ are biharmonic in D because they can be represented as real or imaginary part of the functions like $\tilde{f}_1(z) + \overline{z}\tilde{f}_2(z)$, where $\tilde{f}_j(z)$, $j = 1, 2$, are analytic in D. Note also that the only difference between the analytic coefficients of $\overline{z}$ in u_j and v_j representations, $j = n, n-1$, is the sign.

Let us examine action of the differential operator S from (9.5) on an m-harmonic function. Let

$$u(x, y) = \mathrm{Re}[\sum_{k=0}^{m-1} \overline{z}^k \tilde{f}_k(z)], \quad v(x, y) = \mathrm{Im}[\sum_{k=0}^{m-1} \overline{z}^k \tilde{g}_k(z)]. \tag{9.6}$$

It can be easily verified that

$$\operatorname{Re} S[u+\imath v] = \operatorname{Re}\{\sum_{k=0}^{m-1} \overline{z}^k \frac{k_1-k_2}{4} \int\int [\tilde{f}_k(z) - \tilde{g}_k(z)]dzdz +$$
$$+\sum_{k=0}^{m-1} \overline{z}^{k+1} \frac{k_1+k_2}{2(k+1)} \int \tilde{f}_k(z)dz +$$
$$+\sum_{k=0}^{m-1} \overline{z}^{k+2} \frac{k_1-k_2}{4(k+1)(k+2)} [\tilde{f}_k(z) + \tilde{g}_k(z)]\},$$

$$\operatorname{Im} S[u+\imath v] = \operatorname{Im}\{\sum_{k=0}^{m-1} \overline{z}^k \frac{k_1-k_2}{4} \int\int [\tilde{f}_k(z) - \tilde{g}_k(z)]dzdz +$$
$$+\sum_{k=0}^{m-1} \overline{z}^{k+1} \frac{k_1+k_2}{2(k+1)} \int \tilde{g}_k(z)dz -$$
$$-\sum_{k=0}^{m-1} \overline{z}^{k+2} \frac{k_1-k_2}{4(k+1)(k+2)} [\tilde{f}_k(z) + \tilde{g}_k(z)]\}.$$

The coefficients of $\overline{z}^{m+1}$ in the representations of $\operatorname{Re} S[u+iv]$ and $\operatorname{Im} S[u+iv]$ vanish if $\tilde{f}_{m-1}(z) = -\tilde{g}_{m-1}(z)$ in (9.6), and also for this case the only difference between the analytic coefficients of $\overline{z}^m$ in these representations is the sign.

Now it can be easily calculated that the components of the interpolation solution $(\breve{u}(x,y,t), \breve{v}(x,y,t))$ are $([\frac{n}{2}]+2)$ -harmonic functions.

9.1.3 Interpolation of the initial data

Let us consider the following problem: there should be found the vector of displacements (u,v) in the domain D, where the differentiable in D functions $u(x,y,t)$, $v(x,y,t)$ satisfy the equations (9.1), the boundary conditions (2) and satisfy the initial velocities $\vec{V}(x_k,y_k,t_0) = (V_k^1, V_k^2)$, $k=1,...,m$ and the initial displacements $\vec{a}(x_{k'},y_{k'},t_0) = (a_{k'}^1, a_{k'}^2)$, $k'=1,...,m'$, (x_k,y_k)

and $(x_{k'}, y_{k'})$ being inner points of D.

We search for the solution of the problem with boundary and initial data in the form

$$u(x,y,t) + \imath v(x,y,t) = \breve{u}(x,y,t) + \imath\breve{v}(x,y,t) + \grave{u}(x,y,t) + \imath\grave{v}(x,y,t),$$

where the displacements $(\breve{u}(x,y,t), \breve{v}(x,y,t))$ are constructed as the interpolation solutions of the boundary value problem.

The displacements $(\grave{u}(x,y,t), \grave{v}(x,y,t))$ satisfy the equations (9.1), vanish at $(x,y) \in C$, $t = t_j$, $j = 0, 1, ..., n$, with

$$\begin{gathered} \grave{u}(x_{k'}, y_{k'}, t_0) = a^1_{k'} - \breve{u}(x_{k'}, y_{k'}, t_0), \\ \grave{v}(x_{k'}, y_{k'}, t_0) = a^2_{k'} - \breve{v}(x_{k'}, y_{k'}, t_0), k' = 1, ..., m', \\ \grave{u}'_t(x_k, y_k, t_0) = V^1_k - \breve{u}'_t(x_k, y_k, t_0), \\ \grave{v}'_t(x, y, t_0) = V^2_k - \breve{v}'_t(x_k, y_k, t_0), k = 1, ..., m. \end{gathered} \tag{9.7}$$

We search for the displacements $(\grave{u}, \grave{v})$ in the form

$$\begin{gathered} \grave{u}(x,y,t) = \textstyle\sum_{k=0}^{n} \grave{u}_k(x,y)t^k + \sum_{j=1}^{l} P^1_j(x,y)t^{n+j}, \\ \grave{v}(x,y,t) = \textstyle\sum_{k=0}^{n} \grave{v}_(x,y)t^k + \sum_{j=1}^{l} P^2_j(x,y)t^{n+j}, \end{gathered} \tag{9.8}$$

where $P^k_j(x,y)$, $k = 1, 2$, $j = 1, ..., l$, are algebraic polynomials in two variables x and y with unknown real coefficients. We describe the scheme of these polynomial construction.

We need the polynomials $P^k_j(x,y)$ to satisfy the relations like (4) in order the displacements (9.8) to satisfy the equations (9.1). Therefore we have for $k = l, l-1$

$$P^1_k(x,y) + \imath P^2_k(x,y) = \mathrm{Re}\{\tilde{P}^1_k(z) - \frac{\overline{z}}{2\kappa}[\tilde{P}^1_k(z) + \tilde{P}^2_k(z)]'\} +$$
$$+\imath\,\mathrm{Im}\{\tilde{P}^2_k(z) + \frac{\overline{z}}{2\kappa}[\tilde{P}^1_k(z) + \tilde{P}^2_k(z)]'\},$$

where $\tilde{P}^j_k(z), j = 1, 2$, are algebraic polynomials in the complex variable

$z = x + \imath y$ with arbitrary complex coefficients. For $k < l - 1$ we have

$$P_k^1(x,y) + \imath P_k^2(x,y) = \mathrm{Re}\{\tilde{P}_k^1(z) - \frac{\overline{z}}{2\kappa}[\tilde{P}_k^1(z) + \tilde{P}_k^2(z)]'\} +$$
$$+\imath\,\mathrm{Im}\{\tilde{P}_k^2(z) + \frac{\overline{z}}{2\kappa}[\tilde{P}_k^1(z) + \tilde{P}_k^2(z)]'\} +$$
$$+(k+1)(k+2)S[P_{k+2}^1(x,y) + \imath P_{k+2}^2(x,y)],$$

where S is the operator from (9.5).

The coefficient of every power of t more than n-th in the representation (9.8) brings to this representation 2 additional complex polynomials. So it creates 2 sets of arbitrary complex parameters. Therefore the function $\grave{u}$ $+\imath\grave{v}$ contains $2l$ sets of arbitrary complex parameters or $4l$ sets of arbitrary real parameters.

When we satisfy the null boundary displacements condition at t_j, $j = 0, ..., n$, for $(\grave{u}, \grave{v})$, we express the boundary values of the corresponding coefficients $\grave{u}_k$, $\grave{v}_k$ via the unknown parameters:

$$[\grave{u}_k(x,y) + i\grave{v}_k(x,y)]|_{(x,y)\in C} = -\sum_{p=1}^{l} \alpha_p^k [P_p^1(x,y) + \imath P_p^2(x,y)]|_{(x,y)\in C}, \quad (9.9)$$

where α_p^k is the fraction with Vandermonde determinant which contains the powers of t_j, $j = 0, 1, ..., n$, from 0-th power till n-th power as the denominator, and the same determinant where the column with t_j^k is changed by the column with t_j^{n+p} as the numerator.

Now we restore the coefficients $\grave{u}_k(x,y)$, $\grave{v}_k(x,y)$, $k = 0, 1, ..., n$, following the scheme presented in section 2 for the coefficients $u_k(x,y)$, $v_k(x,y)$, $k = 0, 1, ..., n$, beginning with $k = n$. Note that the polyharmonic functions $\grave{u}_k(x,y)$, $\grave{v}_k(x,y)$, $k = 0, 1, ..., n$, are linearly dependent on the unknown parameters which are contained in $P_p^1(x,y)$ and $P_p^2(x,y)$, $p = 1, ..., l$, due to the boundary conditions (9.9).

Obviously, every polynomial $P(x,y)$ is a polyharmonic function, because there exists a natural n such that

$$\frac{\partial^{2n} P}{\partial z^n \partial \bar{z}^n} = \frac{1}{2^{2n}} \Delta^n P \equiv 0.$$

Therefore the functions $\grave{u}(x, y, t_0)$, $\grave{v}(x, y, t_0)$ from (8) and also the functions $\grave{u}_t'(x, y, t_0)$, $\grave{v}_t'(x, y, t_0)$ are also polyharmonic and they are lineally dependent on the unknown coefficients which are contained in $P_p^1(x, y)$ and $P_p^2(x, y), \quad p = 1, ..., l$.

Let the number of these unknown coefficients be equal to $2m + 2m'$. Now when we satisfy the initial conditions (9.7) we obtain the linear with respect to $2(m+m')$ unknown real constants system of $2(m+m')$ equations. The solvability problem of this system in general case is rather complicated because the value of the determinant of the system depends on the form of the domain and on the form of the polynomials $P_p^k(x, y)$, $k = 1, 2, \quad p = 1, ..., l$.

Suppose that the corresponding system (9.7) is resolvable with respect to the unknown constants. Then we have the interpolation of the initial conditions by the polyharmonic functions.

The class of polyharmonic functions is a generalisation of the class of polynomials. Any continuous in a finite closed domain function can be approximated by a polyharmonic function with arbitrary accuracy in the space C(D). The authors of [8] proved that the class of polyharmonic in a finite closed domain functions is closed under the uniform convergence. They also proved that any uniformly convergent sequence of polyharmonic functions can be differentiated any time, and the sequence of derivatives converges to the derivative of the limit.

9.1.4 Example

Let D be the unit disk. We give the displacements of the boundary of D at tree moments: $t_0 = 0, t_1 = 1, t_2 = 2$. Let

$$\tilde{u}_0(\theta) \equiv 0, \tilde{v}_0(\theta) \equiv \delta,$$
$$\tilde{u}_1(\theta) = (r-1)\cos\theta, \tilde{v}_1(\theta) = (r-1)\sin\theta,$$
$$\tilde{u}_2(\theta) \equiv \epsilon, \tilde{v}_2(\theta) \equiv 0,$$

where $\theta \in [0, 2\pi]$. So the unite disk is displaced in OY direction at the moment $t_0 = 0$, it is pressed uniformly at the moment $t_1 = 1$, and at the moment $t_2 = 2$ the unit disk is displaced in OX direction.

The interpolation solution of this problem which satisfies the equations (9.1) and the given boundary conditions has the following form:

$$\breve{u}(x,y,t) + \imath\breve{v}(x,y,t) = \imath\delta - \left(1-|z|^2\right)\frac{\rho(\epsilon+i\delta)}{2(\lambda+3\mu)} +$$
$$+z\left(1-|z|^2\right)\frac{\rho(r-1)}{4(\lambda+2\mu)} - \frac{\epsilon+3i\delta-4(r-1)z}{2}t +$$
$$+\frac{\epsilon+i\delta-2(r-1)z}{2}t^2, \quad x^2+y^2 \le 1, \quad t \in [0,2].$$

In order to satisfy the additional conditions — the initial displacement at one inner point and the initial velocity at one inner point — we search for the additional displacements $(\grave{u}(x,y,t), \grave{v}(x,y,t))$ in the following form

$$\grave{u}(x,y,t) + \imath\grave{v}(x,y,t) = \grave{u}_0(x,y) + \imath\grave{v}_0(x,y,t) + (\grave{u}_1(x,y) +$$
$$+\imath\grave{v}_1(x,y))t + (\grave{u}_2(x,y) + \imath\grave{v}_2(x,y))t^2 + (a+\imath b)t^3 + (c+\imath d)t^4.$$

When we satisfy the null boundary displacements for $(\grave{u}(x,y,t), \grave{v}(x,y,t))$ at $t = 0, 1, 2,$ we have

$$\grave{u}_2(x,y)+\imath\grave{v}_2(x,y)=-[3(a+ib)+7(c+\imath d)]-\frac{6\rho(c+\imath d)}{\lambda+3\mu}(1-|z|^2),$$
$$\grave{u}_1(x,y)+\imath\grave{v}_1(x,y)=2(a+\imath b)+6(c+\imath d)-\frac{3\rho(a+\imath b)}{\lambda+3\mu}(1-|z|^2),$$
$$\grave{u}_0(x,y)+\imath\grave{v}_0(x,y)=\frac{\rho}{\lambda+3\mu}[3(a+\imath b)+7(c+\imath d)](1-|z|^2)+$$
$$+\frac{\rho^2(\lambda+\mu)(c-\imath d)}{4\mu(\lambda+3\mu)(\lambda+2\mu)}z^2(1-|z|^2)-\frac{3\rho^2(c+\imath d)}{8\mu(\lambda+2\mu)}(1-|z|^4)+$$
$$+\frac{3\rho^2(c+\imath d)}{2\mu(\lambda+2\mu)}(1-|z|^2)-\frac{\rho^2(\lambda+\mu)^2(c+\imath d)}{\mu(\lambda+3\mu)^2(\lambda+2\mu)}(1-|z|^2).$$

Now we obtain the unknown constants a , b , c , d when we satisfy the two initial conditions at two inner points. The determinant of the corresponding linear system depends on the two inner points location and on Lamé coefficients λ and μ . But when we note that the values like $\frac{\rho}{\alpha\lambda+\beta\mu}$ are very small for $\alpha=1,0$, $\beta=1,2,3,$ it becomes evident that this discriminant does not vanish. So the constants a , b , c and d can be uniquely determined for any choice of the two points with initial data.

9.1.5 An estimate of the interpolation crror

We reduce an estimate of the error when the exact solution of the second basic problem of the dynamics of elastic bodies is replaced by the presented interpolation solution, due to the linearity of the problem, to estimating the norm of the solution of the problem which consists of the equations (9.1), boundary conditions

$$(u(x,y,t),v(x,y,t))|_{(x,y)\in C}=(r_1(s,t),r_2(s,t)),\ t\in[t_0,T],\ s\in[0,l],$$

where

$$r_k(s,t_j)=0,\ k=1,2;\ j=0,1,...,n,$$

and initial conditions

$$(u(x,y,t_0), v(x,y,t_0)) = (h_1(x,y), h_2(x,y)),$$
$$(u_t'(x,y,t_0), v_t'(x,y,t_0)) = (V_1(x,y), V_2(x,y)), \quad (x,y) \in D,$$

where

$$h_k(x,y)|_{(x,y)\in C} = V_k(x,y)|_{(x,y)\in C} = 0, \; k = 1, 2.$$

We assume that the functions $h_k(x,y)$ and $V_k(x,y)$ are polyharmonic in D functions bounded in D. This assumption is not very restrictive due to the results of [8]. Let us also assume that

$$\|\sqrt{r_i^2 + r_2^2}\|_{C([0,l]\times[t_0,T])} < \epsilon_0, \; \|\sqrt{V_1^2 + V_2^2}\|_{L_2(D)} < \epsilon_1,$$
$$\|\frac{\partial h_1}{\partial x}\|_{L_2(D)} < \epsilon_2, \; \|\frac{\partial h_2}{\partial y}\|_{L_2(D)} < \epsilon_3, \; \|\frac{\partial h_1}{\partial y} + \frac{\partial h_2}{\partial x}\|_{L_2(D)} < \epsilon_4.$$

We also suppose that the exact solution has second derivatives with respect to x and to y of the components of the displacement vector continuous in D up to the boundary C for all $t \in [t_0, T]$, and that the derivatives with respect to t of these second derivatives are integrable with a modulus along C.

Let us use the well-known relation ([7])

$$\oint_C (\vec{\sigma}_n, \vec{a}_t')ds = \frac{1}{2}\frac{\partial}{\partial t}\left[\iint_D [\rho((u_t')^2 + (v_t')^2) + 2\mu((u_t')^2 + \right.$$
$$\left. + (v_t')^2) + \lambda(u_x' + v_y')^2 + \mu(u_y' + v_x')^2]dxdy\right] \qquad (9.10)$$

where $\vec{a}_t' = (\frac{\partial r_1}{\partial t}, \frac{\partial r_2}{\partial t})$ and $\vec{\sigma}_n$ is the vector of normal stresses.

We integrate equality (9.10) over t from t_0 to t and we obtain the integral on the right-hand side in the form

$$\iint_D [\rho(V_1^2+V_2^2)+2\mu((h'_{1x})^2+(h'_{2y})^2)+\lambda(h'_{1x}+h'_{2y})^2+$$

$$\mu(h'_{1y}+h'_{2x})^2]dxdy+2\oint_C(\vec{\sigma}_n,\vec{a})ds-2\int_{t_0}^{t}\oint(\vec{\sigma}'_{nt},\vec{a})ds.$$

The integral on the right-hand side of (9.10) is a linear combination of squares of norms in the space $L_2(D)$ of several functions, therefore the following estimate holds

$$\rho\|\sqrt{(u'_t)^2+(v'_t)^2}\|^2_{L_2(D)}+2\mu(\|u'_x\|^2_{L_2(D)}+\|v'_y\|^2_{L_2(D)})+\lambda\|u'_x+v'_y\|^2_{L_2(D)}+$$
$$+\mu\|u'_y+v'_x\|^2_{L_2(D)}\leq\rho\epsilon_1^2+2\mu(\epsilon_2^2+\epsilon_3^2)+\lambda(\epsilon_2+\epsilon_3)^2+\mu\epsilon_4^2+$$
$$+2\epsilon_0(\max_{t\in[t_0,T]}\oint|\vec{\sigma}_n|ds+(T-t_0)\max_{t\in[t_0,T]}\oint|\vec{\sigma}'_{nt}|ds)=\delta^2.$$

Now we can estimate the norms of the differences in the components of the stress tensors corresponding to to the true and interpolation solutions of the second basic plane problem of dynamics of elastic bodies, denoting them by σ_{ij} , $i,j=1,2$. We have the following inequalities

$$\|\sigma_{11}\|_{L_2(D)}\leq(\sqrt{\lambda}+\sqrt{2\mu})\delta,\ \|\sigma_{22}\|_{L_2(D)}\leq(\sqrt{\lambda}+\sqrt{2\mu})\delta,$$
$$\|\sigma_{12}\|_{L_2(D)}\leq\sqrt{\mu}\delta.$$

9.1.6 A spline-interpolation solution

The presented method of interpolation solution is the base for spline-interpolation solution construction. We divide the time segment $[t_0,T]$ in n segments $[t_j,t_{j+1}]$. Then we find the interpolation solution for every segment and glue these solutions.

For example, we obtain the continuous in $D\times[t_0,T]$ spline-interpolation solution which satisfies the equation (9.1) in each section $D\times(t_j,t_{j+1})$, and the boundary conditions (9.2) if we construct the linear with respect to t interpolation solution for every section $D \times [t_j, t_{j+1}]$.

We obtain a smooth at the boundary points spline-interpolate solution if we find for every segment $D \times [t_j, t_{j+1}]$ the interpolation solution which takes the given boundary displacements at $t = t_{j+1}$, takes the given boundary displacements and velocities and given displacements and velocities at a finite number of inner points of D at $t = t_j$. We use the given boundary data for $j = 0$. And for $j > 0$ we use at $t = t_j$ the data from the interpolation solution found for the previous segment.

To this end we search for each interpolation solution in the form

$$\breve{u} + \imath\breve{v} + \grave{u} + \imath\grave{v},$$

where

$$\breve{u} + \imath\breve{v} = \sum_{k=0}^{2} (u_k(x, y) + \imath v_k(x, y)) t^k$$

satisfies the equations (9.1) at $D \times (t_j, t_{j+1})$, takes the given boundary displacements at $C \times t_j$ and at $C \times t_{j+1}$ and has the given velocities at $C \times t_{j+1}$. Note that we have to solve the linear system with the determinant $-(t_{j+1} - t_j)^2$ instead of Vandermonde determinant in order to obtain the boundary values of the coefficients $u_k(x, y)$ and $v_k(x, y)$, $k = 0, 1, 2$. The summand

$$\grave{u} + \imath\grave{v} = \sum_{k=0}^{2} (\grave{u}_k(x, y) + \imath\grave{v}_k(x, y)) t^k + \sum_{p=1}^{l} (P_p^1(x, y) + \imath P_p^2(x, y)) t^{2+p}$$

satisfies the equations (9.1) at $D \times (t_j, t_{j+1})$, has the null boundary displacements at $C \times t_j$, at $C \times t_{j+1}$ and null velocities at $C \times t_{j+1}$ and it interpolates the displacements and velocities in D via the given or the found data at some inner points of D at $t = t_j$ due to the presence of the polynomials.

9.2 On the spline-interpolation solution of the 3D dynamic problem

The main system of equations for the dynamic problem of elasticity is the following

$$
\begin{aligned}
(\lambda+\mu)\frac{\partial\theta}{\partial x}+\mu\Delta u&=\partial^2_{tt}u,\\
(\lambda+\mu)\frac{\partial\theta}{\partial y}+\mu\Delta v&=\partial^2_{tt}v,\\
(\lambda+\mu)\frac{\partial\theta}{\partial h}+\mu\Delta w&=\partial^2_{tt}w,
\end{aligned}
\tag{9.11}
$$

where

$$\theta=\frac{\partial u}{\partial x}+\frac{\partial v}{\partial y}+\frac{\partial w}{\partial h}.$$

The formulation of the second dynamic problem of elasticity follows: given displacements

$$(f_1(\sigma,\tau,t),f_2(\sigma,\tau,t),f_3(\sigma,\tau,t))$$

at the surface S of the solid M at any moment $t\geq 0$ and the displacements

$$(g_1(x,y,h),g_2(x,y,h),g_3(x,y,h))$$

and the velocities

$$(p_1(x,y,h),p_2(x,y,h),p_3(x,y,h))$$

in the whole solid M at the initial moment $t=0$, it should be possible to find the displacements

$$(u(x,y,h,t),v(x,y,h,t),w(x,y,h,t))$$

in the elastic solid M for $t>0$ which satisfy system (9.11).

Here we consider some variations of interpolation solution of this problem for the circular cylinder. We search for the interpolation solutions in the form of the polynomials in variables h and t, so we have to change boundary and initial conditions of the second dynamic problem of elasticity.

9.2.1 The linear spline-interpolation solution

The most simple spline for this problem is the spline linear with respect to the variable t.

We formulate the following interpolation problem: given the displacements on the surface S at the moments $t = 0$, $t = t_1$,.... $t = t_n$, given the velocities on the surface S at the moment $t = 0$, it should be possible to find the displacements in M for $t \in [0, t_n]$.

At first we approximate the solution only for the moments close to $t = 0$. We construct the solution in the form

$$u = u_0(x, y, h) + u_1(x, y, h)t, v = v_0(x, y, h) + v_1(x, y, h)t, \quad (9.12)$$
$$w = w_0(x, y, h) + w_1(x, y, h)t.$$

Let us put the components in the form (9.12) into the system (9.11) and compare the coefficients of t^0 and t. The right hand side of (9.11) vanishes. Obviously we obtain the 3D second boundary value problems of elasticity for the displacements (u_0, v_0, w_0) and (u_1, v_1, w_1). We can also construct the spline-interpolation solutions for the last problems, so we should change the data on the whole surface S by the data on a finite number of circles on S and the data at the ends of M.

After we have reconstructed the displacements in M for $t \in [0, t_1/2]$ we apply the linear spline for the time intervals $[t_k, t_k + 1]$, $k = 0, ..., n$, where $t_0 = t_1/2$. We take the spline in the form

$$\begin{aligned} u &= u_{0,k}(x, y, h) + u_{1,k}(x, y, h)(t - t_k), \\ v &= v_{0,k}(x, y, h) + v_{1,k}(x, y, h)(t - t_k), \\ w &= w_{0,k}(x, y, h) + w_{1,k}(x, y, h)(t - t_k). \end{aligned} \quad (9.13)$$

We know the displacements at the points of S at the moments t_k and t_{k+1} (we use the boundary values of the spline constructed in the neighborhood of $t = 0$ at the moment $t = t_1/2$), so we obtain the values at the same points of S of the coefficients $u_{j,k}$, $v_{j,k}$, $w_{j,k}$, $j = 0, 1$. Now we again have the 3D second boundary value problems for the displacements $(u_{0,k}, v_{0,k}, w_{0,k})$ and $(u_{1,k}, v_{1,k}, w_{1,k})$ with $k = 1, .., n$.

This solution is the simplest one. We must fix the given data only at the subset of the points of the boundary surface S at a finite number of moments. Note that in any practical situation we rarely have any information on the processes inside the solid M so we may assume that the only information we know is given only on the side surface S . Hence, we can apply the linear spline construction unless we have to use some additional information on S or inside M .

9.2.2 Reduction to the static problem

Suppose now that we want to find the solution of the second dynamic problem of elasticity: given the displacements at the surface S (or any subset of this surface) of the solid M and the initial displacements and velocities in the whole solid it should be possible to find the solution of (9.11) which satisfies the given data.

Suppose that we try to find the solution in the form of the polynomial in the time variable, so that

$$\begin{aligned} u &= \sum_{k=0}^{n} u_k(x, y, h)t^k, \\ v &= \sum_{k=0}^{n} v_k(x, y, h)t^k, \\ w &= \sum_{k=0}^{n} w_k(x, y, h)t^k. \end{aligned} \tag{9.14}$$

Evidently the coefficients $u_0(x,y,h)$, $v_0(x,y,h)$, $w_0(x,y,h)$ and $u_1(x,y,h)$, $v_1(x,y,h)$, $w_1(x,y,h)$ are known via the initial data.

We substitute the displacement components to (9.11) and compare the coefficients of the powers of t . We get the following relations between u_k and u_{k-2} , $k = 2,\ldots,n$:

$$\begin{aligned}(\lambda+\mu)\frac{\partial\theta_{k-2}}{\partial x}+\mu\Delta u_{k-2}&=k(k-1)u_k,\\(\lambda+\mu)\frac{\partial\theta_{k-2}}{\partial y}+\mu\Delta v_{k-2}&=k(k-1)v_k,\\(\lambda+\mu)\frac{\partial\theta_{k-2}}{\partial h}+\mu\Delta w_{k-2}&=k(k-1)w_k.\end{aligned}\tag{9.15}$$

So in the case when the initial displacements and velocities are given in the whole body the functions u_0 , v_0 , w_0 and u_1 , v_1 , w_1 uniquely determine all the other components u_k , v_k , w_k , $k = 2,\ldots,n$ by formulas (9.15) independently of any boundary data. Moreover, the polynomial representation (9.14) of the solution of the dynamic equations (9.11) is possible only if there exists an $n \in \mathbb{N}$ such that

$$L^n\begin{pmatrix}u_0(x,y,h)\\v_0(x,y,h)\\w_0(x,y,h)\end{pmatrix}=L^n\begin{pmatrix}u_1(x,y,h)\\v_1(x,y,h)\\w_1(x,y,h)\end{pmatrix}=\begin{pmatrix}0\\0\\0\end{pmatrix},\tag{9.16}$$

here

$$L=\begin{pmatrix}(\lambda+\mu)\frac{\partial^2}{\partial x^2}+\mu\Delta & (\lambda+\mu)\frac{\partial^2}{\partial x\partial y} & (\lambda+\mu)\frac{\partial^2}{\partial x\partial h}\\(\lambda+\mu)\frac{\partial^2}{\partial x\partial y} & (\lambda+\mu)\frac{\partial^2}{\partial y^2}+\mu\Delta & (\lambda+\mu)\frac{\partial^2}{\partial h\partial y}\\(\lambda+\mu)\frac{\partial^2}{\partial x\partial h} & (\lambda+\mu)\frac{\partial^2}{\partial y\partial h} & (\lambda+\mu)\frac{\partial^2}{\partial h^2}+\mu\Delta\end{pmatrix}.$$

Let us use the following notation:

$$(\partial^2_{i,j})=\begin{pmatrix}\frac{\partial^2}{\partial x^2} & \frac{\partial^2}{\partial x\partial y} & \frac{\partial^2}{\partial x\partial h}\\\frac{\partial^2}{\partial x\partial y} & \frac{\partial^2}{\partial y^2} & \frac{\partial^2}{\partial h\partial y}\\\frac{\partial^2}{\partial x\partial h} & \frac{\partial^2}{\partial y\partial h} & \frac{\partial^2}{\partial h^2}\end{pmatrix},$$

$$\mathrm{diag}(\Delta) = \begin{pmatrix} \Delta & 0 & 0 \\ 0 & \Delta & 0 \\ 0 & 0 & \Delta \end{pmatrix}.$$

Then

$$L^n = ((\lambda+\mu)(\partial^2_{i,j}) + \mu\,\mathrm{diag}(\Delta))^n =$$

$$= \sum_{k=0}^{n} C_n^k (\lambda+\mu)^k \mu^{n-k} (\Delta^{n-k} \partial^2_{i,j})^k.$$

The relations (9.16) hold true, for example, for polynomial initial conditions.

Therefore we can not apply any polynomial in the time variable t solution of system (9.11) if we want to satisfy the initial displacements and velocities in the whole body and simultaneously satisfy any boundary conditions.

Now we ignore the initial conditions: we search for the solution of the dynamic equations (9.11) which meets given boundary displacements at a finite set of sections of S by the planes parallel to XOY at $n+1$ moments of time in the form (9.14). These boundary data provide the boundary displacements of the coefficients $u_k(x,y,h)$, $v_k(x,y,h)$, $w_k(x,y,h)$ at the given set of sections of S , $k = 0,...,n$.

It seems natural to start reconstruction of the displacement coordinates from the coefficients of the highest degrees of t , i.e. u_n , v_n , w_n and u_{n-1} , v_{n-1} , w_{n-1} . These functions are the solutions of the static 3D second boundary value problems of elasticity. So we can apply spline-interpolation solution presented in the previous chapters if we give the boundary data at the finite number of circles on S and at the points on the ends of M .

Then we reconstruct the coefficients of all degrees of t consequently according to (9.15). This can be done by application of the procedure similar to one used for solution of equilibrium equations (1.1) in the first chapter. Suppose that the highest degree of the components u_k , v_k , w_k with respect to h is m, i.e.

$$u_k = \sum_{q=0}^{m} u_{k,q}(x,y)h^q,$$

$$v_k = \sum_{q=0}^{m} v_{k,q}(x,y)h^q,$$

$$w_k = \sum_{q=0}^{m} w_{k,q}(x,y)h^q,$$

here $k = 0, \dots, n$.

Now for the functions $u_{k,m}$ and $v_{k,m}$, $k = n-2, \dots, 0$ we get the relation

$$\frac{\partial}{\partial \overline{z}}((\lambda+2\mu)(\frac{\partial u_{k,m}}{\partial x} + \frac{\partial v_{k,m}}{\partial y}) + \imath\mu(-\frac{\partial u_{k,m}}{\partial y} + \frac{\partial v_{k,m}}{\partial x})) =$$
$$= (k+2)(k+1)(u_{k+2,m} + \imath v_{k+2,m}).$$

Hence

$$((\lambda+2\mu)(\frac{\partial u_{k,m}}{\partial x} + \frac{\partial v_{k,m}}{\partial y}) + \imath\mu(-\frac{\partial u_{k,m}}{\partial y} + \frac{\partial v_{k,m}}{\partial x})) =$$
$$= (k+2)(k+1)\int (u_{k+2,m} + \imath v_{k+2,m})d\overline{z} + f_{k,m}(z) = F_{k,m}(z,\overline{z}) + f_{k,m}(z),$$

here $F_{k,m}$ is already known function.

Thus

$$\frac{\partial}{\partial z}(u_{k,m} + \imath v_{k,m}) = \frac{F_{k,m}(z,\overline{z}) + \overline{F_{k,m}(z,\overline{z})}}{4(\lambda+2\mu)} + \frac{F_{k,m}(z,\overline{z}) - \overline{F_{k,m}(z,\overline{z})}}{4\mu} +$$
$$+ \frac{f_{k,m}(z) + \overline{f_{k,m}(z)}}{4(\lambda+2\mu)} + \frac{f_{k,m}(z) - \overline{f_{k,m}(z)}}{4\mu}.$$

So

$$-2\mu(u_{k,m} + \imath v_{k,m}) = -\kappa\phi_{k,m}(z) + z\overline{\phi'_{k,m}(z)} + \overline{\psi'_{k,m}(z)} + \Phi_{k,m}(z,\overline{z}),$$

here $\phi_{k,m}(z)$, $\psi_{k,m}(z)$ are unknown analytic functions and $\Phi_{k,m}(z,\overline{z})$ is a known function.

Similarly for the third displacement coordinate $w_{k,m}$, $k = n-2, \ldots, 0$ we get the relation

$$\Delta w_{k,m} = (k+2)(k+1)w_{k+2,m}.$$

So

$$w_{k,m} = \int\int (k+2)(k+1)w_{k+2,m} dz d\bar{z} + \operatorname{Re} \rho_{k,m}(z).$$

In fact since the functions $w_{k+2,m}$ and $u_{k+2,m} + \imath v_{k+2,m}$ are known we get the representations of the coefficients $w_{k,m}$ and $u_{k,m} + \imath v_{k,m}$ via the analytic functions $\phi_{k,m}(z)$, $\psi_{k,m}(z)$ and $\rho_{k,m}(z)$. These analytic functions can be restored via the boundary conditions as in Chapter 2.

The coefficients $u_{k,q} + \imath v_{k,q}$ and $w_{k,q}$ for $q < m$ can be found in the way similar to one presented in the first chapter. We represent the coefficients via the analytic functions $\phi_{k,q}(z)$, $\psi_{k,q}(z)$ and $\rho_{k,q}(z)$, $k = 0, \ldots, n$, $q = 0, \ldots, m$, and we reconstruct these functions using the boundary conditions.

9.2.3 Reduction to the 2D dynamic problem

Let us first search for the solution of system (9.11) in the form

$$u(x, y, h, t) = \sum_{k=0}^{n} u_k(x, y, t)h^k;$$

$$v(x, y, h, t) = \sum_{k=0}^{n} v_k(x, y, t)h^k;$$

$$w(x, y, h, t) = \sum_{k=0}^{n} w_k(x, y, t)h^k.$$

We substitute these expressions into system (9.11)and after equating the coefficients of common degrees of h we obtain the following relations:

$$(\lambda+\mu)(\partial_x(\partial_x u_k + \partial_y v_k) + (k+1)\partial_x w_{k+1}) +$$
$$+\mu(\Delta u_k + (k+2)(k+1)u_{k+2}) = \rho\partial^2_{tt}u_k, \qquad (9.17)$$
$$(\lambda+\mu)(\partial_y(\partial_x u_k + \partial_y v_k) + (k+1)\partial_y w_{k+1}) +$$
$$+\mu(\Delta v_k + (k+2)(k+1)v_{k+2}) = \rho\partial^2_{tt}v_k, \qquad (9.18)$$
$$(\lambda+\mu)((k+1)(\partial_x u_{k+1} + \partial_y v_{k+1}) + (k+2)(k+1)w_{k+2}) +$$
$$+\mu(\Delta w_k + (k+2)(k+1)w_{k+2}) = \rho\partial^2_{tt}w_k. \qquad (9.19)$$

So for the first pair of displacement components $u_n(x,y,t)$, $v_n(x,y,t)$ the equations (9.17) and (9.18) form the system of equations for the 2D dynamic problem presented in the first section of the current chapter. We can apply the methods of the first section if we consider the following representations of the displacement coefficients:

$$u_n(x,y,t) = \sum_{j=1}^{m} u_{n,j}(x,y)t^j,$$

and

$$v_n(x,y,t) = \sum_{j=1}^{m} v_{n,j}(x,y)t^j.$$

The functions $u_{n,j}(x,y)$ and $v_{n,j}(x,y)$, $j=0,\ldots,m$, can be computed consequently using the formula

$$\frac{\partial}{\partial \overline{z}}\left[(\lambda+2\mu)\left(\frac{\partial u_{n,j}}{\partial x}+\frac{\partial v_{n,j}}{\partial y}\right)+\right.$$
$$\left.+\imath\mu\left(\frac{\partial v_{n,j}}{\partial x}-\frac{\partial u_{n,j}}{\partial y}\right)\right] = \frac{\rho(j+1)(j+2)}{2}\left[u_{n,j+2}+\imath v_{n,j+2}\right]. \qquad (9.20)$$

Thus the procedure of finding $u_{n,j}$, $v_{n,j}$ is exactly the same that was used in Section 9.1.

The relation on the third coordinate is the following:

$$\mu\Delta w_n = \rho\partial^2_{tt}w_n.$$

So if we use the representation of this displacement component similar to the representation of the first pair of coordinates, i.e.

$$w_n(x,y,t) = \sum_{j=1}^{m} w_{n,j}(x,y)t^j,$$

we get the following set of relations on $w_{n,j}(x,y)$, $j = 0,\dots,m$:

$$4\mu\frac{\partial}{\partial \overline{z}}\frac{\partial}{\partial z}w_{n,j}(z,\overline{z}) = \rho(j+2)(j+1)w_{n,j+2}(z,\overline{z}). \qquad (9.21)$$

Thus as in the first chapter we have $w_{n,m}(z,\overline{z}) = \operatorname{Re} r_{n,m}(z)$ and $w_{n,m-1}(z,\overline{z}) = \operatorname{Re} r_{n,m-1}(z)$ for some analytic functions $r_{n,m}(z)$, $r_{n,m-1}(z)$. Hence these coefficients also can be computed consequently with the help of the formula

$$w_{n,j-2}(z,\overline{z}) = \rho m(m-1)\int\int w_{n,j}(z,\overline{z})dzd\overline{z} + \operatorname{Re}(r_{n,j-2}(z)).$$

We can reconstruct the analytic functions $r_{n,j}(z)$, $j = 0,\dots,m$, using the given boundary conditions.

Then we find the functions $u_{n-1}(x,y,t)$, $v_{n-1}(x,y,t)$ and $w_{n-1}(x,y,t)$. Again we search for the solutions in the form

$$u_{n-1}(x,y,t) = \sum_{j=1}^{m} u_{n-1,j}(x,y)t^j$$

and

$$v_{n-1}(x,y,t) = \sum_{j=1}^{m} v_{n-1,j}(x,y)t^j.$$

We use formulas (9.17), (9.18) in order to get the relation

$$\frac{\partial}{\partial \overline{z}}[(\lambda+2\mu)(\frac{\partial u_{n-1,j}}{\partial x} + \frac{\partial v_{n-1,j}}{\partial y}) + \imath\mu(\frac{\partial v_{n-1,j}}{\partial x} - \\ -\frac{\partial u_{n-1,j}}{\partial y})] = -n(\lambda+\mu)\frac{\partial}{\partial \overline{z}}w_{n,j} + \\ +\frac{\rho(j+1)(j+2)}{2}[u_{n-1,j+2} + \imath v_{n-1,j+2}]. \qquad (9.22)$$

So we restore the coefficients $u_{n-1,j}+\imath v_{n-1,j}$ consequently using the boundary conditions.

Equation (9.19) provides us with the relations on the third coordinate, represented as

$$w_{n-1}(x,y,t)=\sum_{j=1}^{m} w_{n-1,j}(x,y)t^j,$$

in the following form:

$$4\mu\frac{\partial}{\partial \overline{z}}\frac{\partial}{\partial z}w_{n-1,j}(z,\overline{z})=\rho(j+2)(j+1)w_{n-1,j+2}(z,\overline{z})- \qquad (9.23)$$
$$-n(\lambda+\mu)\operatorname{Re}(\frac{\partial}{\partial z}(u_{n,j}+v_{n,j})).$$

Again the solution of these equations is similar to the solution of relations (9.21).

In the general case we first assume that

$$u_k(x,y,t)=\sum_{j=1}^{m} u_{k,j}(x,y)t^j,$$

$$v_k(x,y,t)=\sum_{j=1}^{m} v_{k,j}(x,y)t^j$$

and

$$w_k(x,y,t)=\sum_{j=1}^{m} w_{k,j}(x,y)t^j.$$

Then we use formulas (9.17), (9.18) in order to get the relation for the function $u_{k,j}+\imath v_{k,j}$ reconstruction.

$$\frac{\partial}{\partial \overline{z}}[(\lambda+2\mu)(\frac{\partial u_{k,j}}{\partial x}+\frac{\partial v_{k,j}}{\partial y})+\imath\mu(\frac{\partial v_{k,j}}{\partial x}-\frac{\partial u_{k,j}}{\partial y})]=$$
$$-(k+1)(\lambda+\mu)\frac{\partial}{\partial \overline{z}}w_{k+1,j}-\mu(k+2)(k+1)(u_{k+2,j}+\imath v_{k+2,j}+ \qquad (9.24)$$
$$+\frac{\rho(j+1)(j+2)}{2}[u_{k,j+2}+\imath v_{k,j+2}].$$

Similarly for $w_{k,j}$ we obtain the relation

$$4\mu\frac{\partial}{\partial \overline{z}}\frac{\partial}{\partial z}w_{k,j}(z,\overline{z}) = \rho(j+2)(j+1)w_{k,j+2}(z,\overline{z}) - \\ -(k+1)(\lambda+\mu)\,\mathrm{Re}(\frac{\partial}{\partial z}(u_{k+1,j}+\imath v_{k+1,j})) - \\ -(\lambda+\mu)(k+2)(k+1)w_{k+2,j}(z,\overline{z}). \qquad (9.25)$$

We descend from n down to 0 and reconstruct all the functions u_k, v_k and w_k.

Suppose now that we have the given data of the boundary values of displacement components at $n+1$ sections of the surface S by the planes parallel to XOY. For different values of h we obtain two linear systems with respect to the boundary values of $u_k + \imath v_k$ and w_k. These systems can be resolved since their determinant is the determinant of the Vandermonde matrix. Now we apply the following procedure: we solve the plane dynamic problem for $u_n(x,y,t)$, $v_n(x,y,t)$, $w_n(x,y,t)$ using the system of boundary values at different moments t_j of these functions and initial displacement and velocity values at some fixed points inside the corresponding sections of M. Then we find $u_{n-1}(x,y,t)$, $v_{n-1}(x,y,t)$, $w_{n-1}(x,y,t)$ from equations (9.21)–(9.23). And so on.

Now we formulate the problem: given the displacements at $n+1$ circles – the sections by the planes $h = h_k$, $k = 0,...,n$, of the surface S of the circular cylinder at $m+1$ separate moments of time beginning with $t = 0$, given the displacements and the velocities at the points (x_l, y_l, h_k), $l = 1,...,N$, $k = 0,...,n$, at the moment $t = 0$ inside the cylinder, it should be possible to find the solution of system (9.11) which meets the given data.

We search for the solution in the form

$$u(x,y,h,t)=\sum_{k=0}^{n}\sum_{j=0}^{m}u_{k,j}(x,y)t^j h^k,$$

$$v(x,y,h,t)=\sum_{k=0}^{n}\sum_{j=0}^{m}u_{k,j}(x,y)t^j h^k,$$

$$w(x,y,h,t)=\sum_{k=0}^{n}\sum_{j=0}^{m}u_{k,j}(x,y)t^j h^k.$$

The given data provide the boundary values of $u_{k,j}(x,y)+\imath v_{k,j}(x,y)$ and $w_{k,j}(x,y)$ and also the values of $u_{k,0}(x_l,y_l)$, $v_{k,0}(x_l,y_l)$, $w_{k,0}(x_l,y_l)$, $u_{k,1}(x_l,y_l)$, $v_{k,1}(x_l,y_l)$ and $w_{k,1}(x_l,y_l)$, $l=1,...,N$, $k=0,...,n$. Now we can apply the construction similar to one presented in Section 9.1.

We can search for the coefficients of h_k in the form

$$u_k=\breve{u}_k+\grave{u}_k,\; v_k=\breve{v}_k+\grave{v}_k,\; w_k=\breve{w}_k+\grave{w}_k.$$

where the displacement components $\breve{u}_k,\breve{v}_k,\breve{w}_k$ meet the boundary conditions at the unit circle at the given separate moments and the components $\grave{u}_k,\grave{v}_k,\grave{w}_k$ simultaneously vanish at the unit circle at any given moment and take given displacements and velocities at the given points at the moment $t=0$. The additional parameters help us to satisfy the given initial data as it is done in Section 9.1.

CHAPTER 10

Appendices

Pyotr N. Ivanshin,

Kazan Federal University

pivanshin@gmail.com

ABSTRACT: In the first appendix we give the thorough method for solution gluing on the common section of adjacent layers presented in general form in Chapter 2. The second appendix contains general solution of the 3D second basic elasticity theory problem for conoid of revolution with one shifted point at one of the bases. The main idea of this solution is given in Chapter 6. The third appendix gives the solution algorythm.

Elena A. Shirokova and Pyotr N. Ivanshin (Eds)

10.1 Appendix 1. General system of equations for the third degree spline

We search for the solutions in the form

$$(u+\imath v)(z,\overline{z},h)=\sum_{k=0}^{n}(u_k(z,\overline{z})+\imath v_k(z,\overline{z}))h^k,$$

$$w(z,\overline{z},h)=\sum_{k=0}^{n}w_k(z,\overline{z})h^k.$$

We suppose that all the given conditions are polynomials in z and $\overline{z}$. So we assume that the analytic functions in the representations of the solution are as follows: $\phi_k=\sum_{l=0}^{n}a_l^k z^l$, $\psi_k=\sum_{l=0}^{n}b_l^k z^l$, $\rho_k=\sum_{l=0}^{n}c_l^k z^l$, $k=0,1,2,3$. Thus we must find the coefficients a_l^k, b_l^k and c_l^k. In order to do this we must use the given boundary conditions. Note that there are two sets of these conditions. The first is the given boundary conditions on one of the ends. The second is the boundary condition on the circle bounding the second end.

Let us consider at first the set of equations on one of the ends. Then we have the relations on the pair of displacement coordinates in the form

$$(u+\imath v)(z,\overline{z},0)=\sum_{k=0}^{\infty}A_k z^k+\sum_{k=1}^{\infty}B_k\overline{z}^k+|z|^2(\sum_{k=0}^{\infty}A_k^1 z^k+$$

$$+\sum_{k=1}^{\infty}B_k^1\overline{z}^k)+|z|^4\sum_{k=0}^{\infty}B_k^2\overline{z}^k.$$

Also we consider the similar relations on the third displacement coordinate:

$$w=\mathrm{Re}(\sum_{k=0}^{\infty}C_k z^k+|z|^2\sum_{k=0}^{\infty}\overline{C_k^1}\overline{z}^k+|z|^4\sum_{k=0}^{\infty}\overline{C_k^2}\overline{z}^k.$$

The constant summand of the first relation gives us the coefficients $a_0^0+\imath\hat{a}_0^0$

which will be determined via the equation

$$\frac{3a_0^0}{2(\lambda+\mu)}+\frac{3\imath\hat{a}_0^0}{2(\lambda+\mu)}+\frac{a_0^0\lambda}{2\mu(\lambda+\mu)}+\frac{\imath\hat{a}_0^0\lambda}{2\mu(\lambda+\mu)}=A_0.$$

The coefficient of $\overline{z}$ of the same equation yields $b_1^0+\imath\hat{b}_1^0$:

$$-\frac{b_1^0\overline{z}}{2\mu}+\frac{\imath\hat{b}_1^0\overline{z}}{2\mu}=B_1.$$

The coefficient of $\overline{z}^2$ provides us with $b_2^0+\imath\hat{b}_2^0$:

$$-\frac{b_2^0\overline{z}^2}{2\mu}+\frac{i\hat{b}_2^0\overline{z}^2}{2\mu}=B_2.$$

Similarly the coefficient of $\overline{z}^n$, gives us $b_n^0+\imath\hat{b}_n^0$:

$$-\frac{b_n^0\overline{z}^n}{2\mu}+\frac{\imath\hat{b}_n^0\overline{z}^n}{2\mu}=B_n, n=3,4\ldots$$

The coefficient of z is $a_1^0+\imath\hat{a}_1^0$.

$$-\frac{a_1^0z}{2\mu}+\frac{\imath\hat{a}_1^0z}{2\mu}+\frac{3a_1^0z}{2(\lambda+\mu)}+\frac{3\imath\hat{a}_1^0z}{2(\lambda+\mu)}+\frac{a_1^0\lambda z}{2\mu(\lambda+\mu)}+\frac{\imath\hat{a}_1^0\lambda z}{2\mu(\lambda+\mu)}-$$

$$-\frac{c_0^1\lambda z}{4(\lambda+2\mu)}+\frac{3\imath\hat{c}_0^1\lambda z}{4(\lambda+2\mu)}+\frac{\imath\hat{c}_0^1\lambda^2z}{4\mu(\lambda+2\mu)}-\frac{c_0^1\mu z}{4(\lambda+2\mu)}+\frac{\imath\hat{c}_0^1\mu z}{2(\lambda+2\mu)}=A_1z$$

The second series of coefficients of the first equation is one with $|z|^2$. So the coefficients of $|z|^2$ and $\overline{z}|z|^2$ give us a_0^2, $\hat{a}_0^2$ and b_0^2, $\hat{b}_0^2$. The first equation is

$$-\frac{a_2^0\overline{z}z}{\mu}+\frac{\imath\hat{a}_2^0\overline{z}z}{\mu}+\frac{3a_0^2\lambda\overline{z}z}{4(\lambda+2\mu)^2}+\frac{3\imath\hat{a}_0^2\lambda\overline{z}z}{4(\lambda+2\mu)^2}+\frac{a_0^2\lambda^2\overline{z}z}{4\mu(\lambda+2\mu)^2}+\frac{\imath\hat{a}_0^2\lambda^2\overline{z}z}{4\mu(\lambda+2\mu)^2}+$$

$$+\frac{a_0^2\mu\overline{z}z}{2(\lambda+2\mu)^2}+\frac{\imath\hat{a}_0^2\mu\overline{z}z}{2(\lambda+2\mu)^2}+\frac{3b_0^2\overline{z}z}{4(\lambda+2\mu)}-\frac{3\imath\hat{b}_0^2\overline{z}z}{4(\lambda+2\mu)}-\frac{c_1^1\lambda\overline{z}z}{\lambda+2\mu}+\frac{\imath\hat{c}_1^1\lambda\overline{z}z}{\lambda+2\mu}+$$

$$+\frac{b_0^2\lambda\overline{z}z}{4\mu(\lambda+2\mu)}-\frac{\imath\hat{b}_0^2\lambda\overline{z}z}{4\mu(\lambda+2\mu)}-\frac{c_1^1\lambda^2\overline{z}z}{4\mu(\lambda+2\mu)}+\frac{\imath\hat{c}_1^1\lambda^2\overline{z}z}{4\mu(\lambda+2\mu)}-\frac{3c_1^1\mu\overline{z}z}{4(\lambda+2\mu)}+$$

$$+\frac{3\imath\hat{c}_1^1\mu\overline{z}z}{4(\lambda+2\mu)}-\frac{a_0^2\lambda\overline{z}z}{(\lambda+\mu)(\lambda+2\mu)}-\frac{\imath\hat{a}_0^2\lambda\overline{z}z}{(\lambda+\mu)(\lambda+2\mu)}-\frac{a_0^2\lambda^2\overline{z}z}{4\mu(\lambda+\mu)(\lambda+2\mu)}-$$

$$-\frac{\imath\hat{a}_0^2\lambda^2\overline{z}z}{4\mu(\lambda+\mu)(\lambda+2\mu)}-\frac{5a_0^2\mu\overline{z}z}{4(\lambda+\mu)(\lambda+2\mu)}-\frac{5\imath\hat{a}_0^2\mu\overline{z}z}{4(\lambda+\mu)(\lambda+2\mu)}=A_0^1|z|^2.$$

The second one—with $\overline{z}|z|^2$ —is as follows:

$$\frac{9b_1^2\overline{z}^2z}{8(\lambda+2\mu)}-\frac{9\imath\hat{b}_1^2\overline{z}^2z}{8(\lambda+2\mu)}-\frac{3c_2^1\lambda\overline{z}^2z}{2(\lambda+2\mu)}+\frac{3\imath\hat{c}_2^1\lambda\overline{z}^2z}{2(\lambda+2\mu)}+\frac{3b_1^2\lambda\overline{z}^2z}{8\mu(\lambda+2\mu)}-\frac{3\imath\hat{b}_1^2\lambda\overline{z}^2z}{8\mu(\lambda+2\mu)}-$$

$$-\frac{3c_2^1\lambda^2\overline{z}^2z}{8\mu(\lambda+2\mu)}+\frac{3\imath\hat{c}_2^1\lambda^2\overline{z}^2z}{8\mu(\lambda+2\mu)}-\frac{9c_2^1\mu\overline{z}^2z}{8(\lambda+2\mu)}+\frac{9\imath\hat{c}_2^1\mu\overline{z}^2z}{8(\lambda+2\mu)}-\frac{a_3^03\overline{z}^2z}{2\mu}+$$

$$+\frac{\imath\hat{a}_3^03\overline{z}^2z}{2\mu}=B_1^1t|z|^2.$$

The coefficients of $\overline{z}^2|z|^2$ and $|z|^4$ yield b_2^2 $\hat{b}_2^2$ and c_3^1, $\hat{c}_3^1$. So we have the system of equation. The first relation is

$$\frac{3b_2^2\overline{z}^3z}{2(\lambda+2\mu)}-\frac{3\imath\hat{b}_2^2\overline{z}^3z}{2(\lambda+2\mu)}+\frac{b_2^2\lambda\overline{z}^3z}{2\mu(\lambda+2\mu)}-\frac{\imath\hat{b}_2^2\lambda\overline{z}^3z}{2\mu(\lambda+2\mu)}-\frac{c_3^1\lambda4\overline{z}^3z}{2(\lambda+2\mu)}+\frac{\imath\hat{c}_3^1\lambda4\overline{z}^3z}{2(\lambda+2\mu)}-$$

$$-\frac{c_3^1\lambda^24\overline{z}^3z}{8\mu(\lambda+2\mu)}+\frac{\imath\hat{c}_3^1\lambda^24\overline{z}^3z}{8\mu(\lambda+2\mu)}-\frac{3c_3^1\mu4\overline{z}^3z}{8(\lambda+2\mu)}+\frac{3\imath\hat{c}_3^1\mu4\overline{z}^3z}{8(\lambda+2\mu)}-$$

$$-\frac{a_4^04\overline{z}^3z}{2\mu}+\frac{\imath\hat{a}_4^04\overline{z}^3z}{2\mu}=B_2^1\overline{z}^2|z|^2.$$

The second relation — the coefficient of $|z|^4$ — gives

$$\frac{9a_2^2\lambda\overline{z}^2z^2}{8(\lambda+2\mu)^2}-\frac{9\imath\hat{a}_2^2\lambda\overline{z}^2z^2}{8(\lambda+2\mu)^2}+\frac{27c_1^3\lambda^2\overline{z}^2z^2}{8(\lambda+2\mu)^2}-\frac{27\imath\hat{c}_1^3\lambda^2\overline{z}^2z^2}{8(\lambda+2\mu)^2}+\frac{3a_2^2\lambda^2\overline{z}^2z^2}{8\mu(\lambda+2\mu)^2}-$$

$$-\frac{3\imath\hat{a}_2^2\lambda^2\overline{z}^2z^2}{8\mu(\lambda+2\mu)^2}+\frac{9c_1^3\lambda^3\overline{z}^2z^2}{16\mu(\lambda+2\mu)^2}-$$

$$-\frac{9\imath\hat{c}_1^3\lambda^3\overline{z}^2z^2}{16\mu(\lambda+2\mu)^2}+\frac{3a_2^2\mu\overline{z}^2z^2}{4(\lambda+2\mu)^2}-\frac{3\imath\hat{a}_2^2\mu\overline{z}^2z^2}{4(\lambda+2\mu)^2}+$$

$$+\frac{99c_1^3\lambda\mu\overline{z}^2z^2}{16(\lambda+2\mu)^2}-\frac{99\imath\hat{c}_1^3\lambda\mu\overline{z}^2z^2}{16(\lambda+2\mu)^2}+\frac{27c_1^3\mu^2\overline{z}^2z^2}{8(\lambda+2\mu)^2}-\frac{27\imath\hat{c}_1^3\mu^2\overline{z}^2z^2}{8(\lambda+2\mu)^2}+\frac{9a_2^2\overline{z}^2z^2}{8(\lambda+2\mu)}-$$

$$-\frac{9\imath\hat{a}_2^2\overline{z}^2z^2}{8(\lambda+2\mu)}+\frac{3a_2^2\lambda\overline{z}^2z^2}{8\mu(\lambda+2\mu)}-\frac{3\imath\hat{a}_2^2\lambda\overline{z}^2z^2}{8\mu(\lambda+2\mu)}=A_0^2|z|^4.$$

And so on, we use the coefficients of $\overline{z}^n|z|^2$ in order to find b_n^2, $\hat{b}_n^2$, $n = 3, 4\ldots$:

$$\frac{3b_n^2\overline{z}^{1+n}(n+2)z}{8(\lambda+2\mu)} - \frac{3\imath\hat{b}_n^2\overline{z}^{1+n}(n+2)z}{8(\lambda+2\mu)} + \frac{b_n^2\lambda\overline{z}^{1+n}(n+2)z}{8\mu(\lambda+2\mu)} - \frac{\imath\hat{b}_n^2\lambda\overline{z}^{1+n}(n+2)z}{8\mu(\lambda+2\mu)} -$$

$$-\frac{c_{n+1}^1\lambda(n+2)\overline{z}^{n+1}z}{2(\lambda+2\mu)} + \frac{\imath\hat{c}_{n+1}^1\lambda(n+2)\overline{z}^{n+1}z}{2(\lambda+2\mu)} - \frac{c_{n+1}^1\lambda^2(n+2)\overline{z}^{n+1}z}{8\mu(\lambda+2\mu)} +$$

$$+\frac{\imath\hat{c}_{n+1}^1\lambda^2(n+2)\overline{z}^{n+1}z}{8\mu(\lambda+2\mu)} - \frac{3c_{n+1}^1\mu(n+2)\overline{z}^{n+1}z}{8(\lambda+2\mu)} + \frac{3\imath\hat{c}_{n+1}^1\mu(n+2)\overline{z}^{n+1}z}{8(\lambda+2\mu)} -$$

$$-\frac{a_{n+2}^0(n+3)\overline{z}^{1+n}z}{2\mu} + \frac{\imath\hat{a}_{n+2}^0(n+3)\overline{z}^{1+n}z}{2\mu} = B_n^1\overline{z}^n|z|^2.$$

At the same time the coefficients of z^2 provide us with a_2^0, $\hat{a}_2^0$:

$$\frac{3a_2^0z^2}{2(\lambda+\mu)} + \frac{3\imath\hat{a}_2^0z^2}{2(\lambda+\mu)} + \frac{a_2^0\lambda z^2}{2\mu(\lambda+\mu)} + \frac{\imath\hat{a}_2^0\lambda z^2}{2\mu(\lambda+\mu)} + \frac{3a_0^2\lambda z^2}{8(\lambda+2\mu)^2} - \frac{3\imath\hat{a}_0^2\lambda z^2}{8(\lambda+2\mu)^2} +$$

$$+\frac{a_0^2\lambda^2z^2}{8\mu(\lambda+2\mu)^2} - \frac{\imath\hat{a}_0^2\lambda^2z^2}{8\mu(\lambda+2\mu)^2} + \frac{a_0^2\mu z^2}{4(\lambda+2\mu)^2} - \frac{\imath\hat{a}_0^2\mu z^2}{4(\lambda+2\mu)^2} + \frac{3a_0^2z^2}{8(\lambda+2\mu)} -$$

$$-\frac{3\imath\hat{a}_0^2z^2}{8(\lambda+2\mu)} - \frac{b_0^2z^2}{8(\lambda+2\mu)} - \frac{\imath\hat{b}_0^2z^2}{8(\lambda+2\mu)} + \frac{c_1^1\lambda z^2}{4(\lambda+2\mu)} + \frac{\imath\hat{c}_1^1\lambda z^2}{4(\lambda+2\mu)} + \frac{a_0^2\lambda z^2}{8\mu(\lambda+2\mu)} -$$

$$-\frac{\imath\hat{a}_0^2\lambda z^2}{8\mu(\lambda+2\mu)} - \frac{b_0^2\lambda z^2}{8\mu(\lambda+2\mu)} - \frac{\imath\hat{b}_0^2\lambda z^2}{8\mu(\lambda+2\mu)} + \frac{c_1^1\lambda^2z^2}{8\mu(\lambda+2\mu)} + \frac{\imath\hat{c}_1^1\lambda^2z^2}{8\mu(\lambda+2\mu)} +$$

$$+\frac{c_1^1\mu z^2}{8(\lambda+2\mu)} + \frac{\imath\hat{c}_1^1\mu z^2}{8(\lambda+2\mu)} = A_2z^2.$$

We use the coefficient of $z|z|^2$ for reconstruction of a_1^2, $\hat{a}_1^2$.

$$\frac{3a_1^2\lambda\overline{z}z^2}{2(\lambda+2\mu)^2} + \frac{21c_0^3\lambda^2\overline{z}z^2}{8(\lambda+2\mu)^2} - \frac{15\imath\hat{c}_0^3\lambda^2\overline{z}z^2}{8(\lambda+2\mu)^2} + \frac{a_1^2\lambda^2\overline{z}z^2}{2\mu(\lambda+2\mu)^2} + \frac{3c_0^3\lambda^3\overline{z}z^2}{8\mu(\lambda+2\mu)^2} -$$

$$-\frac{3\imath\hat{c}_0^3\lambda^3\overline{z}z^2}{8\mu(\lambda+2\mu)^2} + \frac{a_1^2\mu\overline{z}z^2}{(\lambda+2\mu)^2} + \frac{21c_0^3\lambda\mu\overline{z}z^2}{4(\lambda+2\mu)^2} - \frac{3\imath\hat{c}_0^3\lambda\mu\overline{z}z^2}{(\lambda+2\mu)^2} + \frac{3c_0^3\mu^2\overline{z}z^2}{(\lambda+2\mu)^2} - \frac{3\imath\hat{c}_0^3\mu^2\overline{z}z^2}{2(\lambda+2\mu)^2} +$$

$$+\frac{3a_1^2\overline{z}z^2}{4(\lambda+2\mu)} - \frac{3\imath\hat{a}_1^2\overline{z}z^2}{4(\lambda+2\mu)} + \frac{a_1^2\lambda\overline{z}z^2}{4\mu(\lambda+2\mu)} - \frac{\imath\hat{a}_1^2\lambda\overline{z}z^2}{4\mu(\lambda+2\mu)} - \frac{a_1^2\lambda\overline{z}z^2}{(\lambda+\mu)(\lambda+2\mu)} -$$

$$-\frac{\imath\hat{a}_1^2\lambda\overline{z}z^2}{(\lambda+\mu)(\lambda+2\mu)} - \frac{a_1^2\lambda^2\overline{z}z^2}{4\mu(\lambda+\mu)(\lambda+2\mu)} - \frac{\imath\hat{a}_1^2\lambda^2\overline{z}z^2}{4\mu(\lambda+\mu)(\lambda+2\mu)} -$$

$$-\frac{5a_1^2\mu\overline{z}z^2}{4(\lambda+\mu)(\lambda+2\mu)}-\frac{5\imath\hat{a}_1^2\mu\overline{z}z^2}{4(\lambda+\mu)(\lambda+2\mu)}=A_1^1z|z|^2.$$

Then we pass to the series of coefficients of $|z|^4$.

Thus the coefficient of $\overline{z}|z|^4$ yields c_2^3, $\hat{c}_2^3$

$$\frac{9c_2^3\lambda^2\overline{z}^3z^2}{2(\lambda+2\mu)^2}-\frac{9\imath\hat{c}_2^3\lambda^2\overline{z}^3z^2}{2(\lambda+2\mu)^2}+\frac{3c_2^3\lambda^3\overline{z}^3z^2}{4\mu(\lambda+2\mu)^2}-\frac{3\imath\hat{c}_2^3\lambda^3\overline{z}^3z^2}{4\mu(\lambda+2\mu)^2}+\frac{33c_2^3\lambda\mu\overline{z}^3z^2}{4(\lambda+2\mu)^2}-$$

$$-\frac{33\imath\hat{c}_2^3\lambda\mu\overline{z}^3z^2}{4(\lambda+2\mu)^2}+\frac{9c_2^3\mu^2\overline{z}^3z^2}{2(\lambda+2\mu)^2}-\frac{9\imath\hat{c}_2^3\mu^2\overline{z}^3z^2}{2(\lambda+2\mu)^2}+\frac{3a_3^2\lambda 4}{8(\lambda+2\mu)^2}-\frac{3\imath\hat{a}_3^2\lambda 4\overline{z}^3z^2}{8(\lambda+2\mu)^2}+$$

$$+\frac{a_3^2\lambda^2 4\overline{z}^3z^2}{8\mu(\lambda+2\mu)^2}-\frac{\imath\hat{a}_3^2\lambda^2 4\overline{z}^3z^2}{8\mu(\lambda+2\mu)^2}+\frac{a_3^2\mu 4\overline{z}^3z^2}{4(\lambda+2\mu)^2}-\frac{\imath\hat{a}_3^2\mu 4\overline{z}^3z^2}{4(\lambda+2\mu)^2}+\frac{3a_3^2 4\overline{z}^3z^2}{8(\lambda+2\mu)}-$$

$$-\frac{3\imath\hat{a}_3^2 4\overline{z}^3z^2}{8(\lambda+2\mu)}+\frac{a_3^2\lambda 4\overline{z}^3z^2}{8\mu(\lambda+2\mu)}-\frac{\imath\hat{a}_3^2\lambda 4\overline{z}^3z^2}{8\mu(\lambda+2\mu)}=B_1^2\overline{z}|z|^4.$$

In general, the coefficients of $\overline{z}^{n-2}|z|^4$, $n=4,\ldots,$ define c_{n-1}^3, $\hat{c}_{n-1}^3$:

$$\frac{3a_n^2\lambda(n+1)}{8(\lambda+2\mu)^2}-\frac{3\imath\hat{a}_n^2\lambda(n+1)\overline{z}^nz^2}{8(\lambda+2\mu)^2}+\frac{a_n^2\lambda^2(n+1)\overline{z}^nz^2}{8\mu(\lambda+2\mu)^2}-\frac{\imath\hat{a}_n^2\lambda^2(n+1)\overline{z}^nz^2}{8\mu(\lambda+2\mu)^2}+$$

$$+\frac{a_n^2\mu(n+1)\overline{z}^nz^2}{4(\lambda+2\mu)^2}-\frac{\imath\hat{a}_n^2\mu(n+1)\overline{z}^nz^2}{4(\lambda+2\mu)^2}+\frac{3a_n^2(n+1)\overline{z}^nz^2}{8(\lambda+2\mu)}-\frac{3\imath\hat{a}_n^2(n+1)\overline{z}^nz^2}{8(\lambda+2\mu)}+$$

$$+\frac{a_n^2\lambda(n+1)\overline{z}^nz^2}{8\mu(\lambda+2\mu)}-\frac{\imath\hat{a}_n^2\lambda(n+1)\overline{z}^nz^2}{8\mu(\lambda+2\mu)}+\frac{9c_{n-1}^3\lambda^2\overline{z}^n(n+1)z^2}{8(\lambda+2\mu)^2}-$$

$$-\frac{9\imath\hat{c}_{n-1}^3\lambda^2\overline{z}^n(n+1)z^2}{8(\lambda+2\mu)^2}+\frac{3c_{n-1}^3\lambda^3\overline{z}^n(n+1)z^2}{16\mu(\lambda+2\mu)^2}-\frac{3\imath\hat{c}_{n-1}^3\lambda^3\overline{z}^n(n+1)z^2}{16\mu(\lambda+2\mu)^2}+$$

$$+\frac{33c_{n-1}^3\lambda\mu\overline{z}^n(n+1)z^2}{16(\lambda+2\mu)^2}-\frac{33\imath\hat{c}_{n-1}^3\lambda\mu\overline{z}^n(n+1)z^2}{16(\lambda+2\mu)^2}+\frac{9c_{n-1}^3\mu^2\overline{z}^n(n+1)z^2}{8(\lambda+2\mu)^2}-$$

$$-\frac{9\imath\hat{c}_{n-1}^3\mu^2\overline{z}^n(n+1)z^2}{8(\lambda+2\mu)^2}=B_n^2\overline{z}^{n-2}|z|^4.$$

The same with z^3 provides us with a_3^0, $\hat{a}_3^0$:

$$-\frac{b_1^2z^3}{8(\lambda+2\mu)}-\frac{\imath\hat{b}_1^2z^3}{8(\lambda+2\mu)}+\frac{c_2^1\lambda z^3}{4(\lambda+2\mu)}+\frac{\imath\hat{c}_2^1\lambda z^3}{4(\lambda+2\mu)}-\frac{b_1^2\lambda z^3}{8\mu(\lambda+2\mu)}-$$

$$-\frac{\imath\hat{b}_1^2\lambda z^3}{8\mu(\lambda+2\mu)}+\frac{c_2^1\lambda^2z^3}{8\mu(\lambda+2\mu)}+\frac{\imath\hat{c}_2^1\lambda^2z^3}{8\mu(\lambda+2\mu)}+\frac{c_2^1\mu z^3}{8(\lambda+2\mu)}+\frac{\imath\hat{c}_2^1\mu z^3}{8(\lambda+2\mu)}+\frac{3a_3^0z^3}{2(\lambda+\mu)}+$$

$$+\frac{3\imath\hat{a}_3^0 z^3}{2(\lambda+\mu)}+\frac{a_3^0\lambda z^3}{2\mu(\lambda+\mu)}+\frac{\imath\hat{a}_3^0\lambda z^3}{2\mu(\lambda+\mu)}=A_3z^3.$$

We use $z^2|z|^2$ in order to find a_2^2, $\hat{a}_2^2$:

$$\frac{3a_2^2\lambda\overline{z}z^3}{4(\lambda+2\mu)^2}+\frac{3\imath\hat{a}_2^2\lambda\overline{z}z^3}{4(\lambda+2\mu)^2}+\frac{3c_1^3\lambda^2\overline{z}z^3}{8(\lambda+2\mu)^2}+\frac{3\imath\hat{c}_1^3\lambda^2\overline{z}z^3}{8(\lambda+2\mu)^2}+\frac{a_2^2\lambda^2\overline{z}z^3}{4\mu(\lambda+2\mu)^2}+$$

$$+\frac{\imath\hat{a}_2^2\lambda^2\overline{z}z^3}{4\mu(\lambda+2\mu)^2}+\frac{a_2^2\mu\overline{z}z^3}{2(\lambda+2\mu)^2}+\frac{\imath\hat{a}_2^2\mu\overline{z}z^3}{2(\lambda+2\mu)^2}+\frac{9c_1^3\lambda\mu\overline{z}z^3}{8(\lambda+2\mu)^2}+\frac{9\imath\hat{c}_1^3\lambda\mu\overline{z}z^3}{8(\lambda+2\mu)^2}+$$

$$+\frac{3c_1^3\mu^2\overline{z}z^3}{4(\lambda+2\mu)^2}+\frac{3\imath\hat{c}_1^3\mu^2\overline{z}z^3}{4(\lambda+2\mu)^2}-\frac{a_2^2\lambda\overline{z}z^3}{(\lambda+\mu)(\lambda+2\mu)}-\frac{\imath\hat{a}_2^2\lambda\overline{z}z^3}{(\lambda+\mu)(\lambda+2\mu)}-$$

$$-\frac{a_2^2\lambda^2tz^3}{4\mu(\lambda+\mu)(\lambda+2\mu)}-\frac{\imath\hat{a}_2^2\lambda^2\overline{z}z^3}{4\mu(\lambda+\mu)(\lambda+2\mu)}-\frac{5a_2^2\mu\overline{z}z^3}{4(\lambda+\mu)(\lambda+2\mu)}-$$

$$-\frac{5\imath\hat{a}_2^2\mu\overline{z}z^3}{4(\lambda+\mu)(\lambda+2\mu)}=A_2^1z^2|z|^2.$$

The coefficient of z^4 gives us a_4^0, $\hat{a}_4^0$:

$$-\frac{b_2^2z^4}{8(\lambda+2\mu)}-\frac{\imath\hat{b}_2^2z^4}{8(\lambda+2\mu)}-\frac{b_2^2\lambda z^4}{8\mu(\lambda+2\mu)}-\frac{\imath\hat{b}_2^2\lambda z^4}{8\mu(\lambda+2\mu)}+\frac{c_3^1\lambda z^4}{4(\lambda+2\mu)}+\frac{i\hat{c}_3^1\lambda z^4}{4(\lambda+2\mu)}+$$

$$+\frac{c_3^1\lambda^2z^4}{8\mu(\lambda+2\mu)}+\frac{i\hat{c}_3^1\lambda^2z^4}{8\mu(\lambda+2\mu)}+\frac{c_3^1\mu z^4}{8(\lambda+2\mu)}+\frac{\imath\hat{c}_3^1\mu z^4}{8(\lambda+2\mu)}+\frac{3a_4^0z^4}{2(\lambda+\mu)}+\frac{3\imath\hat{a}_4^0z^4}{2(\lambda+\mu)}+$$

$$+\frac{a_4^0\lambda z^4}{2\mu(\lambda+\mu)}+\frac{\imath\hat{a}_4^0\lambda z^4}{2\mu(\lambda+\mu)}=A_4z^4.$$

We use $z^3|z|^2$ to determine a_3^2, $\hat{a}_3^2$

$$\frac{3c_2^3\lambda^2\overline{z}z^4}{8(\lambda+2\mu)^2}+\frac{3\imath\hat{c}_2^3\lambda^2\overline{z}z^4}{8(\lambda+2\mu)^2}+\frac{9c_2^3\lambda\mu\overline{z}z^4}{8(\lambda+2\mu)^2}+\frac{9\imath\hat{c}_2^3\lambda\mu\overline{z}z^4}{8(\lambda+2\mu)^2}+\frac{3c_2^3\mu^2\overline{z}z^4}{4(\lambda+2\mu)^2}+\frac{3\imath\hat{c}_2^3\mu^2\overline{z}z^4}{4(\lambda+2\mu)^2}+$$

$$+\frac{3a_3^2\lambda\overline{z}z^4}{4(\lambda+2\mu)^2}+\frac{3\imath\hat{a}_3^2\lambda\overline{z}z^4}{4(\lambda+2\mu)^2}+\frac{a_3^2\lambda^2\overline{z}z^4}{4\mu(\lambda+2\mu)^2}+\frac{\imath\hat{a}_3^2\lambda^2\overline{z}z^4}{4\mu(\lambda+2\mu)^2}+\frac{a_3^2\mu\overline{z}z^4}{2(\lambda+2\mu)^2}+$$

$$+\frac{\imath\hat{a}_3^2\mu\overline{z}z^4}{2(\lambda+2\mu)^2}-\frac{a_3^2\lambda\overline{z}z^4}{(\lambda+\mu)(\lambda+2\mu)}-\frac{\imath\hat{a}_3^2\lambda\overline{z}z^4}{(\lambda+\mu)(\lambda+2\mu)}-\frac{a_3^2\lambda^2\overline{z}z^4}{4\mu(\lambda+\mu)(\lambda+2\mu)}-$$

$$-\frac{\imath\hat{a}_3^2\lambda^2\overline{z}z^4}{4\mu(\lambda+\mu)(\lambda+2\mu)}-\frac{5a_3^2\mu\overline{z}z^4}{4(\lambda+\mu)(\lambda+2\mu)}-\frac{5\imath\hat{a}_n^2\mu\overline{z}z^4}{4(\lambda+\mu)(\lambda+2\mu)}=A_3^1z^3|z|^2.$$

Again in general the coefficient of $z^n|z|^2$ provides us with a_n^2 , $\hat{a}_n^2$.

$$\frac{3a_n^2\lambda\overline{z}z^{n+1}}{4(\lambda+2\mu)^2}+\frac{31\hat{a}_n^2\lambda\overline{z}z^{n+1}}{4(\lambda+2\mu)^2}+\frac{a_n^2\lambda^2\overline{z}z^{n+1}}{4\mu(\lambda+2\mu)^2}+\frac{1\hat{a}_n^2\lambda^2\overline{z}z^{n+1}}{4\mu(\lambda+2\mu)^2}+\frac{a_n^2\mu\overline{z}z^{n+1}}{2(\lambda+2\mu)^2}+$$

$$+\frac{1\hat{a}_n^2\mu\overline{z}z^{n+1}}{2(\lambda+2\mu)^2}-\frac{a_n^2\lambda\overline{z}z^{n+1}}{(\lambda+\mu)(\lambda+2\mu)}-\frac{1\hat{a}_n^2\lambda\overline{z}z^{n+1}}{(\lambda+\mu)(\lambda+2\mu)}-\frac{a_n^2\lambda^2\overline{z}z^{n+1}}{4\mu(\lambda+\mu)(\lambda+2\mu)}-$$

$$-\frac{1\hat{a}_n^2\lambda^2\overline{z}z^{n+1}}{4\mu(\lambda+\mu)(\lambda+2\mu)}-\frac{5a_n^2\mu\overline{z}z^{n+1}}{4(\lambda+\mu)(\lambda+2\mu)}-\frac{51\hat{a}_n^2\mu\overline{z}z^{n+1}}{4(\lambda+\mu)(\lambda+2\mu)}+\frac{3c_{n-1}^3\lambda^2\overline{z}z^{n+1}}{8(\lambda+2\mu)^2}+$$

$$+\frac{31\hat{c}_{n-1}^3\lambda^2\overline{z}z^{n+1}}{8(\lambda+2\mu)^2}+\frac{9c_{n-1}^3\lambda\mu\overline{z}z^{n+1}}{8(\lambda+2\mu)^2}+\frac{91\hat{c}_{n-1}^3\lambda\mu\overline{z}z^{n+1}}{8(\lambda+2\mu)^2}+\frac{3c_{n-1}^3\mu^2\overline{z}z^{n+1}}{4(\lambda+2\mu)^2}+$$

$$+\frac{31\hat{c}_{n-1}^3\mu^2\overline{z}z^{n+1}}{4(\lambda+2\mu)^2}=A_n^1z^n|z|^2.$$

Also in general the coefficients of z^n , $n=4,\dots$ give us a_n^0 , $\hat{a}_n^0$:

$$\frac{3a_n^0z^n}{2(\lambda+\mu)}+\frac{31\hat{a}_n^0z^n}{2(\lambda+\mu)}+\frac{a_n^0\lambda z^n}{2\mu(\lambda+\mu)}+\frac{1\hat{a}_n^0\lambda z^n}{2\mu(\lambda+\mu)}+\frac{c_{n-1}^1\lambda z^n}{4(\lambda+2\mu)}+\frac{1\hat{c}_{n-1}^1\lambda z^n}{4(\lambda+2\mu)}+$$

$$+\frac{c_{n-1}^1\lambda^2z^n}{8\mu(\lambda+2\mu)}+\frac{1\hat{c}_{n-1}^1\lambda^2z^n}{8\mu(\lambda+2\mu)}+\frac{c_{n-1}^1\mu z^n}{8(\lambda+2\mu)}+\frac{1\hat{c}_{n-1}^1\mu z^n}{8(\lambda+2\mu)}-\frac{b_{n-2}^2z^n}{8(\lambda+2\mu)}-$$

$$-\frac{1\hat{b}_{n-2}^2z^n}{8(\lambda+2\mu)}-\frac{b_{n-2}^2\lambda z^n}{8\mu(\lambda+2\mu)}-\frac{i\hat{b}_{n-2}^2\lambda z^n}{8\mu(\lambda+2\mu)}=A_nz^n.$$

Consider now the second set of relations on displacement coordinates, namely the third coordinate:

$$w(z,\overline{z},0)=\mathrm{Re}(\sum_{k=0}^{\infty}C_kz^k+|z|^2\sum_{k=0}^{\infty}\overline{C_k^1}\overline{z}^k+|z|^4\sum_{k=0}^{\infty}\overline{C_k^2}\overline{z}^k.$$

The constant term defines the unknown c_0^0 :

$$c_0^0=C_0.$$

The coefficient of $\overline{z}$ gives us c_1^0 , $\hat{c}_1^0$:

$$\frac{c_1^0\overline{z}}{2}-\frac{1\hat{c}_1^0\overline{z}}{2}-\frac{a_0^1\overline{z}}{4\mu}-\frac{1\hat{a}_0^1\overline{z}}{4\mu}=\overline{C_1}\overline{z}.$$

Similarly, $\overline{z}^2$ provides c_2^0 and $\hat{c}_2^0$:

$$\frac{c_2^0\overline{z}^2}{2} - \frac{1}{2}\imath\hat{c}_2^0\overline{z}^2 = \overline{C_2}\overline{z}^2.$$

So in general, the coefficients of $\overline{z}^n$ yield c_n^0, $\hat{c}_n^0$

$$\frac{c_n^0\overline{z}^n}{2} - \frac{1}{2}\imath\hat{c}_n^0\overline{z}^n = \overline{C_n}\overline{z}^n.$$

Then we consider the summands containing $|z|^2$. So the coefficient of $|z|^2$ provides c_0^2 by formula:

$$-c_0^2\overline{z}z - \frac{a_1^1\overline{z}z}{2\mu} - \frac{c_0^2\lambda\overline{z}z}{2\mu} + \frac{c_0^2\lambda\overline{z}z}{2(\lambda+2\mu)} + \frac{c_0^2\lambda^2\overline{z}z}{4\mu(\lambda+2\mu)} + \frac{c_0^2\mu\overline{z}z}{4(\lambda+2\mu)} = C_0^1|z|^2.$$

The same with $\overline{z}|z|^2$ can be used to determine c_1^2, $\hat{c}_1^2$:

$$-\frac{1}{2}c_1^2\overline{z}^2z+\frac{1}{2}\imath\hat{c}_1^2\overline{z}^2z+\frac{3a_0^3\lambda\overline{z}^2z}{16\mu^2}+\frac{3\imath\hat{a}_0^3\lambda\overline{z}^2z}{16\mu^2}-\frac{a_2^1\overline{z}^2z}{4\mu}+\frac{3a_0^3\overline{z}^2z}{8\mu}+\frac{\imath\hat{a}_2^1\overline{z}^2z}{4\mu}+\frac{3\imath\hat{a}_0^3\overline{z}^2z}{8\mu}-$$

$$-\frac{c_1^2\lambda\overline{z}^2z}{4\mu}+\frac{\imath\hat{c}_1^2\lambda\overline{z}^2z}{4\mu}-\frac{3b_0^3\overline{z}^2z}{16(\lambda+2\mu)}+\frac{3\imath\hat{b}_0^3\overline{z}^2z}{16(\lambda+2\mu)}+\frac{c_1^2\lambda\overline{z}^2z}{4(\lambda+2\mu)}-\frac{\imath\hat{c}_1^2\lambda\overline{z}^2z}{4(\lambda+2\mu)}-$$

$$-\frac{3a_0^3\lambda^2\overline{z}^2z}{16\mu^2(\lambda+2\mu)}-\frac{3\imath\hat{a}_0^3\lambda^2\overline{z}^2z}{16\mu^2(\lambda+2\mu)}-\frac{3a_0^3\lambda\overline{z}^2z}{8\mu(\lambda+2\mu)}-\frac{3\imath\hat{a}_0^3\lambda\overline{z}^2z}{8\mu(\lambda+2\mu)}-\frac{3b_0^3\lambda\overline{z}^2z}{16\mu(\lambda+2\mu)}+$$

$$+\frac{3\imath\hat{b}_0^3\lambda\overline{z}^2z}{16\mu(\lambda+2\mu)}+\frac{c_1^2\lambda^2\overline{z}^2z}{8\mu(\lambda+2\mu)}-\frac{\imath\hat{c}_1^2\lambda^2\overline{z}^2z}{8\mu(\lambda+2\mu)}+\frac{c_1^2\mu\overline{z}^2z}{8(\lambda+2\mu)}-\frac{\imath\hat{c}_1^2\mu\overline{z}^2z}{8(\lambda+2\mu)}=\overline{C_1^1}\overline{z}|z|^2.$$

Similarly, the coefficient of $\overline{z}^2|z|^2$ gives us c_2^2, $\hat{c}_2^2$

$$-\frac{1}{2}c_2^2\overline{z}^3z+\frac{1}{2}\imath\hat{c}_2^2\overline{z}^3z-\frac{c_2^2\lambda\overline{z}^3z}{4\mu}+\frac{\imath\hat{c}_2^2\lambda\overline{z}^3z}{4\mu}-\frac{3b_1^3\overline{z}^3z}{16(\lambda+2\mu)}+\frac{3\imath\hat{b}_1^3\overline{z}^3z}{16(\lambda+2\mu)}+\frac{c_2^2\lambda\overline{z}^3z}{4(\lambda+2\mu)}-$$

$$-\frac{\imath\hat{c}_2^2\lambda\overline{z}^3z}{4(\lambda+2\mu)}-\frac{3b_1^3\lambda\overline{z}^3z}{16\mu(\lambda+2\mu)}+\frac{3\imath\hat{b}_1^3\lambda\overline{z}^3z}{16\mu(\lambda+2\mu)}+\frac{c_2^2\lambda^2\overline{z}^3z}{8\mu(\lambda+2\mu)}-\frac{\imath\hat{c}_2^2\lambda^2\overline{z}^3z}{8\mu(\lambda+2\mu)}+$$

$$+\frac{c_2^2\mu\overline{z}^3z}{8(\lambda+2\mu)}-\frac{\imath\hat{c}_2^2\mu\overline{z}^3z}{8(\lambda+2\mu)}-\frac{a_3^1\overline{z}^3z}{4\mu}+\frac{\imath\hat{a}_3^1\overline{z}^3z}{4\mu}=\overline{C_2^1}\overline{z}^2|z|^2.$$

Also we consider the coefficient of $\overline{z}^3|z|^2$ in order to obtain c_3^2, $\hat{c}_3^2$:

$$-\frac{3b_2^3\overline{z}^4z}{16(\lambda+2\mu)}+\frac{3\imath\hat{b}_2^3\overline{z}^4z}{16(\lambda+2\mu)}-\frac{3b_2^3\lambda\overline{z}^4z}{16\mu(\lambda+2\mu)}+\frac{3\imath\hat{b}_2^3\lambda\overline{z}^4z}{16\mu(\lambda+2\mu)}-\frac{1}{2}c_3^2\overline{z}^4z+\frac{1}{2}\imath\hat{c}_3^2\overline{z}^4z-$$

$$-\frac{c_3^2\lambda\overline{z}^4 z}{4\mu}+\frac{\imath\hat{c}_3^2\lambda\overline{z}^4 z}{4\mu}+\frac{c_3^2\lambda\overline{z}^4 z}{4(\lambda+2\mu)}-\frac{\imath\hat{c}_3^2\lambda\overline{z}^4 z}{4(\lambda+2\mu)}+\frac{c_3^2\lambda^2\overline{z}^4 z}{8\mu(\lambda+2\mu)}-\frac{\imath\hat{c}_3^2\lambda^2\overline{z}^4 z}{8\mu(\lambda+2\mu)}+$$

$$+\frac{c_3^2\mu\overline{z}^4 z}{8(\lambda+2\mu)}-\frac{\imath\hat{c}_3^2\mu\overline{z}^4 z}{8(\lambda+2\mu)}-\frac{a_4^1\overline{z}^4 z}{4\mu}+\frac{\imath\hat{a}_4^1\overline{z}^4 z}{4\mu}=\overline{C_3^1}\overline{z}^3|z|^2.$$

In general the coefficient of $\overline{z}^n|z|^2$ yields c_n^2, $\hat{c}_n^2$, $n=4,\ldots$:

$$-\frac{1}{2}c_n^2\overline{z}^{n+1}z+\frac{1}{2}\imath\hat{c}_n^2\overline{z}^{n+1}z-\frac{c_n^2\lambda\overline{z}^{n+1}z}{4\mu}+\frac{\imath\hat{c}_n^2\lambda\overline{z}^{n+1}z}{4\mu}+\frac{c_n^2\lambda\overline{z}^{n+1}z}{4(\lambda+2\mu)}-\frac{\imath\hat{c}_n^2\lambda\overline{z}^{n+1}z}{4(\lambda+2\mu)}+$$

$$+\frac{c_n^2\lambda^2\overline{z}^{n+1}z}{8\mu(\lambda+2\mu)}-\frac{\imath\hat{c}_n^2\lambda^2\overline{z}^{n+1}z}{8\mu(\lambda+2\mu)}+\frac{c_n^2\mu\overline{z}^{n+1}z}{8(\lambda+2\mu)}-\frac{\imath\hat{c}_n^2\mu\overline{z}^{n+1}z}{8(\lambda+2\mu)}-\frac{a_{n+1}^1\overline{z}^{n+1}z}{4\mu}+\frac{\imath\hat{a}_{n+1}^1\overline{z}^{n+1}z}{4\mu}-$$

$$-\frac{3b_{n-1}^3\overline{z}^{n+1}z}{16(\lambda+2\mu)}+\frac{3\imath\hat{b}_{n-1}^3\overline{z}^{n+1}z}{16(\lambda+2\mu)}-\frac{3b_{n-1}^3\lambda\overline{z}^{n+1}z}{16\mu(\lambda+2\mu)}+\frac{3\imath\hat{b}_{n-1}^3\lambda\overline{z}^{n+1}z}{16\mu(\lambda+2\mu)}=C_n^1\overline{z}^n|z|^2.$$

Now we consider the coefficients of the summands containing $|z|^4$. Then the initial summand $|z|^4$ provides us with a_1^3:

$$\frac{3a_1^3\lambda\overline{z}^2z^2}{8\mu^2}+\frac{3a_1^3\overline{z}^2z^2}{4\mu}-\frac{3a_1^3\lambda^2\overline{z}^2z^2}{8\mu^2(\lambda+2\mu)}-\frac{3a_1^3\lambda\overline{z}^2z^2}{4\mu(\lambda+2\mu)}=C_0^2|z|^4.$$

The coefficient of $\overline{z}|z|^4$ defines a_2^3, $\hat{a}_2^3$:

$$\frac{3a_2^3\lambda\overline{z}^3z^2}{16\mu^2}-\frac{3\imath\hat{a}_2^3\lambda\overline{z}^3z^2}{16\mu^2}+\frac{3a_2^3\overline{z}^3z^2}{8\mu}-\frac{3\imath\hat{a}_2^3\overline{z}^3z^2}{8\mu}-$$

$$-\frac{3a_2^3\lambda^2\overline{z}^3z^2}{16\mu^2(\lambda+2\mu)}+\frac{3\imath\hat{a}_2^3\lambda^2\overline{z}^3z^2}{16\mu^2(\lambda+2\mu)}-\frac{3a_2^3\lambda\overline{z}^3z^2}{8\mu(\lambda+2\mu)}+\frac{3\imath\hat{a}_2^3\lambda\overline{z}^3z^2}{8\mu(\lambda+2\mu)}=\overline{C_1^2}\overline{z}|z|^4.$$

The general term $\overline{z}^{n-1}|z|^4$ defines a_n^3, $\hat{a}_n^3$, $n=3,4,\ldots$:

$$\frac{3a_n^3\lambda\overline{z}^{n+1}z^2}{16\mu^2}-\frac{3\imath\hat{a}_n^3\lambda\overline{z}^{n+1}z^2}{16\mu^2}+\frac{3a_n^3\overline{z}^{n+1}z^2}{8\mu}-\frac{3\imath\hat{a}_n^3\overline{z}^{n+1}z^2}{8\mu}-\frac{3a_n^3\lambda^2\overline{z}^{n+1}z^2}{16\mu^2(\lambda+2\mu)}+$$

$$+\frac{3\imath\hat{a}_n^3\lambda^2\overline{z}^{n+1}z^2}{16\mu^2(\lambda+2\mu)}-\frac{3a_n^3\lambda\overline{z}^{n+1}z^2}{8\mu(\lambda+2\mu)}+\frac{3\imath\hat{a}_n^3\lambda\overline{z}^{n+1}z^2}{8\mu(\lambda+2\mu)}=\overline{C_{n-1}^2}\overline{z}^{n-1}|z|^4.$$

Then we turn to the second level of the selected layer. First we consider the relations on the upper circle. The first coordinate of the displacement vector is then the real part of $u+\imath v$. Assume that $\mathrm{Re}(u+\imath v)=\mathrm{Re}(\sum_{k=0}^{\infty}\alpha_k z^k)$.

The constant part provides us with the relations on a_0^1, $\hat{a}_0^1$

$$
\begin{aligned}
&-\frac{a_2^0}{\mu}-\frac{a_2^1}{\mu}-\frac{3a_2^2}{\mu}-\frac{3a_2^3}{\mu}+\frac{\imath\hat{a}_2^0}{\mu}+\frac{\imath\hat{a}_2^1}{\mu}+\frac{3\imath\hat{a}_2^2}{\mu}+\frac{3\imath\hat{a}_2^3}{\mu}-\frac{b_0^0}{2\mu}-\frac{b_0^1}{2\mu}-\\
&-\frac{b_0^2}{\mu}-\frac{b_0^3}{\mu}+\frac{\imath\hat{b}_0^0}{2\mu}+\frac{\imath\hat{b}_0^1}{2\mu}+\frac{i\hat{b}_0^2}{\mu}+\frac{\imath\hat{b}_0^3}{\mu}+\frac{3a_0^0}{2(\lambda+\mu)}+\frac{3a_0^1}{2(\lambda+\mu)}+\frac{3a_0^2}{2(\lambda+\mu)}+\frac{3a_0^3}{2(\lambda+\mu)}+\\
&+\frac{3\imath\hat{a}_0^0}{2(\lambda+\mu)}+\frac{3\imath\hat{a}_0^1}{2(\lambda+\mu)}+\frac{3\imath\hat{a}_0^2}{2(\lambda+\mu)}+\frac{3\imath\hat{a}_0^3}{2(\lambda+\mu)}+\frac{a_0^0\lambda}{2\mu(\lambda+\mu)}+\frac{a_0^1\lambda}{2\mu(\lambda+\mu)}+\\
&+\frac{a_0^2\lambda}{2\mu(\lambda+\mu)}+\frac{a_0^3\lambda}{2\mu(\lambda+\mu)}+\frac{\imath\hat{a}_0^0\lambda}{2\mu(\lambda+\mu)}+\frac{\imath\hat{a}_0^1\lambda}{2\mu(\lambda+\mu)}+\frac{\imath\hat{a}_0^2\lambda}{2\mu(\lambda+\mu)}+\frac{\imath\hat{a}_0^3\lambda}{2\mu(\lambda+\mu)}+\\
&+\frac{3a_0^2\lambda}{4(\lambda+2\mu)^2}+\frac{9a_2^2\lambda}{8(\lambda+2\mu)^2}+\frac{3\imath\hat{a}_0^2\lambda}{4(\lambda+2\mu)^2}-\frac{9\imath\hat{a}_2^2\lambda}{8(\lambda+2\mu)^2}+\frac{27c_1^3\lambda^2}{8(\lambda+2\mu)^2}-\\
&-\frac{27\imath\hat{c}_1^3\lambda^2}{8(\lambda+2\mu)^2}+\frac{a_0^2\lambda^2}{4\mu(\lambda+2\mu)^2}+\frac{3a_2^2\lambda^2}{8\mu(\lambda+2\mu)^2}+\frac{\imath\hat{a}_0^2\lambda^2}{4\mu(\lambda+2\mu)^2}-\frac{3\imath\hat{a}_2^2\lambda^2}{8\mu(\lambda+2\mu)^2}+\\
&+\frac{9c_1^3\lambda^3}{16\mu(\lambda+2\mu)^2}-\frac{9\imath\hat{c}_1^3\lambda^3}{16\mu(\lambda+2\mu)^2}+\frac{a_0^2\mu}{2(\lambda+2\mu)^2}+\frac{3a_2^2\mu}{4(\lambda+2\mu)^2}+\frac{\imath\hat{a}_0^2\mu}{2(\lambda+2\mu)^2}-\\
&-\frac{3\imath\hat{a}_2^2\mu}{4(\lambda+2\mu)^2}+\frac{99c_1^3\lambda\mu}{16(\lambda+2\mu)^2}-\frac{99\imath\hat{c}_1^3\lambda\mu}{16(\lambda+2\mu)^2}+\frac{27c_1^3\mu^2}{8(\lambda+2\mu)^2}-\frac{27\imath\hat{c}_1^3\mu^2}{8(\lambda+2\mu)^2}+\\
&+\frac{9a_2^2}{8(\lambda+2\mu)}+\frac{3a_0^3}{4(\lambda+2\mu)}+\frac{9a_2^3}{2(\lambda+2\mu)}-\frac{9\imath\hat{a}_2^2}{8(\lambda+2\mu)}+\frac{3\imath\hat{a}_0^3}{4(\lambda+2\mu)}-\frac{9\imath\hat{a}_2^3}{2(\lambda+2\mu)}+\\
&+\frac{3b_0^2}{4(\lambda+2\mu)}+\frac{9b_0^3}{4(\lambda+2\mu)}-\frac{3\imath\hat{b}_0^2}{4(\lambda+2\mu)}-\frac{9\imath\hat{b}_0^3}{4(\lambda+2\mu)}-\frac{c_1^1\lambda}{\lambda+2\mu}-\frac{2c_1^2\lambda}{\lambda+2\mu}-\frac{9c_1^3\lambda}{\lambda+2\mu}+\\
&+\frac{\imath\hat{c}_1^1\lambda}{\lambda+2\mu}+\frac{2\imath\hat{c}_1^2\lambda}{\lambda+2\mu}+\frac{9\imath\hat{c}_1^3\lambda}{\lambda+2\mu}+\frac{3a_2^2\lambda}{8\mu(\lambda+2\mu)}+\frac{3a_0^3\lambda}{4\mu(\lambda+2\mu)}+\frac{9a_2^3\lambda}{4\mu(\lambda+2\mu)}-\\
&-\frac{3\imath\hat{a}_2^2\lambda}{8\mu(\lambda+2\mu)}+\frac{3\imath\hat{a}_0^3\lambda}{4\mu(\lambda+2\mu)}-\frac{9\imath\hat{a}_2^3\lambda}{4\mu(\lambda+2\mu)}+\frac{b_0^2\lambda}{4\mu(\lambda+2\mu)}+\frac{3b_0^3\lambda}{4\mu(\lambda+2\mu)}-\\
&-\frac{\imath\hat{b}_0^2\lambda}{4\mu(\lambda+2\mu)}-\frac{3\imath\hat{b}_0^3\lambda}{4\mu(\lambda+2\mu)}-\frac{c_1^1\lambda^2}{4\mu(\lambda+2\mu)}-\frac{c_1^2\lambda^2}{2\mu(\lambda+2\mu)}-\frac{9c_1^3\lambda^2}{4\mu(\lambda+2\mu)}+\\
&+\frac{\imath\hat{c}_1^1\lambda^2}{4\mu(\lambda+2\mu)}+\frac{\imath\hat{c}_1^2\lambda^2}{2\mu(\lambda+2\mu)}+\frac{9\imath\hat{c}_1^3\lambda^2}{4\mu(\lambda+2\mu)}-\frac{3c_1^1\mu}{4(\lambda+2\mu)}-\frac{3c_1^2\mu}{2(\lambda+2\mu)}-\frac{27c_1^3\mu}{4(\lambda+2\mu)}+
\end{aligned}
$$

$$+\frac{3\imath\hat{c}_1^1\mu}{4(\lambda+2\mu)}+\frac{3\imath\hat{c}_1^2\mu}{2(\lambda+2\mu)}+\frac{27\imath\hat{c}_1^3\mu}{4(\lambda+2\mu)}-\frac{a_0^2\lambda}{(\lambda+\mu)(\lambda+2\mu)}-\frac{3a_0^3\lambda}{(\lambda+\mu)(\lambda+2\mu)}-$$

$$-\frac{\imath\hat{a}_0^2\lambda}{(\lambda+\mu)(\lambda+2\mu)}-\frac{3\imath\hat{a}_0^3\lambda}{(\lambda+\mu)(\lambda+2\mu)}-$$

$$-\frac{a_0^2\lambda^2}{4\mu(\lambda+\mu)(\lambda+2\mu)}-\frac{3a_0^3\lambda^2}{4\mu(\lambda+\mu)(\lambda+2\mu)}-$$

$$-\frac{\imath\hat{a}_0^2\lambda^2}{4\mu(\lambda+\mu)(\lambda+2\mu)}-\frac{3\imath\hat{a}_0^3\lambda^2}{4\mu(\lambda+\mu)(\lambda+2\mu)}-\frac{5a_0^2\mu}{4(\lambda+\mu)(\lambda+2\mu)}-$$

$$-\frac{15a_0^3\mu}{4(\lambda+\mu)(\lambda+2\mu)}-\frac{5\imath\hat{a}_0^2\mu}{4(\lambda+\mu)(\lambda+2\mu)}-\frac{15\imath\hat{a}_0^3\mu}{4(\lambda+\mu)(\lambda+2\mu)}=\alpha_0.$$

The first degree of z provides us with the relation on a_1^1, $\hat{a}_1^1$:

$$-\frac{a_1^0z}{2\mu}-\frac{a_1^1z}{2\mu}-\frac{a_1^2z}{\mu}-\frac{a_1^3z}{\mu}+\frac{\imath\hat{a}_1^0z}{2\mu}+\frac{\imath\hat{a}_1^1z}{2\mu}+\frac{\imath\hat{a}_1^2z}{\mu}+\frac{\imath\hat{a}_1^3z}{\mu}-\frac{b_1^0z}{2\mu}-$$

$$-\frac{b_1^1z}{2\mu}-\frac{3b_1^2z}{\mu}-\frac{3b_1^3z}{\mu}-\frac{\imath\hat{b}_1^0z}{2\mu}-\frac{\imath\hat{b}_1^1z}{2\mu}-\frac{3\imath\hat{b}_1^2z}{\mu}-\frac{3\imath\hat{b}_1^3z}{\mu}+\frac{3a_1^0z}{2(\lambda+\mu)}+\frac{3a_1^1z}{2(\lambda+\mu)}+$$

$$+\frac{3a_1^2z}{\lambda+\mu}+\frac{3a_1^3z}{\lambda+\mu}+\frac{3\imath\hat{a}_1^0z}{2(\lambda+\mu)}+\frac{3\imath\hat{a}_1^1z}{2(\lambda+\mu)}+\frac{3\imath\hat{a}_1^2z}{\lambda+\mu}+\frac{3\imath\hat{a}_1^3z}{\lambda+\mu}+\frac{a_1^0\lambda z}{2\mu(\lambda+\mu)}+$$

$$+\frac{a_1^1\lambda z}{2\mu(\lambda+\mu)}+\frac{a_1^2\lambda z}{\mu(\lambda+\mu)}+\frac{a_1^3\lambda z}{\mu(\lambda+\mu)}+\frac{\imath\hat{a}_1^0\lambda z}{2\mu(\lambda+\mu)}+\frac{\imath\hat{a}_1^1\lambda z}{2\mu(\lambda+\mu)}+\frac{\imath\hat{a}_1^2\lambda z}{\mu(\lambda+\mu)}+$$

$$+\frac{\imath\hat{a}_1^3\lambda z}{\mu(\lambda+\mu)}+\frac{3a_1^2\lambda z}{2(\lambda+2\mu)^2}+\frac{21c_0^3\lambda^2z}{8(\lambda+2\mu)^2}+\frac{9c_2^3\lambda^2z}{2(\lambda+2\mu)^2}-\frac{15\imath\hat{c}_0^3\lambda^2z}{8(\lambda+2\mu)^2}+$$

$$+\frac{9\imath\hat{c}_2^3\lambda^2z}{2(\lambda+2\mu)^2}+\frac{a_1^2\lambda^2z}{2\mu(\lambda+2\mu)^2}+\frac{3c_0^3\lambda^3z}{8\mu(\lambda+2\mu)^2}+\frac{3c_2^3\lambda^3z}{4\mu(\lambda+2\mu)^2}-\frac{3\imath\hat{c}_0^3\lambda^3z}{8\mu(\lambda+2\mu)^2}+$$

$$+\frac{3\imath\hat{c}_2^3\lambda^3z}{4\mu(\lambda+2\mu)^2}+\frac{a_1^2\mu z}{(\lambda+2\mu)^2}+\frac{21c_0^3\lambda\mu z}{4(\lambda+2\mu)^2}+\frac{33c_2^3\lambda\mu z}{4(\lambda+2\mu)^2}-\frac{3\imath\hat{c}_0^3\lambda\mu z}{(\lambda+2\mu)^2}+$$

$$+\frac{33\imath\hat{c}_2^3\lambda\mu z}{4(\lambda+2\mu)^2}+\frac{3c_0^3\mu^2z}{(\lambda+2\mu)^2}+\frac{9c_2^3\mu^2z}{2(\lambda+2\mu)^2}-\frac{3\imath\hat{c}_0^3\mu^2z}{2(\lambda+2\mu)^2}+\frac{9\imath\hat{c}_2^3\mu^2z}{2(\lambda+2\mu)^2}+$$

$$+\frac{3a_1^2z}{4(\lambda+2\mu)}+\frac{15a_1^3z}{4(\lambda+2\mu)}-\frac{3\imath\hat{a}_1^2z}{4(\lambda+2\mu)}-\frac{9\imath\hat{a}_1^3z}{4(\lambda+2\mu)}+\frac{9b_1^2z}{8(\lambda+2\mu)}+\frac{27b_1^3z}{8(\lambda+2\mu)}+$$

$$+\frac{9\imath\hat{b}_1^2z}{8(\lambda+2\mu)}+\frac{27\imath\hat{b}_1^3z}{8(\lambda+2\mu)}-\frac{c_0^1\lambda z}{4(\lambda+2\mu)}-\frac{3c_2^1\lambda z}{2(\lambda+2\mu)}-\frac{c_0^2\lambda z}{2(\lambda+2\mu)}-\frac{3c_2^2\lambda z}{\lambda+2\mu}-$$

$$-\frac{3c_0^3\lambda z}{2(\lambda+2\mu)} - \frac{18c_2^3\lambda z}{\lambda+2\mu} + \frac{3\imath\hat{c}_0^1\lambda z}{4(\lambda+2\mu)} - \frac{3\imath\hat{c}_2^1\lambda z}{2(\lambda+2\mu)} + \frac{3\imath\hat{c}_0^2\lambda z}{2(\lambda+2\mu)} - \frac{3\imath\hat{c}_2^2\lambda z}{\lambda+2\mu} +$$

$$+\frac{9\imath\hat{c}_0^3\lambda z}{2(\lambda+2\mu)} - \frac{18\imath\hat{c}_2^3\lambda z}{\lambda+2\mu} + \frac{a_1^2\lambda z}{4\mu(\lambda+2\mu)} + \frac{9a_1^3\lambda z}{4\mu(\lambda+2\mu)} - \frac{\imath\hat{a}_1^2\lambda z}{4\mu(\lambda+2\mu)} - \frac{3\imath\hat{a}_1^3\lambda z}{4\mu(\lambda+2\mu)} +$$

$$+\frac{3b_1^2\lambda z}{8\mu(\lambda+2\mu)} + \frac{9b_1^3\lambda z}{8\mu(\lambda+2\mu)} + \frac{3\imath\hat{b}_1^2\lambda z}{8\mu(\lambda+2\mu)} + \frac{9\imath\hat{b}_1^3\lambda z}{8\mu(\lambda+2\mu)} - \frac{3c_2^1\lambda^2 z}{8\mu(\lambda+2\mu)} -$$

$$-\frac{3c_2^2\lambda^2 z}{4\mu(\lambda+2\mu)} - \frac{9c_2^3\lambda^2 z}{2\mu(\lambda+2\mu)} + \frac{\imath\hat{c}_0^1\lambda^2 z}{4\mu(\lambda+2\mu)} - \frac{3\imath\hat{c}_2^1\lambda^2 z}{8\mu(\lambda+2\mu)} + \frac{\imath\hat{c}_0^2\lambda^2 z}{2\mu(\lambda+2\mu)} -$$

$$-\frac{3\imath\hat{c}_2^2\lambda^2 z}{4\mu(\lambda+2\mu)} + \frac{3\imath\hat{c}_0^3\lambda^2 z}{2\mu(\lambda+2\mu)} - \frac{9\imath\hat{c}_2^3\lambda^2 z}{2\mu(\lambda+2\mu)} - \frac{c_0^1\mu z}{4(\lambda+2\mu)} - \frac{9c_2^1\mu z}{8(\lambda+2\mu)} - \frac{c_0^2\mu z}{2(\lambda+2\mu)} -$$

$$-\frac{9c_2^2\mu z}{4(\lambda+2\mu)} - \frac{3c_0^3\mu z}{2(\lambda+2\mu)} - \frac{27c_2^3\mu z}{2(\lambda+2\mu)} + \frac{\imath\hat{c}_0^1\mu z}{2(\lambda+2\mu)} - \frac{9\imath\hat{c}_2^1\mu z}{8(\lambda+2\mu)} + \frac{\imath\hat{c}_0^2\mu z}{\lambda+2\mu} -$$

$$-\frac{9\imath\hat{c}_2^2\mu z}{4(\lambda+2\mu)} + \frac{3\imath\hat{c}_0^3\mu z}{\lambda+2\mu} - \frac{27\imath\hat{c}_2^3\mu z}{2(\lambda+2\mu)} - \frac{a_1^2\lambda z}{(\lambda+\mu)(\lambda+2\mu)} - \frac{3a_1^3\lambda z}{(\lambda+\mu)(\lambda+2\mu)} -$$

$$-\frac{\imath\hat{a}_1^2\lambda z}{(\lambda+\mu)(\lambda+2\mu)} - \frac{3\imath\hat{a}_1^3\lambda z}{(\lambda+\mu)(\lambda+2\mu)} - \frac{a_1^2\lambda^2 z}{4\mu(\lambda+\mu)(\lambda+2\mu)} -$$

$$-\frac{3a_1^3\lambda^2 z}{4\mu(\lambda+\mu)(\lambda+2\mu)} - \frac{\imath\hat{a}_1^2\lambda^2 z}{4\mu(\lambda+\mu)(\lambda+2\mu)} - \frac{3\imath\hat{a}_1^3\lambda^2 z}{4\mu(\lambda+\mu)(\lambda+2\mu)} -$$

$$-\frac{5a_1^2\mu z}{4(\lambda+\mu)(\lambda+2\mu)} - \frac{15a_1^3\mu z}{4(\lambda+\mu)(\lambda+2\mu)} - \frac{5\imath\hat{a}_1^2\mu z}{4(\lambda+\mu)(\lambda+2\mu)} -$$

$$-\frac{15\imath\hat{a}_1^3\mu z}{4(\lambda+\mu)(\lambda+2\mu)} = \alpha_1 z.$$

The coefficient of z^2 gives us a_2^1, $\hat{a}_2^1$:

$$-\frac{b_2^0 z^2}{2\mu} - \frac{b_2^1 z^2}{2\mu} - \frac{6b_2^2 z^2}{\mu} - \frac{6b_2^3 z^2}{\mu} - \frac{\imath\hat{b}_2^0 z^2}{2\mu} - \frac{\imath\hat{b}_2^1 z^2}{2\mu} - \frac{6\imath\hat{b}_2^2 z^2}{\mu} - \frac{6\imath\hat{b}_2^3 z^2}{\mu} +$$

$$+\frac{3a_2^0 z^2}{2(\lambda+\mu)} + \frac{3a_2^1 z^2}{2(\lambda+\mu)} + \frac{9a_2^2 z^2}{2(\lambda+\mu)} + \frac{9a_2^3 z^2}{2(\lambda+\mu)} + \frac{3\imath\hat{a}_2^0 z^2}{2(\lambda+\mu)} +$$

$$+\frac{3\imath\hat{a}_2^1 z^2}{2(\lambda+\mu)} + \frac{9\imath\hat{a}_2^2 z^2}{2(\lambda+\mu)} + \frac{9\imath\hat{a}_2^3 z^2}{2(\lambda+\mu)} + \frac{a_2^0\lambda z^2}{2\mu(\lambda+\mu)} + \frac{a_2^1\lambda z^2}{2\mu(\lambda+\mu)} + \frac{3a_2^2\lambda z^2}{2\mu(\lambda+\mu)} +$$

$$+\frac{3a_2^3\lambda z^2}{2\mu(\lambda+\mu)} + \frac{\imath\hat{a}_2^0\lambda z^2}{2\mu(\lambda+\mu)} + \frac{\imath\hat{a}_2^1\lambda z^2}{2\mu(\lambda+\mu)} + \frac{3\imath\hat{a}_2^2\lambda z^2}{2\mu(\lambda+\mu)} + \frac{3\imath\hat{a}_2^3\lambda z^2}{2\mu(\lambda+\mu)} +$$

$$+\frac{3a_0^2\lambda z^2}{8(\lambda+2\mu)^2}+\frac{3a_2^2\lambda z^2}{4(\lambda+2\mu)^2}-\frac{3\imath\hat{a}_0^2\lambda z^2}{8(\lambda+2\mu)^2}+\frac{3\imath\hat{a}_2^2\lambda z^2}{4(\lambda+2\mu)^2}+\frac{3c_1^3\lambda^2 z^2}{8(\lambda+2\mu)^2}+$$

$$+\frac{3\imath\hat{c}_1^3\lambda^2 z^2}{8(\lambda+2\mu)^2}+\frac{a_0^2\lambda^2 z^2}{8\mu(\lambda+2\mu)^2}+\frac{a_2^2\lambda^2 z^2}{4\mu(\lambda+2\mu)^2}-\frac{\imath\hat{a}_0^2\lambda^2 z^2}{8\mu(\lambda+2\mu)^2}+\frac{\imath\hat{a}_2^2\lambda^2 z^2}{4\mu(\lambda+2\mu)^2}+$$

$$+\frac{a_0^2\mu z^2}{4(\lambda+2\mu)^2}+\frac{a_2^2\mu z^2}{2(\lambda+2\mu)^2}-\frac{\imath\hat{a}_0^2\mu z^2}{4(\lambda+2\mu)^2}+\frac{\imath\hat{a}_2^2\mu z^2}{2(\lambda+2\mu)^2}+\frac{9c_1^3\lambda\mu z^2}{8(\lambda+2\mu)^2}+$$

$$+\frac{9\imath\hat{c}_1^3\lambda\mu z^2}{8(\lambda+2\mu)^2}+\frac{3c_1^3\mu^2 z^2}{4(\lambda+2\mu)^2}+\frac{3\imath\hat{c}_1^3\mu^2 z^2}{4(\lambda+2\mu)^2}+\frac{3a_0^2 z^2}{8(\lambda+2\mu)}+\frac{3a_0^3 z^2}{2(\lambda+2\mu)}+$$

$$+\frac{3a_2^3 z^2}{4(\lambda+2\mu)}-\frac{3\imath\hat{a}_0^2 z^2}{8(\lambda+2\mu)}-\frac{3\imath\hat{a}_0^3 z^2}{2(\lambda+2\mu)}+\frac{3\imath\hat{a}_2^3 z^2}{4(\lambda+2\mu)}-\frac{b_0^2 z^2}{8(\lambda+2\mu)}+$$

$$+\frac{3b_2^2 z^2}{2(\lambda+2\mu)}-\frac{3b_0^3 z^2}{8(\lambda+2\mu)}+\frac{9b_2^3 z^2}{2(\lambda+2\mu)}-\frac{\imath\hat{b}_0^2 z^2}{8(\lambda+2\mu)}+\frac{3\imath\hat{b}_2^2 z^2}{2(\lambda+2\mu)}-\frac{3\imath\hat{b}_0^3 z^2}{8(\lambda+2\mu)}+$$

$$+\frac{9\imath\hat{b}_2^3 z^2}{2(\lambda+2\mu)}+\frac{c_1^1\lambda z^2}{4(\lambda+2\mu)}+\frac{c_1^2\lambda z^2}{2(\lambda+2\mu)}+\frac{9c_1^3\lambda z^2}{4(\lambda+2\mu)}+\frac{\imath\hat{c}_1^1\lambda z^2}{4(\lambda+2\mu)}+\frac{\imath\hat{c}_1^2\lambda z^2}{2(\lambda+2\mu)}+$$

$$+\frac{9\imath\hat{c}_1^3\lambda z^2}{4(\lambda+2\mu)}+\frac{a_0^2\lambda z^2}{8\mu(\lambda+2\mu)}+\frac{3a_0^3\lambda z^2}{4\mu(\lambda+2\mu)}+\frac{3a_2^3\lambda z^2}{4\mu(\lambda+2\mu)}-\frac{\imath\hat{a}_0^2\lambda z^2}{8\mu(\lambda+2\mu)}-$$

$$-\frac{3\imath\hat{a}_0^3\lambda z^2}{4\mu(\lambda+2\mu)}+\frac{3\imath\hat{a}_2^3\lambda z^2}{4\mu(\lambda+2\mu)}-\frac{b_0^2\lambda z^2}{8\mu(\lambda+2\mu)}+\frac{b_2^2\lambda z^2}{2\mu(\lambda+2\mu)}-\frac{3b_0^3\lambda z^2}{8\mu(\lambda+2\mu)}+$$

$$+\frac{3b_2^3\lambda z^2}{2\mu(\lambda+2\mu)}-\frac{\imath\hat{b}_0^2\lambda z^2}{8\mu(\lambda+2\mu)}+\frac{\imath\hat{b}_2^2\lambda z^2}{2\mu(\lambda+2\mu)}-\frac{3\imath\hat{b}_0^3\lambda z^2}{8\mu(\lambda+2\mu)}+\frac{3\imath\hat{b}_2^3\lambda z^2}{2\mu(\lambda+2\mu)}+$$

$$+\frac{c_1^1\lambda^2 z^2}{8\mu(\lambda+2\mu)}+\frac{c_1^2\lambda^2 z^2}{4\mu(\lambda+2\mu)}+\frac{9c_1^3\lambda^2 z^2}{8\mu(\lambda+2\mu)}+\frac{\imath\hat{c}_1^1\lambda^2 z^2}{8\mu(\lambda+2\mu)}+\frac{\imath\hat{c}_1^2\lambda^2 z^2}{4\mu(\lambda+2\mu)}+$$

$$+\frac{9\imath\hat{c}_1^3\lambda^2 z^2}{8\mu(\lambda+2\mu)}+\frac{c_1^1\mu z^2}{8(\lambda+2\mu)}+\frac{c_1^2\mu z^2}{4(\lambda+2\mu)}+\frac{9c_1^3\mu z^2}{8(\lambda+2\mu)}+\frac{\imath\hat{c}_1^1\mu z^2}{8(\lambda+2\mu)}+\frac{\imath\hat{c}_1^2\mu z^2}{4(\lambda+2\mu)}+$$

$$+\frac{9\imath\hat{c}_1^3\mu z^2}{8(\lambda+2\mu)}-\frac{a_2^2\lambda z^2}{(\lambda+\mu)(\lambda+2\mu)}-\frac{3a_2^3\lambda z^2}{(\lambda+\mu)(\lambda+2\mu)}-\frac{\imath\hat{a}_2^2\lambda z^2}{(\lambda+\mu)(\lambda+2\mu)}-$$

$$-\frac{3\imath\hat{a}_2^3\lambda z^2}{(\lambda+\mu)(\lambda+2\mu)}-\frac{a_2^2\lambda^2 z^2}{4\mu(\lambda+\mu)(\lambda+2\mu)}-\frac{3a_2^3\lambda^2 z^2}{4\mu(\lambda+\mu)(\lambda+2\mu)}-$$

$$-\frac{\imath\hat{a}_2^2\lambda^2 z^2}{4\mu(\lambda+\mu)(\lambda+2\mu)}-\frac{3\imath\hat{a}_2^3\lambda^2 z^2}{4\mu(\lambda+\mu)(\lambda+2\mu)}-\frac{5a_2^2\mu z^2}{4(\lambda+\mu)(\lambda+2\mu)}-$$

$$-\frac{15a_2^3\mu z^2}{4(\lambda+\mu)(\lambda+2\mu)}-\frac{5\imath\hat{a}_2^2\mu z^2}{4(\lambda+\mu)(\lambda+2\mu)}-\frac{15\imath\hat{a}_2^3\mu z^2}{4(\lambda+\mu)(\lambda+2\mu)}-\frac{c_3^1\lambda(4)z^2}{2(\lambda+2\mu)}-$$

$$-\frac{c_3^2\lambda(4)z^2}{\lambda+2\mu}-\frac{\imath\hat{c}_3^1\lambda(4)z^2}{2(\lambda+2\mu)}-\frac{\imath\hat{c}_3^2\lambda(4)z^2}{\lambda+2\mu}-\frac{c_3^1\lambda^2 4z^2}{8\mu(\lambda+2\mu)}-\frac{c_3^2\lambda^2 4z^2}{4\mu(\lambda+2\mu)}-\frac{\imath\hat{c}_3^1\lambda^2 4z^2}{8\mu(\lambda+2\mu)}-$$

$$-\frac{\imath\hat{c}_3^2\lambda^2 4z^2}{4\mu(\lambda+2\mu)}-\frac{3c_3^1\mu 4z^2}{8(\lambda+2\mu)}-\frac{3c_3^2\mu 4z^2}{4(\lambda+2\mu)}-\frac{3\imath\hat{c}_3^1\mu 4z^2}{8(\lambda+2\mu)}-\frac{3\imath\hat{c}_3^2\mu 4z^2}{4(\lambda+2\mu)}=\alpha_2 z^2.$$

The coefficient of z^n gives us a_n^1, $\hat{a}_n^1$, $n=3,4,\dots$:

$$\frac{3a_n^2\lambda z^n}{4(\lambda+2\mu)^2}+\frac{3\imath\hat{a}_n^2\lambda z^n}{4(\lambda+2\mu)^2}+\frac{a_n^2\lambda^2 z^n}{4\mu(\lambda+2\mu)^2}+\frac{\imath\hat{a}_n^2\lambda^2 z^n}{4\mu(\lambda+2\mu)^2}+\frac{a_n^2\mu z^n}{2(\lambda+2\mu)^2}+$$

$$+\frac{\imath\hat{a}_n^2\mu z^n}{2(\lambda+2\mu)^2}+\frac{3a_n^3 z^n}{4(\lambda+2\mu)}+\frac{3\imath\hat{a}_n^3 z^n}{4(\lambda+2\mu)}+\frac{3a_n^3\lambda z^n}{4\mu(\lambda+2\mu)}+\frac{3\imath\hat{a}_n^3\lambda z^n}{4\mu(\lambda+2\mu)}-$$

$$-\frac{a_n^2\lambda z^n}{(\lambda+\mu)(\lambda+2\mu)}-\frac{3a_n^3\lambda z^n}{(\lambda+\mu)(\lambda+2\mu)}-$$

$$-\frac{\imath\hat{a}_n^2\lambda z^n}{(\lambda+\mu)(\lambda+2\mu)}-\frac{3\imath\hat{a}_n^3\lambda z^n}{(\lambda+\mu)(\lambda+2\mu)}-$$

$$-\frac{a_n^2\lambda^2 z^n}{4\mu(\lambda+\mu)(\lambda+2\mu)}-\frac{3a_n^3\lambda^2 z^n}{4\mu(\lambda+\mu)(\lambda+2\mu)}-\frac{\imath\hat{a}_n^2\lambda^2 z^n}{4\mu(\lambda+\mu)(\lambda+2\mu)}-$$

$$-\frac{3\imath\hat{a}_n^3\lambda^2 z^n}{4\mu(\lambda+\mu)(\lambda+2\mu)}-\frac{5a_n^2\mu z^n}{4(\lambda+\mu)(\lambda+2\mu)}-\frac{15a_n^3\mu z^n}{4(\lambda+\mu)(\lambda+2\mu)}-$$

$$-\frac{5\imath\hat{a}_n^2\mu z^n}{4(\lambda+\mu)(\lambda+2\mu)}-\frac{15\imath\hat{a}_n^3\mu z^n}{4(\lambda+\mu)(\lambda+2\mu)}+$$

$$+\frac{3a_n^2(n+1)z^n}{2(\lambda+\mu)}+\frac{3a_n^3(n+1)z^n}{2(\lambda+\mu)}+$$

$$+\frac{3\imath\hat{a}_n^2(n+1)z^n}{2(\lambda+\mu)}+\frac{3\imath\hat{a}_n^3(n+1)z^n}{2(\lambda+\mu)}+\frac{a_n^2\lambda(n+1)z^n}{2\mu(\lambda+\mu)}+\frac{a_n^3\lambda(n+1)z^n}{2\mu(\lambda+\mu)}+$$

$$+\frac{\imath\hat{a}_n^2\lambda(n+1)z^n}{2\mu(\lambda+\mu)}+\frac{\imath\hat{a}_n^3\lambda(n+1)z^n}{2\mu(\lambda+\mu)}-\frac{c_{n+1}^1\lambda nz^n}{2(\lambda+2\mu)}-\frac{c_{n+1}^2\lambda nz^n}{\lambda+2\mu}-$$

$$-\frac{\imath\hat{c}_{n+1}^1\lambda nz^n}{2(\lambda+2\mu)}-\frac{\imath\hat{c}_{n+1}^2\lambda nz^n}{\lambda+2\mu}-\frac{c_{n+1}^1\lambda^2 nz^n}{8\mu(\lambda+2\mu)}-\frac{c_{n+1}^2\lambda^2 nz^n}{4\mu(\lambda+2\mu)}-\frac{i\hat{c}_{n+1}^1\lambda^2 nz^n}{8\mu(\lambda+2\mu)}-$$

$$-\frac{\imath\hat{c}^2_{n+1}\lambda^2 n z^n}{4\mu(\lambda+2\mu)}-\frac{3c^1_{n+1}\mu n z^n}{8(\lambda+2\mu)}-\frac{3c^2_{n+1}\mu n z^n}{4(\lambda+2\mu)}-\frac{3\imath\hat{c}^1_{n+1}\mu n z^n}{8(\lambda+2\mu)}-\frac{3\imath\hat{c}^2_{n+1}\mu n z^n}{4(\lambda+2\mu)}+$$

$$+\frac{c^1_{n-1}\lambda z^n}{4(\lambda+2\mu)}+\frac{c^2_{n-1}\lambda z^n}{2(\lambda+2\mu)}+\frac{\imath\hat{c}^1_{n-1}\lambda z^n}{4(\lambda+2\mu)}+\frac{i\hat{c}^2_{n-1}\lambda z^n}{2(\lambda+2\mu)}+\frac{c^1_{n-1}\lambda^2 z^n}{8\mu(\lambda+2\mu)}+$$

$$+\frac{c^2_{n-1}\lambda^2 z^n}{4\mu(\lambda+2\mu)}+\frac{\imath\hat{c}^1_{n-1}\lambda^2 z^n}{8\mu(\lambda+2\mu)}+\frac{\imath\hat{c}^2_{n-1}\lambda^2 z^n}{4\mu(\lambda+2\mu)}+\frac{c^1_{n-1}\mu z^n}{8(\lambda+2\mu)}+\frac{c^2_{n-1}\mu z^n}{4(\lambda+2\mu)}+$$

$$+\frac{\imath\hat{c}^1_{n-1}\mu z^n}{8(\lambda+2\mu)}+\frac{\imath\hat{c}^2_{n-1}\mu z^n}{4(\lambda+2\mu)}-$$

$$-\frac{a^0_{n+2}(n+2)z^n}{2\mu}-\frac{a^1_{n+2}(n+2)z^n}{2\mu}-\frac{\imath\hat{a}^0_{n+2}(n+2)z^n}{2\mu}-$$

$$-\frac{\imath\hat{a}^1_{n+2}(n+2)z^n}{2\mu}-\frac{b^0_n z^n}{2\mu}-\frac{b^1_n z^n}{2\mu}-\frac{i\hat{b}^0_n z^n}{2\mu}-\frac{\imath\hat{b}^1_n z^n}{2\mu}+\frac{3a^0_n z^n}{2(\lambda+\mu)}+$$

$$+\frac{3a^1_n z^n}{2(\lambda+\mu)}+\frac{3\imath\hat{a}^0_n z^n}{2(\lambda+\mu)}+\frac{3\imath\hat{a}^1_n z^n}{2(\lambda+\mu)}+\frac{a^0_n\lambda z^n}{2\mu(\lambda+\mu)}+\frac{a^1_n\lambda z^n}{2\mu(\lambda+\mu)}+$$

$$+\frac{\imath\hat{a}^0_n\lambda z^n}{2\mu(\lambda+\mu)}+\frac{\imath\hat{a}^1_n\lambda z^n}{2\mu(\lambda+\mu)}+\frac{9c^3_{n+1}\lambda^2(n+3)z^n}{8(\lambda+2\mu)^2}+\frac{9\imath\hat{c}^3_{n+1}\lambda^2(n+3)z^n}{8(\lambda+2\mu)^2}+$$

$$+\frac{3c^3_{n+1}\lambda^3(n+3)z^n}{16\mu(\lambda+2\mu)^2}+\frac{3\imath\hat{c}^3_{n+1}\lambda^3(n+3)z^{-2+n}}{16\mu(\lambda+2\mu)^2}+\frac{33c^3_{n+1}\lambda\mu(n+3)z^n}{16(\lambda+2\mu)^2}+$$

$$+\frac{33\imath\hat{c}^3_{n+1}\lambda\mu(n+3)z^n}{16(\lambda+2\mu)^2}+\frac{9c^3_{n+1}\mu^2(n+3)z^n}{8(\lambda+2\mu)^2}+\frac{9\imath\hat{c}^3_{n+1}\mu^2(n+3)z^n}{8(\lambda+2\mu)^2}+$$

$$+\frac{3c^3_{n+1}\lambda(n+3)z^n}{2(\lambda+2\mu)}+\frac{3i\hat{c}^3_{n+1}\lambda(n+3)z^n}{2(\lambda+2\mu)}+\frac{3c^3_{n+1}\lambda^2(n+3)z^n}{8\mu(\lambda+2\mu)}+$$

$$+\frac{3\imath\hat{c}^3_{n+1}\lambda^2(n+3)z^n}{8\mu(\lambda+2\mu)}+\frac{9c^3_{n+1}\mu(n+3)z^n}{8(\lambda+2\mu)}+\frac{9\imath\hat{c}^3_{n+1}\mu(n+3)z^n}{8(\lambda+2\mu)}-$$

$$-\frac{3c^3_{n+1}\lambda(n+3)^2 z^n}{2(\lambda+2\mu)}-\frac{3\imath\hat{c}^3_{n+1}\lambda(n+3)^2 z^n}{2(\lambda+2\mu)}-\frac{3c^3_{n+1}\lambda^2(n+3)^2 z^n}{8\mu(\lambda+2\mu)}-$$

$$-\frac{3\imath\hat{c}^3_{n+1}\lambda^2(n+3)^2 z^n}{8\mu(\lambda+2\mu)}-\frac{9c^3_{n+1}\mu(n+3)^2 z^n}{8(\lambda+2\mu)}-$$

$$-\frac{9\imath\hat{c}^3_{n+1}\mu(n+3)^2 z^n}{8(\lambda+2\mu)}+\frac{b^2_n(n+2)z^n}{2\mu}+$$

$$+\frac{b_n^3(n+2)z^n}{2\mu}+\frac{\imath\hat{b}_n^2(n+2)z^n}{2\mu}+\frac{\imath\hat{b}_n^3(n+2)z^n}{2\mu}+\frac{3b_n^2(n+2)z^n}{8(\lambda+2\mu)}+\frac{9b_n^3(n+2)z^n}{8(\lambda+2\mu)}+$$

$$+\frac{3\imath\hat{b}_n^2(n+2)z^n}{8(\lambda+2\mu)}+\frac{9\imath\hat{b}_n^3(n+2)z^n}{8(\lambda+2\mu)}+\frac{b_n^2\lambda(n+2)z^n}{8\mu(\lambda+2\mu)}+\frac{3b_n^3\lambda(n+2)z^n}{8\mu(\lambda+2\mu)}+$$

$$+\frac{\imath\hat{b}_n^2\lambda(n+2)z^n}{8\mu(\lambda+2\mu)}+\frac{3\imath\hat{b}_n^3\lambda(n+2)z^n}{8\mu(\lambda+2\mu)}-\frac{b_n^2(n+2)^2z^n}{2\mu}-\frac{b_n^3(n+2)^2z^n}{2\mu}-$$

$$-\frac{\imath\hat{b}_n^2(n+2)^2z^n}{2\mu}-\frac{\imath\hat{b}_n^3(n+2)^2z^n}{2\mu}+\frac{3c_{n-1}^3\lambda^2z^n}{8(\lambda+2\mu)^2}+\frac{3\imath\hat{c}_{n-1}^3\lambda^2z^n}{8(\lambda+2\mu)^2}+\frac{9c_{n-1}^3\lambda\mu z^n}{8(\lambda+2\mu)^2}+$$

$$+\frac{9\imath\hat{c}_{n-1}^3\lambda\mu z^n}{8(\lambda+2\mu)^2}+\frac{3c_{n-1}^3\mu^2z^n}{4(\lambda+2\mu)^2}+\frac{3\imath\hat{c}_{n-1}^3\mu^2z^n}{4(\lambda+2\mu)^2}+\frac{3c_{n-1}^3\lambda(n+1)z^n}{4(\lambda+2\mu)}+$$

$$+\frac{3\imath\hat{c}_{n-1}^3\lambda(n+1)z^n}{4(\lambda+2\mu)}+\frac{3c_{n-1}^3\lambda^2(n+1)z^n}{8\mu(\lambda+2\mu)}+\frac{3\imath\hat{c}_{n-1}^3\lambda^2(n+1)z^n}{8\mu(\lambda+2\mu)}+$$

$$+\frac{3c_{n-1}^3\mu(n+1)z^n}{8(\lambda+2\mu)}+\frac{3\imath\hat{c}_{n-1}^3\mu nz^n}{8(\lambda+2\mu)}-\frac{b_{n-2}^2z^n}{8(\lambda+2\mu)}-\frac{3b_{n-2}^3z^n}{8(\lambda+2\mu)}-\frac{\imath\hat{b}_{n-2}^2z^n}{8(\lambda+2\mu)}-$$

$$-\frac{3\imath\hat{b}_{n-2}^3z^n}{8(\lambda+2\mu)}-\frac{b_{n-2}^2\lambda z^n}{8\mu(\lambda+2\mu)}-\frac{3b_{n-2}^3\lambda z^n}{8\mu(\lambda+2\mu)}-\frac{\imath\hat{b}_{n-2}^2\lambda z^n}{8\mu(\lambda+2\mu)}-\frac{3\imath\hat{b}_{n-2}^3\lambda z^n}{8\mu(\lambda+2\mu)}=\alpha_nz^n.$$

Let us consider the relations on the imaginary part of $u+\imath v$, i.e. $\mathrm{Im}(u+\imath v)=\mathrm{Im}(\sum_{k=1}^{\infty}\beta_kz^k)$. The coefficient of z yields the unknown b_1^1, $\hat{b}_1^1$:

$$-\frac{a_1^0z}{2\mu}-\frac{a_1^1z}{2\mu}-\frac{a_1^2z}{\mu}-\frac{a_1^3z}{\mu}+\frac{\imath\hat{a}_1^0z}{2\mu}+\frac{\imath a_1^1z}{2\mu}+\frac{\imath\hat{a}_1^2z}{\mu}+\frac{\imath\hat{a}_1^3z}{\mu}+\frac{b_1^0z}{2\mu}+$$

$$+\frac{b_1^1z}{2\mu}+\frac{3b_1^2z}{\mu}+\frac{3b_1^3z}{\mu}+\frac{\imath\hat{b}_1^0z}{2\mu}+\frac{\imath\hat{b}_1^1z}{2\mu}+\frac{3\imath\hat{b}_1^2z}{\mu}+\frac{3\imath\hat{b}_1^3z}{\mu}+\frac{3a_1^0z}{2(\lambda+\mu)}+\frac{3a_1^1z}{2(\lambda+\mu)}+$$

$$+\frac{3a_1^2z}{\lambda+\mu}+\frac{3a_1^3z}{\lambda+\mu}+\frac{3\imath\hat{a}_1^0z}{2(\lambda+\mu)}+\frac{3\imath a_1^1z}{2(\lambda+\mu)}+\frac{3\imath\hat{a}_1^2z}{\lambda+\mu}+\frac{3\imath\hat{a}_1^3z}{\lambda+\mu}+\frac{a_1^0\lambda z}{2\mu(\lambda+\mu)}+\frac{a_1^1\lambda z}{2\mu(\lambda+\mu)}+$$

$$+\frac{a_1^2\lambda z}{\mu(\lambda+\mu)}+\frac{a_1^3\lambda z}{\mu(\lambda+\mu)}+\frac{\imath\hat{a}_1^0\lambda z}{2\mu(\lambda+\mu)}+\frac{\imath a_1^1\lambda z}{2\mu(\lambda+\mu)}+\frac{\imath\hat{a}_1^2\lambda z}{\mu(\lambda+\mu)}+\frac{\imath\hat{a}_1^3\lambda z}{\mu(\lambda+\mu)}+$$

$$+\frac{3a_1^2\lambda z}{2(\lambda+2\mu)^2}+\frac{21c_0^3\lambda^2z}{8(\lambda+2\mu)^2}-\frac{9c_2^3\lambda^2z}{2(\lambda+2\mu)^2}-\frac{15\imath\hat{c}_0^3\lambda^2z}{8(\lambda+2\mu)^2}-\frac{9\imath\hat{c}_2^3\lambda^2z}{2(\lambda+2\mu)^2}+$$

$$+\frac{a_1^2\lambda^2 z}{2\mu(\lambda+2\mu)^2}+\frac{3c_0^3\lambda^3 z}{8\mu(\lambda+2\mu)^2}-\frac{3c_2^3\lambda^3 z}{4\mu(\lambda+2\mu)^2}-\frac{3\imath\hat{c}_0^3\lambda^3 z}{8\mu(\lambda+2\mu)^2}-\frac{3\imath\hat{c}_2^3\lambda^3 z}{4\mu(\lambda+2\mu)^2}+$$

$$+\frac{a_1^2\mu z}{(\lambda+2\mu)^2}+\frac{21c_0^3\lambda\mu z}{4(\lambda+2\mu)^2}-\frac{33c_2^3\lambda\mu z}{4(\lambda+2\mu)^2}-\frac{3\imath\hat{c}_0^3\lambda\mu z}{(\lambda+2\mu)^2}-\frac{33\imath\hat{c}_2^3\lambda\mu z}{4(\lambda+2\mu)^2}+$$

$$+\frac{3c_0^3\mu^2 z}{(\lambda+2\mu)^2}-\frac{9c_2^3\mu^2 z}{2(\lambda+2\mu)^2}-\frac{3\imath\hat{c}_0^3\mu^2 z}{2(\lambda+2\mu)^2}-\frac{9\imath\hat{c}_2^3\mu^2 z}{2(\lambda+2\mu)^2}+\frac{3a_1^2 z}{4(\lambda+2\mu)}+\frac{15a_1^3 z}{4(\lambda+2\mu)}-$$

$$-\frac{3\imath\hat{a}_1^2 z}{4(\lambda+2\mu)}-\frac{9\imath\hat{a}_1^3 z}{4(\lambda+2\mu)}-\frac{9b_1^2 z}{8(\lambda+2\mu)}-\frac{27b_1^3 z}{8(\lambda+2\mu)}-\frac{9\imath\hat{b}_1^2 z}{8(\lambda+2\mu)}-\frac{27\imath\hat{b}_1^3 z}{8(\lambda+2\mu)}-$$

$$-\frac{c_0^1\lambda z}{4(\lambda+2\mu)}+\frac{3c_2^1\lambda z}{2(\lambda+2\mu)}-\frac{c_0^2\lambda z}{2(\lambda+2\mu)}+\frac{3c_2^2\lambda z}{\lambda+2\mu}-\frac{3c_0^3\lambda z}{2(\lambda+2\mu)}+\frac{18c_2^3\lambda z}{\lambda+2\mu}+$$

$$+\frac{3\imath\hat{c}_0^1\lambda z}{4(\lambda+2\mu)}+\frac{3\imath\hat{c}_2^1\lambda z}{2(\lambda+2\mu)}+\frac{3\imath\hat{c}_0^2\lambda z}{2(\lambda+2\mu)}+\frac{3\imath\hat{c}_2^2\lambda z}{\lambda+2\mu}+\frac{9\imath\hat{c}_0^3\lambda z}{2(\lambda+2\mu)}+\frac{18\imath\hat{c}_2^3\lambda z}{\lambda+2\mu}+$$

$$+\frac{a_1^2\lambda z}{4\mu(\lambda+2\mu)}+\frac{9a_1^3\lambda z}{4\mu(\lambda+2\mu)}-\frac{\imath\hat{a}_1^2\lambda z}{4\mu(\lambda+2\mu)}-\frac{3\imath\hat{a}_1^3\lambda z}{4\mu(\lambda+2\mu)}-\frac{3b_1^2\lambda z}{8\mu(\lambda+2\mu)}-$$

$$-\frac{9b_1^3\lambda z}{8\mu(\lambda+2\mu)}-\frac{3\imath\hat{b}_1^2\lambda z}{8\mu(\lambda+2\mu)}-\frac{9\imath\hat{b}_1^3\lambda z}{8\mu(\lambda+2\mu)}+\frac{3c_2^1\lambda^2 z}{8\mu(\lambda+2\mu)}+\frac{3c_2^2\lambda^2 z}{4\mu(\lambda+2\mu)}+$$

$$+\frac{9c_2^3\lambda^2 z}{2\mu(\lambda+2\mu)}+\frac{\imath\hat{c}_0^1\lambda^2 z}{4\mu(\lambda+2\mu)}+\frac{3\imath\hat{c}_2^1\lambda^2 z}{8\mu(\lambda+2\mu)}+\frac{\imath\hat{c}_0^2\lambda^2 z}{2\mu(\lambda+2\mu)}+\frac{3\imath\hat{c}_2^2\lambda^2 z}{4\mu(\lambda+2\mu)}+$$

$$+\frac{3\imath\hat{c}_0^3\lambda^2 z}{2\mu(\lambda+2\mu)}+\frac{9i\hat{c}_2^3\lambda^2 z}{2\mu(\lambda+2\mu)}-\frac{c_0^1\mu z}{4(\lambda+2\mu)}+\frac{9c_2^1\mu z}{8(\lambda+2\mu)}-\frac{c_0^2\mu z}{2(\lambda+2\mu)}+\frac{9c_2^2\mu z}{4(\lambda+2\mu)}-$$

$$-\frac{3c_0^3\mu z}{2(\lambda+2\mu)}+\frac{27c_2^3\mu z}{2(\lambda+2\mu)}+\frac{\imath\hat{c}_0^1\mu z}{2(\lambda+2\mu)}+\frac{9\imath\hat{c}_2^1\mu z}{8(\lambda+2\mu)}+\frac{\imath\hat{c}_0^2\mu z}{\lambda+2\mu}+\frac{9\imath\hat{c}_2^2\mu z}{4(\lambda+2\mu)}+$$

$$+\frac{3\imath\hat{c}_0^3\mu z}{\lambda+2\mu}+\frac{27\imath\hat{c}_2^3\mu z}{2(\lambda+2\mu)}-\frac{a_1^2\lambda z}{(\lambda+\mu)(\lambda+2\mu)}-\frac{3a_1^3\lambda z}{(\lambda+\mu)(\lambda+2\mu)}-\frac{\imath\hat{a}_1^2\lambda z}{(\lambda+\mu)(\lambda+2\mu)}-$$

$$-\frac{3\imath\hat{a}_1^3\lambda z}{(\lambda+\mu)(\lambda+2\mu)}-\frac{a_1^2\lambda^2 z}{4\mu(\lambda+\mu)(\lambda+2\mu)}-\frac{3a_1^3\lambda^2 z}{4\mu(\lambda+\mu)(\lambda+2\mu)}-$$

$$-\frac{\imath\hat{a}_1^2\lambda^2 z}{4\mu(\lambda+\mu)(\lambda+2\mu)}-\frac{3\imath\hat{a}_1^3\lambda^2 z}{4\mu(\lambda+\mu)(\lambda+2\mu)}-\frac{5a_1^2\mu z}{4(\lambda+\mu)(\lambda+2\mu)}-$$

$$-\frac{15a_1^3\mu z}{4(\lambda+\mu)(\lambda+2\mu)}-\frac{5\imath\hat{a}_1^2\mu z}{4(\lambda+\mu)(\lambda+2\mu)}-\frac{15i\hat{a}_1^3\mu z}{4(\lambda+\mu)(\lambda+2\mu)}=\beta_1 z.$$

Similarly the coefficient of z^2 provides us with b_2^1, $\hat{b}_2^1$:

$$\frac{b_2^0 z^2}{2\mu}+\frac{b_2^1 z^2}{2\mu}+\frac{6b_2^2 z^2}{\mu}+\frac{6b_2^3 z^2}{\mu}+\frac{\imath\hat{b}_2^0 z^2}{2\mu}+\frac{\imath\hat{b}_2^1 z^2}{2\mu}+\frac{6\imath\hat{b}_2^2 z^2}{\mu}+\frac{6\imath\hat{b}_2^3 z^2}{\mu}+$$

$$+\frac{3a_2^0 z^2}{2(\lambda+\mu)}+\frac{3a_2^1 z^2}{2(\lambda+\mu)}+\frac{9a_2^2 z^2}{2(\lambda+\mu)}+\frac{9a_2^3 z^2}{2(\lambda+\mu)}+\frac{3\imath\hat{a}_2^0 z^2}{2(\lambda+\mu)}+\frac{3\imath\hat{a}_2^1 z^2}{2(\lambda+\mu)}+$$

$$+\frac{9\imath\hat{a}_2^2 z^2}{2(\lambda+\mu)}+\frac{9\imath\hat{a}_2^3 z^2}{2(\lambda+\mu)}+\frac{a_2^0\lambda z^2}{2\mu(\lambda+\mu)}+\frac{a_2^1\lambda z^2}{2\mu(\lambda+\mu)}+\frac{3a_2^2\lambda z^2}{2\mu(\lambda+\mu)}+$$

$$+\frac{3a_2^3\lambda z^2}{2\mu(\lambda+\mu)}+\frac{\imath\hat{a}_2^0\lambda z^2}{2\mu(\lambda+\mu)}+\frac{\imath\hat{a}_2^1\lambda z^2}{2\mu(\lambda+\mu)}+\frac{3\imath\hat{a}_2^2\lambda z^2}{2\mu(\lambda+\mu)}+\frac{3\imath\hat{a}_2^3\lambda z^2}{2\mu(\lambda+\mu)}+$$

$$+\frac{3a_0^2\lambda z^2}{8(\lambda+2\mu)^2}+\frac{3a_2^2\lambda z^2}{4(\lambda+2\mu)^2}-\frac{3\imath\hat{a}_0^2\lambda z^2}{8(\lambda+2\mu)^2}+\frac{3\imath\hat{a}_2^2\lambda z^2}{4(\lambda+2\mu)^2}+\frac{3c_1^3\lambda^2 z^2}{8(\lambda+2\mu)^2}+$$

$$+\frac{3\imath\hat{c}_1^3\lambda^2 z^2}{8(\lambda+2\mu)^2}+\frac{a_0^2\lambda^2 z^2}{8\mu(\lambda+2\mu)^2}+\frac{a_2^2\lambda^2 z^2}{4\mu(\lambda+2\mu)^2}-\frac{\imath\hat{a}_0^2\lambda^2 z^2}{8\mu(\lambda+2\mu)^2}+\frac{i\hat{a}_2^2\lambda^2 z^2}{4\mu(\lambda+2\mu)^2}+$$

$$+\frac{a_0^2\mu z^2}{4(\lambda+2\mu)^2}+\frac{a_2^2\mu z^2}{2(\lambda+2\mu)^2}-\frac{\imath\hat{a}_0^2\mu z^2}{4(\lambda+2\mu)^2}+\frac{\imath\hat{a}_2^2\mu z^2}{2(\lambda+2\mu)^2}+\frac{9c_1^3\lambda\mu z^2}{8(\lambda+2\mu)^2}+$$

$$+\frac{9\imath\hat{c}_1^3\lambda\mu z^2}{8(\lambda+2\mu)^2}+\frac{3c_1^3\mu^2 z^2}{4(\lambda+2\mu)^2}+\frac{3\imath\hat{c}_1^3\mu^2 z^2}{4(\lambda+2\mu)^2}+\frac{3a_0^2 z^2}{8(\lambda+2\mu)}+\frac{3a_0^3 z^2}{2(\lambda+2\mu)}+\frac{3a_2^3 z^2}{4(\lambda+2\mu)}-$$

$$-\frac{3\imath\hat{a}_0^2 z^2}{8(\lambda+2\mu)}-\frac{3\imath\hat{a}_0^3 z^2}{2(\lambda+2\mu)}+\frac{3\imath\hat{a}_2^3 z^2}{4(\lambda+2\mu)}-\frac{b_0^2 z^2}{8(\lambda+2\mu)}-\frac{3b_2^2 z^2}{2(\lambda+2\mu)}-\frac{3b_0^3 z^2}{8(\lambda+2\mu)}-$$

$$-\frac{9b_2^3 z^2}{2(\lambda+2\mu)}-\frac{\imath\hat{b}_0^2 z^2}{8(\lambda+2\mu)}-\frac{3\imath\hat{b}_2^2 z^2}{2(\lambda+2\mu)}-\frac{3\imath\hat{b}_0^3 z^2}{8(\lambda+2\mu)}-\frac{9\imath\hat{b}_2^3 z^2}{2(\lambda+2\mu)}+\frac{c_1^1\lambda z^2}{4(\lambda+2\mu)}+$$

$$+\frac{c_1^2\lambda z^2}{2(\lambda+2\mu)}+\frac{9c_1^3\lambda z^2}{4(\lambda+2\mu)}+\frac{\imath\hat{c}_1^1\lambda z^2}{4(\lambda+2\mu)}+\frac{\imath\hat{c}_1^2\lambda z^2}{2(\lambda+2\mu)}+\frac{9\imath\hat{c}_1^3\lambda z^2}{4(\lambda+2\mu)}+\frac{a_0^2\lambda z^2}{8\mu(\lambda+2\mu)}+$$

$$+\frac{3a_0^3\lambda z^2}{4\mu(\lambda+2\mu)}+\frac{3a_2^3\lambda z^2}{4\mu(\lambda+2\mu)}-\frac{\imath\hat{a}_0^2\lambda z^2}{8\mu(\lambda+2\mu)}-\frac{3\imath\hat{a}_0^3\lambda z^2}{4\mu(\lambda+2\mu)}+\frac{3\imath\hat{a}_2^3\lambda z^2}{4\mu(\lambda+2\mu)}-$$

$$-\frac{b_0^2\lambda z^2}{8\mu(\lambda+2\mu)}-\frac{b_2^2\lambda z^2}{2\mu(\lambda+2\mu)}-\frac{3b_0^3\lambda z^2}{8\mu(\lambda+2\mu)}-\frac{3b_2^3\lambda z^2}{2\mu(\lambda+2\mu)}-\frac{\imath\hat{b}_0^2\lambda z^2}{8\mu(\lambda+2\mu)}-$$

$$-\frac{\imath\hat{b}_2^2\lambda z^2}{2\mu(\lambda+2\mu)}-\frac{3\imath\hat{b}_0^3\lambda z^2}{8\mu(\lambda+2\mu)}-\frac{3\imath\hat{b}_2^3\lambda z^2}{2\mu(\lambda+2\mu)}+\frac{c_1^1\lambda^2 z^2}{8\mu(\lambda+2\mu)}+\frac{c_1^2\lambda^2 z^2}{4\mu(\lambda+2\mu)}+$$

$$+\frac{9c_1^3\lambda^2z^2}{8\mu(\lambda+2\mu)}+\frac{\imath\hat{c}_1^1\lambda^2z^2}{8\mu(\lambda+2\mu)}+\frac{\imath\hat{c}_1^2\lambda^2z^2}{4\mu(\lambda+2\mu)}+$$

$$+\frac{9\imath\hat{c}_1^3\lambda^2z^2}{8\mu(\lambda+2\mu)}+\frac{c_1^1\mu z^2}{8(\lambda+2\mu)}+\frac{c_1^2\mu z^2}{4(\lambda+2\mu)}+$$

$$+\frac{9c_1^3\mu z^2}{8(\lambda+2\mu)}+\frac{\imath\hat{c}_1^1\mu z^2}{8(\lambda+2\mu)}+\frac{\imath\hat{c}_1^2\mu z^2}{4(\lambda+2\mu)}+\frac{9\imath\hat{c}_1^3\mu z^2}{8(\lambda+2\mu)}-\frac{a_2^2\lambda z^2}{(\lambda+\mu)(\lambda+2\mu)}-$$

$$-\frac{3a_2^3\lambda z^2}{(\lambda+\mu)(\lambda+2\mu)}-\frac{\imath\hat{a}_2^2\lambda z^2}{(\lambda+\mu)(\lambda+2\mu)}-\frac{3\imath\hat{a}_2^3\lambda z^2}{(\lambda+\mu)(\lambda+2\mu)}-\frac{a_2^2\lambda^2z^2}{4\mu(\lambda+\mu)(\lambda+2\mu)}-$$

$$-\frac{3a_2^3\lambda^2z^2}{4\mu(\lambda+\mu)(\lambda+2\mu)}-\frac{\imath\hat{a}_2^2\lambda^2z^2}{4\mu(\lambda+\mu)(\lambda+2\mu)}-\frac{3\imath\hat{a}_2^3\lambda^2z^2}{4\mu(\lambda+\mu)(\lambda+2\mu)}-$$

$$-\frac{5a_2^2\mu z^2}{4(\lambda+\mu)(\lambda+2\mu)}-\frac{15a_2^3\mu z^2}{4(\lambda+\mu)(\lambda+2\mu)}-\frac{5\imath\hat{a}_2^2\mu z^2}{4(\lambda+\mu)(\lambda+2\mu)}-$$

$$-\frac{15\imath\hat{a}_2^3\mu z^2}{4(\lambda+\mu)(\lambda+2\mu)}=\beta_2z^2.$$

The coefficient of z^n gives the relation on b_n^1, $\hat{b}_n^1$ for $n=3,4,\ldots$:

$$\frac{3a_n^2\lambda z^n}{4(\lambda+2\mu)^2}+\frac{3\imath\hat{a}_n^2\lambda z^n}{4(\lambda+2\mu)^2}+\frac{a_n^2\lambda^2z^n}{4\mu(\lambda+2\mu)^2}+\frac{\imath\hat{a}_n^2\lambda^2z^n}{4\mu(\lambda+2\mu)^2}+\frac{a_n^2\mu z^n}{2(\lambda+2\mu)^2}+$$

$$+\frac{\imath\hat{a}_n^2\mu z^n}{2(\lambda+2\mu)^2}+\frac{3a_n^3z^n}{4(\lambda+2\mu)}+\frac{3\imath\hat{a}_n^3z^n}{4(\lambda+2\mu)}+\frac{3a_n^3\lambda z^n}{4\mu(\lambda+2\mu)}+\frac{3\imath\hat{a}_n^3\lambda z^n}{4\mu(\lambda+2\mu)}-$$

$$-\frac{a_n^2\lambda z^n}{(\lambda+\mu)(\lambda+2\mu)}-\frac{3a_n^3\lambda z^n}{(\lambda+\mu)(\lambda+2\mu)}-\frac{\imath\hat{a}_n^2\lambda z^n}{(\lambda+\mu)(\lambda+2\mu)}-\frac{3\imath\hat{a}_n^3\lambda z^n}{(\lambda+\mu)(\lambda+2\mu)}-$$

$$-\frac{a_n^2\lambda^2z^n}{4\mu(\lambda+\mu)(\lambda+2\mu)}-\frac{3a_n^3\lambda^2z^n}{4\mu(\lambda+\mu)(\lambda+2\mu)}-\frac{\imath\hat{a}_n^2\lambda^2z^n}{4\mu(\lambda+\mu)(\lambda+2\mu)}-$$

$$-\frac{3\imath\hat{a}_n^3\lambda^2z^n}{4\mu(\lambda+\mu)(\lambda+2\mu)}-\frac{5a_n^2\mu z^n}{4(\lambda+\mu)(\lambda+2\mu)}-\frac{15a_n^3\mu z^n}{4(\lambda+\mu)(\lambda+2\mu)}-$$

$$-\frac{5\imath\hat{a}_n^2\mu z^n}{4(\lambda+\mu)(\lambda+2\mu)}-\frac{15\imath\hat{a}_n^3\mu z^n}{4(\lambda+\mu)(\lambda+2\mu)}+\frac{3a_n^2(n+1)z^n}{2(\lambda+\mu)}+\frac{3a_n^3(n+1)z^n}{2(\lambda+\mu)}+$$

$$+\frac{3\imath\hat{a}_n^2(n+1)z^n}{2(\lambda+\mu)}+\frac{3\imath\hat{a}_n^3(n+1)z^n}{2(\lambda+\mu)}+\frac{a_n^2\lambda(n+1)z^n}{2\mu(\lambda+\mu)}+\frac{a_n^3\lambda(n+1)z^n}{2\mu(\lambda+\mu)}+$$

$$+\frac{\imath\hat{a}_n^2\lambda(n+1)z^n}{2\mu(\lambda+\mu)}+\frac{\imath\hat{a}_n^3\lambda(n+1)z^n}{2\mu(\lambda+\mu)}-\frac{a_{n+2}^2(n+3)z^n}{2\mu}-\frac{a_{n+2}^3(n+3)z^n}{2\mu}-$$

$$-\frac{\imath\hat{a}_{n+2}^2(n+3)z^n}{2\mu}-\frac{\imath\hat{a}_{n+2}^3(n+3)z^n}{2\mu}-\frac{3a_{n+2}^2\lambda(n+3)z^n}{8(\lambda+2\mu)^2}-\frac{3\imath\hat{a}_{n+2}^2\lambda(n+3)z^n}{8(\lambda+2\mu)^2}-$$

$$-\frac{a_{n+2}^2\lambda^2(n+3)z^n}{8\mu(\lambda+2\mu)^2}-\frac{\imath\hat{a}_{n+2}^2\lambda^2(n+3)z^n}{8\mu(\lambda+2\mu)^2}-\frac{a_{n+2}^2\mu(n+3)z^n}{4(\lambda+2\mu)^2}-\frac{\imath\hat{a}_{n+2}^2\mu(n+3)z^n}{4(\lambda+2\mu)^2}-$$

$$-\frac{3a_{n+2}^2(n+3)z^n}{8(\lambda+2\mu)}-\frac{3a_{n+2}^3(n+3)z^n}{2(\lambda+2\mu)}-\frac{3\imath\hat{a}_{n+2}^2(n+3)z^n}{8(\lambda+2\mu)}-\frac{3\imath\hat{a}_{n+2}^3(n+3)z^n}{2(\lambda+2\mu)}-$$

$$-\frac{a_{n+2}^2\lambda(n+3)z^n}{8\mu(\lambda+2\mu)}-\frac{3a_{n+2}^3\lambda(n+3)z^n}{4\mu(\lambda+2\mu)}-\frac{\imath\hat{a}_{n+2}^2\lambda(n+3)z^n}{8\mu(\lambda+2\mu)}-\frac{3\imath\hat{a}_{n+2}^3\lambda(n+3)z^n}{4\mu(\lambda+2\mu)}+$$

$$+\frac{a_{n+2}^2(n+3)^2z^n}{2\mu}+\frac{a_{n+2}^3(n+3)^2z^n}{2\mu}+\frac{i\hat{a}_{n+2}^2(n+3)^2z^n}{2\mu}+\frac{\imath\hat{a}_{n+2}^3(n+3)^2z^n}{2\mu}+$$

$$+\frac{c_{n+1}^1\lambda(n+2)z^n}{2(\lambda+2\mu)}+\frac{c_{n+1}^2\lambda(n+2)z^n}{\lambda+2\mu}+\frac{\imath\hat{c}_{n+1}^1\lambda(n+2)z^n}{2(\lambda+2\mu)}+\frac{\imath\hat{c}_{n+1}^2\lambda(n+2)z^n}{\lambda+2\mu}+$$

$$+\frac{c_{n+1}^1\lambda^2(n+2)z^n}{8\mu(\lambda+2\mu)}+\frac{c_{n+1}^2\lambda^2(n+2)z^n}{4\mu(\lambda+2\mu)}+\frac{\imath\hat{c}_{n+1}^1\lambda^2(n+2)z^n}{8\mu(\lambda+2\mu)}+\frac{\imath\hat{c}_{n+1}^2\lambda^2(n+2)z^n}{4\mu(\lambda+2\mu)}+$$

$$+\frac{3c_{n+1}^1\mu(n+2)z^n}{8(\lambda+2\mu)}+\frac{3c_{n+1}^2\mu(n+2)z^n}{4(\lambda+2\mu)}+\frac{3\imath\hat{c}_{n+1}^1\mu(n+2)z^n}{8(\lambda+2\mu)}+$$

$$+\frac{3\imath\hat{c}_{n+1}^2\mu(n+2)z^n}{4(\lambda+2\mu)}+\frac{c_{n-1}^1\lambda z^n}{4(\lambda+2\mu)}+\frac{c_{n-1}^2\lambda z^n}{2(\lambda+2\mu)}+\frac{\imath\hat{c}_{n-1}^1\lambda z^n}{4(\lambda+2\mu)}+\frac{\imath\hat{c}_{n-1}^2\lambda z^n}{2(\lambda+2\mu)}+$$

$$+\frac{c_{n-1}^1\lambda^2z^n}{8\mu(\lambda+2\mu)}+\frac{c_{n-1}^2\lambda^2z^n}{4\mu(\lambda+2\mu)}+\frac{\imath\hat{c}_{n-1}^1\lambda^2z^n}{8\mu(\lambda+2\mu)}+\frac{\imath\hat{c}_{n-1}^2\lambda^2z^n}{4\mu(\lambda+2\mu)}+\frac{c_{n-1}^1\mu z^n}{8(\lambda+2\mu)}+$$

$$+\frac{c_{n-1}^2\mu z^n}{4(\lambda+2\mu)}+\frac{\imath\hat{c}_{n-1}^1\mu z^n}{8(\lambda+2\mu)}+\frac{\imath\hat{c}_{n-1}^2\mu z^n}{4(\lambda+2\mu)}+\frac{a_{n+2}^0(n+2)z^n}{2\mu}+\frac{a_{n+2}^1(n+2)z^n}{2\mu}+$$

$$+\frac{\imath\hat{a}_{n+2}^0(n+2)z^n}{2\mu}+\frac{i\hat{a}_{n+2}^1(n+2)z^n}{2\mu}+\frac{b_n^0z^n}{2\mu}+\frac{b_n^1z^n}{2\mu}+\frac{\imath\hat{b}_n^0z^n}{2\mu}+\frac{i\hat{b}_n^1z^n}{2\mu}+$$

$$+\frac{3a_n^0z^n}{2(\lambda+\mu)}+\frac{3a_n^1z^n}{2(\lambda+\mu)}+\frac{3\imath\hat{a}_n^0z^n}{2(\lambda+\mu)}+\frac{3\imath\hat{a}_n^1z^n}{2(\lambda+\mu)}+\frac{a_n^0\lambda z^n}{2\mu(\lambda+\mu)}+\frac{a_n^1\lambda z^n}{2\mu(\lambda+\mu)}+$$

$$+\frac{\imath\hat{a}_n^0\lambda z^n}{2\mu(\lambda+\mu)}+\frac{\imath\hat{a}_n^1\lambda z^n}{2\mu(\lambda+\mu)}-\frac{9c_{n+1}^3\lambda^2(n+3)z^n}{8(\lambda+2\mu)^2}-\frac{9\imath\hat{c}_{n+1}^3\lambda^2(n+3)z^n}{8(\lambda+2\mu)^2}-$$

$$
\begin{aligned}
&-\frac{3c_{n+1}^{3}\lambda^{3}(n+3)z^{n}}{16\mu(\lambda+2\mu)^{2}}-\frac{3\imath\hat{c}_{n+1}^{3}\lambda^{3}(n+3)z^{n}}{16\mu(\lambda+2\mu)^{2}}-\frac{33c_{n+1}^{3}\lambda\mu(n+3)z^{n}}{16(\lambda+2\mu)^{2}}-\\
&-\frac{33\imath\hat{c}_{n+1}^{3}\lambda\mu(n+3)z^{n}}{16(\lambda+2\mu)^{2}}-\frac{9c_{n+1}^{3}\mu^{2}(n+3)z^{n}}{8(\lambda+2\mu)^{2}}-\frac{9\imath\hat{c}_{n+1}^{3}\mu^{2}(n+3)z^{n}}{8(\lambda+2\mu)^{2}}-\\
&-\frac{3c_{n+1}^{3}\lambda(n+3)z^{n}}{2(\lambda+2\mu)}-\frac{3\imath\hat{c}_{n+1}^{3}\lambda(n+3)z^{n}}{2(\lambda+2\mu)}-\frac{3c_{n+1}^{3}\lambda^{2}(n+3)z^{n}}{8\mu(\lambda+2\mu)}-\\
&-\frac{3\imath\hat{c}_{n+1}^{3}\lambda^{2}(n+3)z^{n}}{8\mu(\lambda+2\mu)}-\frac{9c_{n+1}^{3}\mu(n+3)z^{n}}{8(\lambda+2\mu)}-\frac{9\imath\hat{c}_{n+1}^{3}\mu(n+3)z^{n}}{8(\lambda+2\mu)}+\\
&+\frac{3c_{n+1}^{3}\lambda(n+3)^{2}z^{n}}{2(\lambda+2\mu)}+\frac{3\imath\hat{c}_{n+1}^{3}\lambda(n+3)^{2}z^{n}}{2(\lambda+2\mu)}+\frac{3c_{n+1}^{3}\lambda^{2}(n+3)^{2}z^{n}}{8\mu(\lambda+2\mu)}+\\
&+\frac{3\imath\hat{c}_{n+1}^{3}\lambda^{2}(n+3)^{2}z^{n}}{8\mu(\lambda+2\mu)}+\frac{9c_{n+1}^{3}\mu(n+3)^{2}z^{n}}{8(\lambda+2\mu)}+\frac{9\imath\hat{c}_{n+1}^{3}\mu(n+3)^{2}z^{n}}{8(\lambda+2\mu)}-\frac{b_{n}^{2}(n+2)z^{n}}{2\mu}-\\
&-\frac{b_{n}^{3}(n+2)z^{n}}{2\mu}-\frac{\imath\hat{b}_{n}^{2}(n+2)z^{n}}{2\mu}-\frac{\imath\hat{b}_{n}^{3}(n+2)z^{n}}{2\mu}-\frac{3b_{n}^{2}(n+2)z^{n}}{8(\lambda+2\mu)}-\frac{9b_{n}^{3}(n+2)z^{n}}{8(\lambda+2\mu)}-\\
&-\frac{3\imath\hat{b}_{n}^{2}(n+2)z^{n}}{8(\lambda+2\mu)}-\frac{9\imath\hat{b}_{n}^{3}(n+2)z^{n}}{8(\lambda+2\mu)}-\frac{b_{n}^{2}\lambda(n+2)z^{n}}{8\mu(\lambda+2\mu)}-\frac{3b_{n}^{3}\lambda(n+2)z^{n}}{8\mu(\lambda+2\mu)}-\\
&-\frac{\imath\hat{b}_{n}^{2}\lambda(n+2)z^{n}}{8\mu(\lambda+2\mu)}-\frac{3\imath\hat{b}_{n}^{3}\lambda(n+2)z^{n}}{8\mu(\lambda+2\mu)}+\frac{b_{n}^{2}(n+2)^{2}z^{n}}{2\mu}+\frac{b_{n}^{3}(n+2)^{2}z^{n}}{2\mu}+\\
&+\frac{\imath\hat{b}_{n}^{2}(n+2)^{2}z^{n}}{2\mu}+\frac{\imath\hat{b}_{n}^{3}(n+2)^{2}z^{n}}{2\mu}+\frac{3c_{n-1}^{3}\lambda^{2}z^{n}}{8(\lambda+2\mu)^{2}}+\frac{3\imath\hat{c}_{n-1}^{3}\lambda^{2}z^{n}}{8(\lambda+2\mu)^{2}}+\frac{9c_{n-1}^{3}\lambda\mu z^{n}}{8(\lambda+2\mu)^{2}}+\\
&+\frac{9\imath\hat{c}_{n-1}^{3}\lambda\mu z^{n}}{8(\lambda+2\mu)^{2}}+\frac{3c_{n-1}^{3}\mu^{2}z^{n}}{4(\lambda+2\mu)^{2}}+\frac{3\imath\hat{c}_{n-1}^{3}\mu^{2}z^{n}}{4(\lambda+2\mu)^{2}}+\frac{3c_{n-1}^{3}\lambda(n+1)z^{n}}{4(\lambda+2\mu)}+\\
&+\frac{3\imath\hat{c}_{n-1}^{3}\lambda(n+1)z^{n}}{4(\lambda+2\mu)}+\frac{3c_{n-1}^{3}\lambda^{2}(n+1)z^{n}}{8\mu(\lambda+2\mu)}+\frac{3\imath\hat{c}_{n-1}^{3}\lambda^{2}(n+1)z^{n}}{8\mu(\lambda+2\mu)}+\\
&+\frac{3c_{n-1}^{3}\mu(n+1)z^{n}}{8(\lambda+2\mu)}+\frac{3\imath\hat{c}_{n-1}^{3}\mu(n+1)z^{n}}{8(\lambda+2\mu)}-\frac{b_{n-2}^{2}z^{n}}{8(\lambda+2\mu)}-\frac{3b_{n-2}^{3}z^{n}}{8(\lambda+2\mu)}-\frac{\imath\hat{b}_{n-2}^{2}z^{n}}{8(\lambda+2\mu)}-\\
&-\frac{3\imath\hat{b}_{n-2}^{3}z^{n}}{8(\lambda+2\mu)}-\frac{b_{n-2}^{2}\lambda z^{n}}{8\mu(\lambda+2\mu)}-\frac{3b_{n-2}^{3}\lambda z^{n}}{8\mu(\lambda+2\mu)}-\frac{i\hat{b}_{n-2}^{2}\lambda z^{n}}{8\mu(\lambda+2\mu)}-\frac{3\imath\hat{b}_{n-2}^{3}\lambda z^{n}}{8\mu(\lambda+2\mu)}=\beta_{n}z^{n}.
\end{aligned}
$$

Let us now consider the relations on w on the second level circle. $w = \mathrm{Re}(\sum\limits_{k=0}^{\infty}\gamma_k z^k)$.

The constant summand gives us the relation on c_0^1:

$$c_0^0 + c_0^1 - 4c_0^3 + \frac{3a_1^3\lambda}{8\mu^2} - \frac{9a_1^3}{4\mu} - \frac{c_0^2\lambda}{2\mu} - \frac{3c_0^3\lambda}{\mu} - \frac{a_1^1}{2(\lambda+\mu)} - \frac{2a_1^2}{\lambda+\mu} - \frac{a_1^1\lambda}{2\mu(\lambda+\mu)} -$$

$$- \frac{2a_1^2\lambda}{\mu(\lambda+\mu)} + \frac{3c_0^3\lambda}{\lambda+2\mu} + \frac{3c_0^3\lambda^2}{2\mu(\lambda+2\mu)} + \frac{3c_0^3\mu}{2(\lambda+2\mu)} - \frac{3a_1^3\lambda}{4(\lambda+\mu)(\lambda+2\mu)} +$$

$$+ \frac{3c_0^2\lambda^2}{4(\lambda+\mu)(\lambda+2\mu)} - \frac{3a_1^3\lambda^3}{8\mu^2(\lambda+\mu)(\lambda+2\mu)} - \frac{9a_1^3\lambda^2}{8\mu(\lambda+\mu)(\lambda+2\mu)} +$$

$$+ \frac{c_0^2\lambda^3}{4\mu(\lambda+\mu)(\lambda+2\mu)} + \frac{3c_0^2\lambda\mu}{4(\lambda+\mu)(\lambda+2\mu)} + \frac{c_0^2\mu^2}{4(\lambda+\mu)(\lambda+2\mu)} = \gamma_0.$$

The term with z yields c_1^1 and $\hat{c}_1^1$ via the following equation:

$$c_1^0 z + 2c_1^1 z + c_1^2 z - 3c_1^3 z + \imath\hat{c}_1^0 z + 2\imath\hat{c}_1^1 z + \imath\hat{c}_1^2 z - 3\imath\hat{c}_1^3 z + \frac{3a_0^3\lambda z}{8\mu^2} + \frac{3a_2^3\lambda z}{8\mu^2} - \frac{3\imath\hat{a}_0^3\lambda z}{8\mu^2} +$$

$$+ \frac{3\imath\hat{a}_2^3\lambda z}{8\mu^2} - \frac{3a_0^3 z}{4\mu} - \frac{15a_2^3 z}{4\mu} + \frac{3\imath\hat{a}_0^3 z}{4\mu} - \frac{15\imath\hat{a}_2^3 z}{4\mu} - \frac{c_1^2\lambda z}{2\mu} - \frac{9c_1^3\lambda z}{2\mu} - \frac{i\hat{c}_1^2\lambda z}{2\mu} - \frac{9\imath\hat{c}_1^3\lambda z}{2\mu} -$$

$$- \frac{a_0^1 z}{2(\lambda+\mu)} - \frac{a_2^1 z}{2(\lambda+\mu)} - \frac{a_0^2 z}{\lambda+\mu} - \frac{3a_2^2 z}{\lambda+\mu} + \frac{\imath\hat{a}_0^1 z}{2(\lambda+\mu)} - \frac{\imath\hat{a}_2^1 z}{2(\lambda+\mu)} + \frac{\imath\hat{a}_0^2 z}{\lambda+\mu} - \frac{3\imath\hat{a}_2^2 z}{\lambda+\mu} -$$

$$- \frac{a_0^1\lambda z}{2\mu(\lambda+\mu)} - \frac{a_2^1\lambda z}{2\mu(\lambda+\mu)} - \frac{a_0^2\lambda z}{\mu(\lambda+\mu)} - \frac{3a_2^2\lambda z}{\mu(\lambda+\mu)} + \frac{\imath\hat{a}_0^1\lambda z}{2\mu(\lambda+\mu)} - \frac{\imath\hat{a}_2^1\lambda z}{2\mu(\lambda+\mu)} +$$

$$+ \frac{\imath\hat{a}_0^2\lambda z}{\mu(\lambda+\mu)} - \frac{3\imath\hat{a}_2^2\lambda z}{\mu(\lambda+\mu)} + \frac{9c_1^3\lambda z}{2(\lambda+2\mu)} + \frac{9\imath\hat{c}_1^3\lambda z}{2(\lambda+2\mu)} + \frac{9c_1^3\lambda^2 z}{4\mu(\lambda+2\mu)} + \frac{9\imath\hat{c}_1^3\lambda^2 z}{4\mu(\lambda+2\mu)} +$$

$$+ \frac{9c_1^3\mu z}{4(\lambda+2\mu)} + \frac{9i\hat{c}_1^3\mu z}{4(\lambda+2\mu)} - \frac{3a_0^3\lambda z}{4(\lambda+\mu)(\lambda+2\mu)} - \frac{3a_2^3\lambda z}{4(\lambda+\mu)(\lambda+2\mu)} +$$

$$+ \frac{3\imath\hat{a}_0^3\lambda z}{4(\lambda+\mu)(\lambda+2\mu)} - \frac{3\imath\hat{a}_2^3\lambda z}{4(\lambda+\mu)(\lambda+2\mu)} - \frac{3b_0^3\lambda z}{4(\lambda+\mu)(\lambda+2\mu)} -$$

$$- \frac{3\imath\hat{b}_0^3\lambda z}{4(\lambda+\mu)(\lambda+2\mu)} + \frac{3c_1^2\lambda^2 z}{4(\lambda+\mu)(\lambda+2\mu)} + \frac{3\imath\hat{c}_1^2\lambda^2 z}{4(\lambda+\mu)(\lambda+2\mu)} -$$

$$- \frac{3a_0^3\lambda^3 z}{8\mu^2(\lambda+\mu)(\lambda+2\mu)} - \frac{3a_2^3\lambda^3 z}{8\mu^2(\lambda+\mu)(\lambda+2\mu)} + \frac{3\imath\hat{a}_0^3\lambda^3 z}{8\mu^2(\lambda+\mu)(\lambda+2\mu)} -$$

$$- \frac{3\imath\hat{a}_2^3\lambda^3 z}{8\mu^2(\lambda+\mu)(\lambda+2\mu)} - \frac{9a_0^3\lambda^2 z}{8\mu(\lambda+\mu)(\lambda+2\mu)} - \frac{9a_2^3\lambda^2 z}{8\mu(\lambda+\mu)(\lambda+2\mu)} +$$

$$+\frac{9\imath\hat{a}_0^3\lambda^2 z}{8\mu(\lambda+\mu)(\lambda+2\mu)}-\frac{9\imath\hat{a}_2^3\lambda^2 z}{8\mu(\lambda+\mu)(\lambda+2\mu)}-\frac{3b_0^3\lambda^2 z}{8\mu(\lambda+\mu)(\lambda+2\mu)}-$$

$$-\frac{3\imath\hat{b}_0^3\lambda^2 z}{8\mu(\lambda+\mu)(\lambda+2\mu)}+\frac{c_1^2\lambda^3 z}{4\mu(\lambda+\mu)(\lambda+2\mu)}+\frac{\imath\hat{c}_1^2\lambda^3 z}{4\mu(\lambda+\mu)(\lambda+2\mu)}-$$

$$-\frac{3b_0^3\mu z}{8(\lambda+\mu)(\lambda+2\mu)}-\frac{3\imath\hat{b}_0^3\mu z}{8(\lambda+\mu)(\lambda+2\mu)}+\frac{3c_1^2\lambda\mu z}{4(\lambda+\mu)(\lambda+2\mu)}+$$

$$+\frac{3\imath\hat{c}_1^2\lambda\mu z}{4(\lambda+\mu)(\lambda+2\mu)}+\frac{c_1^2\mu^2 z}{4(\lambda+\mu)(\lambda+2\mu)}+\frac{\imath\hat{c}_1^2\mu^2 z}{4(\lambda+\mu)(\lambda+2\mu)}=\gamma_1 z.$$

The term with z^2 lets us to determine c_2^1, $\hat{c}_2^1$:

$$c_2^0 z^2+3c_2^1 z^2+2c_2^2 z^2+\imath\hat{c}_2^0 z^2+3\imath\hat{c}_2^1 z^2+2\imath\hat{c}_2^2 z^2-\frac{c_2^2\lambda z^2}{2\mu}-\frac{6c_2^3\lambda z^2}{\mu}-\frac{\imath\hat{c}_2^2\lambda z^2}{2\mu}-\frac{6\imath\hat{c}_2^3\lambda z^2}{\mu}+$$

$$+\frac{6c_2^3\lambda z^2}{\lambda+2\mu}+\frac{6\imath\hat{c}_2^3\lambda z^2}{\lambda+2\mu}+\frac{3c_2^3\lambda^2 z^2}{\mu(\lambda+2\mu)}+\frac{3\imath\hat{c}_2^3\lambda^2 z^2}{\mu(\lambda+2\mu)}+\frac{3c_2^3\mu z^2}{\lambda+2\mu}+\frac{3\imath\hat{c}_2^3\mu z^2}{\lambda+2\mu}-$$

$$-\frac{3b_1^3\lambda z^2}{4(\lambda+\mu)(\lambda+2\mu)}-\frac{3\imath\hat{b}_1^3\lambda z^2}{4(\lambda+\mu)(\lambda+2\mu)}+\frac{3c_2^2\lambda^2 z^2}{4(\lambda+\mu)(\lambda+2\mu)}+\frac{3\imath\hat{c}_2^2\lambda^2 z^2}{4(\lambda+\mu)(\lambda+2\mu)}-$$

$$-\frac{3b_1^3\lambda^2 z^2}{8\mu(\lambda+\mu)(\lambda+2\mu)}-\frac{3\imath\hat{b}_1^3\lambda^2 z^2}{8\mu(\lambda+\mu)(\lambda+2\mu)}+\frac{c_2^2\lambda^3 z^2}{4\mu(\lambda+\mu)(\lambda+2\mu)}+$$

$$+\frac{\imath\hat{c}_2^2\lambda^3 z^2}{4\mu(\lambda+\mu)(\lambda+2\mu)}-\frac{3b_1^3\mu z^2}{8(\lambda+\mu)(\lambda+2\mu)}-\frac{3\imath\hat{b}_1^3\mu z^2}{8(\lambda+\mu)(\lambda+2\mu)}+$$

$$+\frac{3c_2^2\lambda\mu z^2}{4(\lambda+\mu)(\lambda+2\mu)}+\frac{3\imath\hat{c}_2^2\lambda\mu z^2}{4(\lambda+\mu)(\lambda+2\mu)}+\frac{c_2^2\mu^2 z^2}{4(\lambda+\mu)(\lambda+2\mu)}+$$

$$+\frac{\imath\hat{c}_2^2\mu^2 z^2}{4(\lambda+\mu)(\lambda+2\mu)}=\gamma_2 z^2.$$

The coefficient of z^n provides us with the relation on c_n^1, $\hat{c}_n^1$ for $n = 3, 4, \ldots$:

$$-c_n^2 z^n-\imath\hat{c}_n^2 z^n-\frac{c_n^2\lambda z^{-1+n}}{2\mu}-\frac{i\hat{c}_n^2\lambda z^n}{2\mu}+\frac{3c_n^2\lambda^2 z^n}{4(\lambda+\mu)(\lambda+2\mu)}+\frac{3\imath\hat{c}_n^2\lambda^2 z^{-1+n}}{4(\lambda+\mu)(\lambda+2\mu)}+$$

$$+\frac{c_n^2\lambda^3 z^n}{4\mu(\lambda+\mu)(\lambda+2\mu)}+\frac{\imath\hat{c}_n^2\lambda^3 z^n}{4\mu(\lambda+\mu)(\lambda+2\mu)}+\frac{3c_n^2\lambda\mu z^n}{4(\lambda+\mu)(\lambda+2\mu)}+$$

$$+\frac{3\imath\hat{c}_n^2\lambda\mu z^n}{4(\lambda+\mu)(\lambda+2\mu)}+\frac{c_n^2\mu^2 z^n}{4(\lambda+\mu)(\lambda+2\mu)}+\frac{\imath\hat{c}_n^2\mu^2 z^{-1+n}}{4(\lambda+\mu)(\lambda+2\mu)}+c_n^1 n z^n+c_n^2 n z^n+$$

$$+\imath\hat{c}_n^1 n z^n+\imath\hat{c}_n^2 n z^n+\frac{3a_{n+1}^3\lambda z^n}{8\mu^2}+\frac{3\imath\hat{a}_{n+1}^3\lambda z^n}{8\mu^2}+\frac{3a_{n+1}^3 z^n}{4\mu}+\frac{3\imath\hat{a}_{n+1}^3 z^n}{4\mu}-$$

$$-\frac{3a_{n+1}^3\lambda z^n}{4(\lambda+\mu)(\lambda+2\mu)}-\frac{3\imath\hat{a}_{n+1}^3\lambda z^n}{4(\lambda+\mu)(\lambda+2\mu)}-\frac{3a_{n+1}^3\lambda^3 z^n}{8\mu^2(\lambda+\mu)(\lambda+2\mu)}-$$

$$-\frac{3\imath\hat{a}_{n+1}^3\lambda^3 z^n}{8\mu^2(\lambda+\mu)(\lambda+2\mu)}-\frac{9a_{n+1}^3\lambda^2 z^n}{8\mu(\lambda+\mu)(\lambda+2\mu)}-\frac{9\imath\hat{a}_{n+1}^3\lambda^2 z^n}{8\mu(\lambda+\mu)(\lambda+2\mu)}-$$

$$-\frac{3a_{n+1}^3(n+2)z^n}{2\mu}-\frac{3\imath\hat{a}_{n+1}^3(n+2)z^n}{2\mu}-\frac{a_{n+1}^2(n+2)z^n}{\lambda+\mu}-\frac{\imath\hat{a}_{n+1}^2(n+2)z^n}{\lambda+\mu}-$$

$$-\frac{a_{n+1}^2\lambda(n+2)z^n}{\mu(\lambda+\mu)}-\frac{\imath\hat{a}_{n+1}^2\lambda(n+2)z^n}{\mu(\lambda+\mu)}-\frac{a_{n+1}^1 z^n}{2(\lambda+\mu)}-\frac{\imath\hat{a}_{n+1}^1 z^n}{2(\lambda+\mu)}-\frac{a_{n+1}^1\lambda z^n}{2\mu(\lambda+\mu)}-$$

$$-\frac{\imath\hat{a}_{n+1}^1\lambda z^n}{2\mu(\lambda+\mu)}+c_n^0 z^n+\imath c_n^0 z^n-4c_n^3(n+2)z^n-4\imath\hat{c}_n^3(n+2)z^n-\frac{3c_n^3\lambda(n+2)z^n}{2\mu}-$$

$$-\frac{3\imath\hat{c}_n^3\lambda(n+2)z^n}{2\mu}+\frac{3c_n^3\lambda(n+2)z^n}{2(\lambda+2\mu)}+\frac{3\imath\hat{c}_n^3\lambda(n+2)z^n}{2(\lambda+2\mu)}+\frac{3c_n^3\lambda^2(n+2)z^n}{4\mu(\lambda+2\mu)}+$$

$$+\frac{3\imath\hat{c}_n^3\lambda^2(n+2)z^n}{4\mu(\lambda+2\mu)}+\frac{3c_n^3\mu(n+2)z^n}{4(\lambda+2\mu)}+\frac{3\imath\hat{c}_n^3\mu(n+2)z^n}{4(\lambda+2\mu)}+c_n^3(n+2)^2 z^n-$$

$$-\frac{3b_{n-1}^3\lambda z^n}{4(\lambda+\mu)(\lambda+2\mu)}-\frac{3\imath\hat{b}_{n-1}^3\lambda z^n}{4(\lambda+\mu)(\lambda+2\mu)}-\frac{3b_{n-1}^3\lambda^2 z^n}{8\mu(\lambda+\mu)(\lambda+2\mu)}-$$

$$-\frac{3\imath\hat{b}_{n-1}^3\lambda^2 z^n}{8\mu(\lambda+\mu)(\lambda+2\mu)}-\frac{3b_{n-1}^3\mu z^n}{8(\lambda+\mu)(\lambda+2\mu)}-\frac{3\imath\hat{b}_{n-1}^3\mu z^n}{8(\lambda+\mu)(\lambda+2\mu)}=\gamma_n z^n.$$

10.2 Solution of the first problem for the symmetric cone with one point at the base of the cone

As in Chapter 6 we solve the second basic problem of elasticity for symmetric conoid within two planes $h=a$ and $h=b$, with radii R and r, respectively.

Let us consider the boundary conditions

$$u(x,y,h)|_{h=a,x^2+y^2=R} = \sum_{j=0}^{n}(d_{1j}\cos j\theta - q_{1j}\sin j\theta),$$

$$v(x,y,h)|_{h=a,x^2+y^2=R} = \sum_{j=0}^{n}(e_{1j}\cos j\theta + f_{1j}\sin j\theta),$$

$$w(x,y,h)|_{h=a,x^2+y^2=R} = \sum_{j=0}^{n}(g_{1j}\cos j\theta - h_{1j}\sin j\theta),$$

$$u(x,y,h)|_{h=b,x^2+y^2=r} = \sum_{j=0}^{n}(d_{2j}\cos j\theta - q_{2j}\sin j\theta),$$

$$v(x,y,h)|_{h=b,x^2+y^2=r} = \sum_{j=0}^{n}(e_{2j}\cos j\theta + f_{2j}\sin j\theta),$$

$$w(x,y,h)|_{h=b,x^2+y^2=r} = \sum_{j=0}^{n}(g_{2j}\cos j\theta) - h_{2j}\sin j\theta).$$

These boundary conditions define the system of equations. The equations on real and imaginary parts of $u + \imath v$ and real part of w are as follows:

$$a_2a^2 + a_3a^3 - \frac{(a_2)(\lambda+3\mu)R^2}{4(\lambda+2\mu)} + \frac{(a_2-\imath b_2)(\lambda+\mu)z^2}{8(\lambda+2\mu)} + \frac{3c_3(\lambda+\mu)zR^2}{8\mu} -$$

$$-\frac{(\lambda+\mu)h_1(z)(R^2/z-\imath)}{8\mu} - \frac{(\lambda+\mu)h_1(z)(R^2/z+\imath)}{8(\lambda+2\mu)} -$$

$$-\frac{R^2\phi_0'(z)/z+\psi_0(z)-\kappa\phi_0(z)}{2\mu} - \frac{1}{2\mu}(a(\frac{3\mu(\lambda+\mu)(\kappa(a_3)R^2-(a_3-\imath b_3)z^2/2)}{2(\lambda+2\mu)} +$$

$$+R^2\phi_1'(z)/z+\psi_1(z)-\kappa\phi_1(z))) = \sum_{j=2}^{n}(d_{1j}+q_{1j}\imath)z^j/R^j + (d_{11}+\imath q_{11})z/R +$$

$$+(d_{10}+\imath q_{10}) + (d_1-\imath q_1)R^2/z - (d_1+\imath q_1)z,$$

$$(\imath b_2)a^2 + (\imath b_3)a^3 - \frac{(\imath b_2)(\lambda+3\mu)R^2}{4(\lambda+2\mu)} + \frac{(a_2-\imath b_2)(\lambda+\mu)z^2}{8(\lambda+2\mu)} + \frac{3c_3(\lambda+\mu)zR^2}{8\mu} -$$

$$-\frac{(\lambda+\mu)h_1(z)(-R^2/z-\imath)}{8\mu} - \frac{(\lambda+\mu)h_1(z)(-R^2/z+\imath)}{8(\lambda+2\mu)} -$$

$$-\frac{-R^2\phi_0'(z)/z - \psi_0(z) - \kappa\phi_0(z)}{2\mu} - \frac{1}{2\mu}(a(\frac{3\mu(\lambda+\mu)(\kappa(\imath b_3)R^2-(a_3-\imath b_3)z^2/2)}{2(\lambda+2\mu)} -$$

$$-R^2\phi_1'(z)/z - \psi_1(z) - \kappa\phi_1(z))) = \sum_{j=2}^{n}(e_{1j}+f_{1j}\imath)z^j/R^j + (e_{11}+\imath f_{11})z/R +$$

$$+(e_{10}+\imath f_{10}) - (d_1-\imath q_1)R^2/z - (d_1+\imath q_1)z,$$

$$\frac{1}{4\mu(\lambda+2\mu)}(4a^2(a_2+\imath b_2+a(a_3+\imath b_3))\mu(\lambda+2\mu) - \mu(3a(a_3+\imath b_3)\kappa(\lambda+\mu)+$$

$$+a_2(\lambda+3\mu)+\imath b_2(\lambda+3\mu))R^2 + 3c_3(\lambda+\mu)(\lambda+2\mu)R^2z + (a_2-\imath b_2+$$

$$+3a(a_3-\imath b_3))\mu(\lambda+\mu)z^2 + (\lambda+\mu)^2\imath h_1(z) + 4k(\lambda+2\mu)(\phi_0(z)+a\phi_1(z))) =$$

$$= \sum_{j=2}^{n} R^{-1-j}((d_{1j}+e_{1j}+\imath(f_{1j}+q_{1j}))Rz^j + R^j((d_{10}+e_{10}+\imath(f_{10}+q_{10}))R+$$

$$+(d_{11}+e_{11}+\imath(f_{11}+q_{11}+2\imath d_1R-2q_1R))z)),$$

$$(a_2)b^2 + (a_3)b^3 - \frac{(a_2)(\lambda+3\mu)r^2}{4(\lambda+2\mu)} + \frac{(a_2-\imath b_2)(\lambda+\mu)z^2}{8(\lambda+2\mu)} + \frac{3c_3(\lambda+\mu)zr^2}{8\mu} -$$

$$-\frac{(\lambda+\mu)h_1(z)(r^2/z-\imath)}{8\mu} - \frac{(\lambda+\mu)h_1(z)(r^2/z+\imath)}{8(\lambda+2\mu)} -$$

$$-\frac{\phi_0'(z)r^2/z + \psi_0(z) - \kappa\phi_0(z)}{2\mu} - \frac{1}{2\mu}(b(\frac{3\mu(\lambda+\mu)(\kappa(a_3)r^2-(a_3-\imath b_3)z^2/2)}{2(\lambda+2\mu)} +$$

$$+\phi_1'(z)r^2/z + \psi_1(z) - \kappa\phi_1(z))) = \sum_{j=2}^{n}(d_{2j}+q_{2j}\imath)z^j/r^j + (d_{21}+\imath q_{21})z/r +$$

$$+(d_{20}+\imath q_{20})+(d_2-\imath q_2)r^2/z-(d_2+\imath q_2)z,$$

$$(\imath b_2)b^2+(\imath b_3)b^3-\frac{(\imath b_2)(\lambda+3\mu)r^2}{4(\lambda+2\mu)}+\frac{(a_2-\imath b_2)(\lambda+\mu)z^2}{8(\lambda+2\mu)}+\frac{3c_3(\lambda+\mu)zr^2}{8\mu}-$$

$$-\frac{(\lambda+\mu)h_1(z)(-r^2/z-\imath)}{8\mu}-\frac{(\lambda+\mu)h_1(z)(-r^2/z+\imath)}{8(\lambda+2\mu)}-$$

$$-\frac{-\phi_0'(z)r^2/z-\psi_0(z)-\kappa\phi_0(z)}{2\mu}-\frac{1}{2\mu}(b(\frac{3\mu(\lambda+\mu)(\kappa(\imath b_3)r^2-(a_3-\imath b_3)z^2/2)}{2(\lambda+2\mu)}-$$

$$-r^2/z\phi_1'(z)-\psi_1(z)-\kappa\phi_1(z)))=\sum_{j=2}^{n}(e_{2j}+f_{2j}\imath)z^j/r^j+(e_{21}+\imath f_{21})z/r+$$

$$+(e_{20}+\imath f_{20})-(d_2-\imath q_2)r^2/z-(d_2+\imath q_2)z,$$

$$\frac{1}{4\mu(\lambda+2\mu)}(4b^2(a_2+a_3b+\imath(b_2+bb_3))\mu(\lambda+2\mu)-\mu(3b(a_3+\imath b_3)\kappa(\lambda+\mu)+$$

$$+a_2(\lambda+3\mu)+\imath b_2(\lambda+3\mu))r^2+3c_3(\lambda+\mu)(\lambda+2\mu)r^2z+$$

$$+(a_2+3a_3b-\imath(b_2+3bb_3))\mu(\lambda+\mu)z^2+(\lambda+\mu)^2ih_1(z)+4\kappa(\lambda+2\mu)(\phi_0(z)+b\phi_1(z)))=$$

$$=\sum_{j=2}^{n}r^{-1-j}((d_{2j}+e_{2j}+\imath(f_{2j}+q_{2j}))rz^j+r^j((d_{20}+e_{20}+\imath(f_{20}+q_{20}))r+$$

$$+(d_{21}+e_{21}+\imath(f_{21}+q_{21}+2\imath d_2r-2q_2r))z)),$$

$$-1/(4\mu)(2\phi_1(z)R^2/z)-(\lambda+2\mu)c_2R^2/(2\mu)+$$

$$+3/16(\lambda+\mu)/(\lambda+2\mu)(R^4(a_3+\imath b_3)/z+R^2z(a_3-\imath b_3))+1/2(2h_0(z))+$$

$$+a(-(\lambda+2\mu)3c_3R^2/2\mu+(2h_1(z)))/2+a^2c_2+a^3c_3=\sum_{j=2}^{n}(g_{1j}+\imath h_{1j})z^j/R^j+$$

$$+(g_{11}+\imath h_{11})z/R+(g_{10}+\imath h_{10})+\overline{d_3}R^2/z-d_3z,$$

$$-1/(4\mu)(2r^2\phi_1(z)/z)-(\lambda+2\mu)c_2r^2/(2\mu)+$$

$$+3/16(\lambda+\mu)/(\lambda+2\mu)(r^4/z(a_3+\imath b_3)+r^2z(a_3-\imath b_3))+(2h_0(z))1/2+$$

$$+b(-(\lambda+2\mu)3c_3\mu r^2/2+(2h_1(z)))/2+b^2c_2+b^3c_3=\sum_{j=2}^{n}(g_{2j}+h_{2j}\imath)z^j/r^j+$$

$$+(g_{21}+\imath h_{21})z/r+(g_{20}+\imath h_{20})+\overline{d_4}r^2/z-d_4z.$$

Thus

$$\phi_1(z)=-\frac{1}{4(a-b)\kappa(\lambda+2\mu)}(-4a_3b^3\lambda\mu-4\imath b^2b_2\lambda\mu-4\imath b^3b_3\lambda\mu-4d_{10}\lambda\mu+$$

$$+4d_{20}\lambda\mu-4e_{10}\lambda\mu+4e_{20}\lambda\mu-4\imath f_{10}\lambda\mu+4\imath f_{20}\lambda\mu-8a_3b^3\mu^2-8\imath b^2b_2\mu^2-8\imath b^3b_3\mu^2-$$

$$-8d_{10}\mu^2+8d_{20}\mu^2-8e_{10}\mu^2+8e_{20}\mu^2-8\imath f_{10}\mu^2+8\imath f_{20}\mu^2+4a^2(a_2+ib_2)\mu(\lambda+2\mu)+$$

$$+4a^3(a_3+\imath b_3)\mu(\lambda+2\mu)-4\imath\lambda\mu q_{10}-8\imath\mu^2q_{10}+4\imath\lambda\mu q_{20}+8\imath\mu^2q_{20}+\imath b_2\lambda\mu r^2+$$

$$+3a_3b\kappa\lambda\mu r^2+3\imath bb_3\kappa\lambda\mu r^2+3\imath b_2\mu^2r^2+3a_3b\kappa\mu^2r^2+3\imath bb_3\kappa\mu^2r^2-\imath b_2\lambda\mu R^2-$$

$$-3\imath b_2\mu^2R^2+a_2\mu(-4b^2(\lambda+2\mu)+(\lambda+3\mu)(r-R)(r+R))+8d_1\lambda\mu z-8d_2\lambda\mu z+$$

$$+16d_1\mu^2z-16d_2\mu^2z+8\imath\lambda\mu q_1z+16\imath\mu^2q_1z-8\imath\lambda\mu q_2z-16\imath\mu^2q_2z+\frac{4d_{21}\lambda\mu z}{r}+$$

$$+\frac{4e_{21}\lambda\mu z}{r}+\frac{4\imath f_{21}\lambda\mu z}{r}+\frac{8d_{21}\mu^2z}{r}+\frac{8e_{21}\mu^2z}{r}+\frac{8\imath f_{21}\mu^2z}{r}+\frac{4\imath\lambda\mu q_{21}z}{r}+\frac{8\imath\mu^2q_{21}z}{r}-$$

$$-3c_3\lambda^2r^2z-9c_3\lambda\mu r^2z-6c_3\mu^2r^2z-\frac{4d_{11}\lambda\mu z}{R}-\frac{4e_{11}\lambda\mu z}{R}-\frac{4\imath f_{11}\lambda\mu z}{R}-\frac{8d_{11}\mu^2z}{R}-$$

$$-\frac{8e_{11}\mu^2z}{R}-\frac{8\imath f_{11}\mu^2z}{R}-\frac{4\imath\lambda\mu q_{11}z}{R}-\frac{8\imath\mu^2q_{11}z}{R}+3c_3\lambda^2R^2z+9c_3\lambda\mu R^2z+6c_3\mu^2R^2z-$$

$$-3a_3b\lambda\mu z^2+3\imath bb_3\lambda\mu z^2-3a_3b\mu^2z^2+3\imath bb_3\mu^2z^2-\sum_{j=2}^{n}4\mu(\lambda+2\mu)r^{-j}R^{-j}((d_{1j}+e_{1j}+$$

$$+\imath(f_{1j}+q_{1j}))r^j-\sum_{j=2}^{n}(d_{2j}+e_{2j}+\imath(f_{2j}+q_{2j}))R^j)z^j+$$

$$+3a\mu(\lambda+\mu)(-(a_3+\imath b_3)\kappa R^2+(a_3-\imath b_3)z^2)).$$

Since $h_0(z)$ and $h_1(z)$ are analytic we must set

$$\overline{d_3}=\frac{a^2a_2\lambda}{2(a-b)\kappa(\lambda+2\mu)}+\frac{a^3a_3\lambda}{2(a-b)\kappa(\lambda+2\mu)}-\frac{a_2b^2\lambda}{2(a-b)\kappa(\lambda+2\mu)}-$$

$$-\frac{a_3b^3\lambda}{2(a-b)\kappa(\lambda+2\mu)}+\frac{1a^2b_2\lambda}{2(a-b)\kappa(\lambda+2\mu)}-\frac{1b^2b_2\lambda}{2(a-b)\kappa(\lambda+2\mu)}+$$

$$+\frac{1a^3b_3\lambda}{2(a-b)\kappa(\lambda+2\mu)}-\frac{1b^3b_3\lambda}{2(a-b)\kappa(\lambda+2\mu)}-\frac{d_{10}\lambda}{2(a-b)\kappa(\lambda+2\mu)}+$$

$$+\frac{d_{20}\lambda}{2(a-b)\kappa(\lambda+2\mu)}-\frac{e_{10}\lambda}{2(a-b)\kappa(\lambda+2\mu)}+\frac{e_{20}\lambda}{2(a-b)\kappa(\lambda+2\mu)}-$$

$$-\frac{1f_{10}\lambda}{2(a-b)\kappa(\lambda+2\mu)}+\frac{1f_{20}\lambda}{2(a-b)\kappa(\lambda+2\mu)}+\frac{a^2a_2\mu}{(a-b)\kappa(\lambda+2\mu)}+$$

$$+\frac{a^3a_3\mu}{(a-b)\kappa(\lambda+2\mu)}-\frac{a_2b^2\mu}{(a-b)\kappa(\lambda+2\mu)}-\frac{a_3b^3\mu}{(a-b)\kappa(\lambda+2\mu)}+$$

$$+\frac{1a^2b_2\mu}{(a-b)\kappa(\lambda+2\mu)}-\frac{1b^2b_2\mu}{(a-b)\kappa(\lambda+2\mu)}+\frac{1a^3b_3\mu}{(a-b)\kappa(\lambda+2\mu)}-$$

$$-\frac{1b^3b_3\mu}{(a-b)\kappa(\lambda+2\mu)}-\frac{d_{10}\mu}{(a-b)\kappa(\lambda+2\mu)}+\frac{d_{20}\mu}{(a-b)\kappa(\lambda+2\mu)}-$$

$$-\frac{e_{10}\mu}{(a-b)\kappa(\lambda+2\mu)}+\frac{e_{20}\mu}{(a-b)\kappa(\lambda+2\mu)}-\frac{1f_{10}\mu}{(a-b)\kappa(\lambda+2\mu)}+$$

$$+\frac{1f_{20}\mu}{(a-b)\kappa(\lambda+2\mu)}-\frac{1\lambda q_{10}}{2(a-b)\kappa(\lambda+2\mu)}-\frac{1\mu q_{10}}{(a-b)\kappa(\lambda+2\mu)}+$$

$$+\frac{1\lambda q_{20}}{2(a-b)\kappa(\lambda+2\mu)}+\frac{1\mu q_{20}}{(a-b)\kappa(\lambda+2\mu)}+\frac{3a_3b\lambda r^2}{8(a-b)(\lambda+2\mu)}+$$

$$+\frac{31bb_3\lambda r^2}{8(a-b)(\lambda+2\mu)}+\frac{a_2\lambda r^2}{8(a-b)\kappa(\lambda+2\mu)}+\frac{1b_2\lambda r^2}{8(a-b)\kappa(\lambda+2\mu)}+$$

$$+\frac{3a_3b\mu r^2}{8(a-b)(\lambda+2\mu)}+\frac{31bb_3\mu r^2}{8(a-b)(\lambda+2\mu)}+\frac{3a_2\mu r^2}{8(a-b)\kappa(\lambda+2\mu)}+$$

$$+\frac{31b_2\mu r^2}{8(a-b)\kappa(\lambda+2\mu)}+\frac{3a_3\lambda R^2}{16(\lambda+2\mu)}-\frac{3aa_3\lambda R^2}{8(a-b)(\lambda+2\mu)}+\frac{31b_3\lambda R^2}{16(\lambda+2\mu)}-$$

$$-\frac{31ab_3\lambda R^2}{8(a-b)(\lambda+2\mu)}-\frac{a_2\lambda R^2}{8(a-b)\kappa(\lambda+2\mu)}-\frac{1b_2\lambda R^2}{8(a-b)\kappa(\lambda+2\mu)}+\frac{3a_3\mu R^2}{16(\lambda+2\mu)}-$$

$$-\frac{3aa_3\mu R^2}{8(a-b)(\lambda+2\mu)}+\frac{31b_3\mu R^2}{16(\lambda+2\mu)}-\frac{31ab_3\mu R^2}{8(a-b)(\lambda+2\mu)}-\frac{3a_2\mu R^2}{8(a-b)\kappa(\lambda+2\mu)}-$$

$$-\frac{3\imath b_2\mu R^2}{8(a-b)\kappa(\lambda+2\mu)},$$

$$\overline{d_4} = \frac{a^2a_2\lambda}{2(a-b)\kappa(\lambda+2\mu)} + \frac{a^3a_3\lambda}{2(a-b)\kappa(\lambda+2\mu)} - \frac{a_2b^2\lambda}{2(a-b)\kappa(\lambda+2\mu)} -$$

$$-\frac{a_3b^3\lambda}{2(a-b)\kappa(\lambda+2\mu)} + \frac{\imath a^2b_2\lambda}{2(a-b)\kappa(\lambda+2\mu)} - \frac{\imath b^2b_2\lambda}{2(a-b)\kappa(\lambda+2\mu)} +$$

$$+\frac{\imath a^3b_3\lambda}{2(a-b)\kappa(\lambda+2\mu)} - \frac{\imath b^3b_3\lambda}{2(a-b)\kappa(\lambda+2\mu)} - \frac{d_{10}\lambda}{2(a-b)\kappa(\lambda+2\mu)} +$$

$$+\frac{d_{20}\lambda}{2(a-b)\kappa(\lambda+2\mu)} - \frac{e_{10}\lambda}{2(a-b)\kappa(\lambda+2\mu)} + \frac{e_{20}\lambda}{2(a-b)\kappa(\lambda+2\mu)} -$$

$$-\frac{\imath f_{10}\lambda}{2(a-b)\kappa(\lambda+2\mu)} + \frac{\imath f_{20}\lambda}{2(a-b)\kappa(\lambda+2\mu)} + \frac{a^2a_2\mu}{(a-b)\kappa(\lambda+2\mu)} +$$

$$+\frac{a^3a_3\mu}{(a-b)\kappa(\lambda+2\mu)} - \frac{a_2b^2\mu}{(a-b)\kappa(\lambda+2\mu)} - \frac{a_3b^3\mu}{(a-b)\kappa(\lambda+2\mu)} +$$

$$+\frac{\imath a^2b_2\mu}{(a-b)\kappa(\lambda+2\mu)} - \frac{\imath b^2b_2\mu}{(a-b)\kappa(\lambda+2\mu)} + \frac{\imath a^3b_3\mu}{(a-b)\kappa(\lambda+2\mu)} -$$

$$-\frac{\imath b^3b_3\mu}{(a-b)\kappa(\lambda+2\mu)} - \frac{d_{10}\mu}{(a-b)\kappa(\lambda+2\mu)} + \frac{d_{20}\mu}{(a-b)\kappa(\lambda+2\mu)} -$$

$$-\frac{e_{10}\mu}{(a-b)\kappa(\lambda+2\mu)} + \frac{e_{20}\mu}{(a-b)\kappa(\lambda+2\mu)} - \frac{\imath f_{10}\mu}{(a-b)\kappa(\lambda+2\mu)} +$$

$$+\frac{\imath f_{20}\mu}{(a-b)\kappa(\lambda+2\mu)} - \frac{\imath\lambda q_{10}}{2(a-b)\kappa(\lambda+2\mu)} - \frac{\imath\mu q_{10}}{(a-b)\kappa(\lambda+2\mu)} +$$

$$+\frac{\imath\lambda q_{20}}{2(a-b)\kappa(\lambda+2\mu)} + \frac{\imath\mu q_{20}}{(a-b)\kappa(\lambda+2\mu)} + \frac{3a_3\lambda r^2}{16(\lambda+2\mu)} + \frac{3a_3b\lambda r^2}{8(a-b)(\lambda+2\mu)} +$$

$$+\frac{3\imath b_3\lambda r^2}{16(\lambda+2\mu)} + \frac{3\imath bb_3\lambda r^2}{8(a-b)(\lambda+2\mu)} + \frac{a_2\lambda r^2}{8(a-b)\kappa(\lambda+2\mu)} + \frac{\imath b_2\lambda r^2}{8(a-b)\kappa(\lambda+2\mu)} +$$

$$+\frac{3a_3\mu r^2}{16(\lambda+2\mu)} + \frac{3a_3b\mu r^2}{8(a-b)(\lambda+2\mu)} + \frac{3\imath b_3\mu r^2}{16(\lambda+2\mu)} + \frac{3\imath bb_3\mu r^2}{8(a-b)(\lambda+2\mu)} +$$

$$+\frac{3a_2\mu r^2}{8(a-b)\kappa(\lambda+2\mu)} + \frac{3\imath b_2\mu r^2}{8(a-b)\kappa(\lambda+2\mu)} - \frac{3aa_3\lambda R^2}{8(a-b)(\lambda+2\mu)} -$$

$$-\frac{3\imath a b_3\lambda R^2}{8(a-b)(\lambda+2\mu)}-\frac{a_2\lambda R^2}{8(a-b)\kappa(\lambda+2\mu)}-\frac{\imath b_2\lambda R^2}{8(a-b)\kappa(\lambda+2\mu)}-$$

$$-\frac{3aa_3\mu R^2}{8(a-b)(\lambda+2\mu)}-\frac{3\imath ab_3\mu R^2}{8(a-b)(\lambda+2\mu)}-\frac{3a_2\mu R^2}{8(a-b)\kappa(\lambda+2\mu)}-$$

$$-\frac{3\imath b_2\mu R^2}{8(a-b)\kappa(\lambda+2\mu)}.$$

Hence

$$h_0(z)=\frac{1}{8(a-b)^2\kappa\mu(\lambda+2\mu)z}(-4a^4\mu(\lambda+2\mu)z(-2bc_3\kappa+(a_3-\imath b_3)z)+$$

$$+4a^3\mu(\lambda+2\mu)z(2bc_2k-2b^2c_3k-a_2z+a_3bz+\imath b_2z-\imath bb_3z)-$$

$$-\sum_{j=2}^{n}a^2r^{-1-j}z(-8(g_{2j}+\imath h_{2j})\kappa\mu(\lambda+2\mu)rz^j-r^j(-16b^2c_2\kappa\mu(\lambda+2\mu)r-$$

$$-8b^3c_3\kappa\mu(\lambda+2\mu)r+4\kappa(\lambda+2\mu)r(2g_{20}\mu+2\imath h_{20}\mu+c_2(\lambda+2\mu)r^2)+\mu(8g_{21}\kappa(\lambda+2\mu)+$$

$$+8\imath h_{21}\kappa(\lambda+2\mu)+3(a_3-\imath b_3)(\lambda+\mu)r(-(1+\kappa)r^2+\kappa R^2))z+$$

$$+4b(\lambda+2\mu)r(3c_3\kappa(\lambda+2\mu)(r-R)(r+R)+(a_2-\imath b_2)\mu z)))+a(8b^4c_3\kappa\mu(\lambda+2\mu)z-$$

$$-4e_{21}\lambda\mu rz-4\imath f_{21}\lambda\mu rz-8e_{21}\mu^2rz-8\imath f_{21}\mu^2rz-4d_{21}\mu(\lambda+2\mu)rz-4\imath\lambda\mu q_{21}rz-$$

$$-8\imath\mu^2q_{21}rz-8d_1\lambda\mu r^2z+8d_2\lambda\mu r^2z-16d_1\mu^2r^2z+16d_2\mu^2r^2z-8\imath\lambda\mu q_1r^2z-$$

$$-16\imath\mu^2q_1r^2z+8\imath\lambda\mu q_2r^2z+16\imath\mu^2q_2r^2z+3c_3\lambda^2r^4z+9c_3\lambda\mu r^4z+6c_3\mu^2r^4z+$$

$$+\frac{4d_{11}\lambda\mu r^2z}{R}+\frac{4e_{11}\lambda\mu r^2z}{R}+\frac{4\imath f_{11}\lambda\mu r^2z}{R}+\frac{8d_{11}\mu^2r^2z}{R}+\frac{8e_{11}\mu^2r^2z}{R}+\frac{8\imath f_{11}\mu^2r^2z}{R}+$$

$$+\frac{4\imath\lambda\mu q_{11}r^2z}{R}+\frac{8\imath\mu^2q_{11}r^2z}{R}-3c_3\lambda^2r^2R^2z-9c_3\lambda\mu r^2R^2z-6c_3\mu^2r^2R^2z+4d_{10}\lambda\mu z^2-$$

$$-4d_{20}\lambda\mu z^2+4e_{10}\lambda\mu z^2-4e_{20}\lambda\mu z^2-4\imath f_{10}\lambda\mu z^2+4\imath f_{20}\lambda\mu z^2+8d_{10}\mu^2z^2-8d_{20}\mu^2z^2+$$

$$+8e_{10}\mu^2z^2-8e_{20}\mu^2z^2-8\imath f_{10}\mu^2z^2+8\imath f_{20}\mu^2z^2-4\imath\lambda\mu q_{10}z^2-8\imath\mu^2q_{10}z^2+4\imath\lambda\mu q_{20}z^2+$$

$$+8\imath\mu^2q_{20}z^2-a_2\lambda\mu r^2z^2+\imath b_2\lambda\mu r^2z^2-3a_2\mu^2r^2z^2+3\imath b_2\mu^2r^2z^2+a_2\lambda\mu R^2z^2-$$

$$-\imath b_2\lambda\mu R^2z^2+3a_2\mu^2R^2z^2-3\imath b_2\mu^2R^2z^2-4d_{2j}\lambda\mu r^{2-j}z^j-4e_{2j}\lambda\mu r^{2-j}z^j-$$

$$-4\imath f_{2j}\lambda\mu r^{2-j}z^j-8d_{2j}\mu^2r^{2-j}z^j-8e_{2j}\mu^2r^{2-j}z^j-8\imath f_{2j}\mu^2r^{2-j}z^j-4\imath\lambda\mu q_{2j}r^{2-j}z^j-$$

$$-8\imath\mu^2 q_{2j} r^{2-j} z^j + 4b^3\mu(\lambda+2\mu)z(2c_2\kappa+(a_3-\imath b_3)z) + 4\mu(\lambda+2\mu)R^{-j}z^j((d_{1j}+e_{1j}+$$
$$+\imath(f_{1j}+q_{1j}))r^2 - 2b(g_{1j}+\imath h_{1j})\kappa z) + 4b^2(\lambda+2\mu)z(3c_3\kappa(\lambda+2\mu)(-r^2+R^2)+$$
$$+(a_2-\imath b_2)\mu z) + \frac{1}{R}(br^{-1-j}z(-8(g_{2j}+\imath h_{2j})\kappa\mu(\lambda+2\mu)rRz^j-$$
$$-r^j(4\kappa(\lambda+2\mu)rR(c_2\lambda r^2+2\mu(g_{10}+g_{20}+\imath(h_{10}+h_{20})+c_2r^2)+c_2(\lambda+2\mu)R^2)+$$
$$+\mu(8(g_{11}+\imath h_{11})\kappa(\lambda+2\mu)r+(8(g_{21}+\imath h_{21})\kappa(\lambda+2\mu)-$$
$$-3(a_3-\imath b_3)(\lambda+\mu)r^3)R - 3(a_3-\imath b_3)(\lambda+\mu)rR^3)z)))) + b(-4e_{11}\lambda\mu Rz-$$
$$-4\imath f_{11}\lambda\mu Rz - 8e_{11}\mu^2 Rz - 8\imath f_{11}\mu^2 Rz - 4d_{11}\mu(\lambda+2\mu)Rz - 4\imath\lambda\mu q_{11}Rz-$$
$$-8\imath\mu^2 q_{11}Rz + 8d_1\lambda\mu R^2 z - 8d_2\lambda\mu R^2 z + 16d_1\mu^2R^2z - 16d_2\mu^2R^2z + 8\imath\lambda\mu q_1R^2z+$$
$$+16\imath\mu^2q_1R^2z - 8\imath\lambda\mu q_2R^2z - 16\imath\mu^2q_2R^2z + \frac{4d_{21}\lambda\mu R^2z}{r} + \frac{4e_{21}\lambda\mu R^2z}{r}+$$
$$+\frac{4\imath f_{21}\lambda\mu R^2z}{r} + \frac{8d_{21}\mu^2R^2z}{r} + \frac{8e_{21}\mu^2R^2z}{r} + \frac{8\imath f_{21}\mu^2R^2z}{r} + \frac{4\imath\lambda\mu q_{21}R^2z}{r}+$$
$$+\frac{8\imath\mu^2q_{21}R^2z}{r} - 3c_3\lambda^2r^2R^2z - 9c_3\lambda\mu r^2R^2z - 6c_3\mu^2r^2R^2z + 3c_3\lambda^2R^4z+$$
$$+9c_3\lambda\mu R^4z + 6c_3\mu^2R^4z - 4d_{10}\lambda\mu z^2 + 4d_{20}\lambda\mu z^2 - 4e_{10}\lambda\mu z^2 + 4e_{20}\lambda\mu z^2+$$
$$+4\imath f_{10}\lambda\mu z^2 - 4\imath f_{20}\lambda\mu z^2 - 8d_{10}\mu^2z^2 + 8d_{20}\mu^2z^2 - 8e_{10}\mu^2z^2 + 8e_{20}\mu^2z^2+$$
$$+8\imath f_{10}\mu^2z^2 - 8\imath f_{20}\mu^2z^2 - 4b^2(a_2-ib_2)\mu(\lambda+2\mu)z^2 - 4b^3(a_3-\imath b_3)\mu(\lambda+2\mu)z^2+$$
$$+4\imath\lambda\mu q_{10}z^2 + 8\imath\mu^2q_{10}z^2 - 4\imath\lambda\mu q_{20}z^2 - 8\imath\mu^2q_{20}z^2 + a_2\lambda\mu r^2z^2 - ib_2\lambda\mu r^2z^2+$$
$$+3a_2\mu^2r^2z^2 - 3\imath b_2\mu^2r^2z^2 - a_2\lambda\mu R^2z^2 + \imath b_2\lambda\mu R^2z^2 - 3a_2\mu^2R^2z^2 + 3\imath b_2\mu^2R^2z^2+$$
$$+4\mu(\lambda+2\mu)(d_{2j}+e_{2j}+\imath(f_{2j}+q_{2j}))r^{-j}R^2z^j - 4d_{1j}\lambda\mu R^{2-j}z^j - 4e_{1j}\lambda\mu R^{2-j}z^j-$$
$$-4\imath f_{1j}\lambda\mu R^{2-j}z^j - 8d_{1j}\mu^2R^{2-j}z^j - 8e_{1j}\mu^2R^{2-j}z^j - 8\imath f_{1j}\mu^2R^{2-j}z^j-$$
$$-4\imath\lambda\mu q_{1j}R^{2-j}z^j - 8\imath\mu^2q_{1j}R^{2-j}z^j + R^{-1-j}z(8b(g_{1j}+\imath h_{1j})\kappa\mu(\lambda+2\mu)Rz^j+$$
$$+R^j(4b\kappa(\lambda+2\mu)R(2g_{10}\mu + 2\imath h_{10}\mu + c_2(\lambda+2\mu)R^2) + b\mu(8g_{11}\kappa(\lambda+2\mu)+$$
$$+8\imath h_{11}\kappa(\lambda+2\mu) + 3(a_3-\imath b_3)(\lambda+\mu)R(\kappa r^2 - (1+\kappa)R^2))z)))),$$

$$h_1(z) = -\frac{1}{8(a-b)^2\kappa\mu(\lambda+2\mu)z}(8b^3c_2\kappa\mu(\lambda+2\mu)z + 8a^4c_3\kappa\mu(\lambda+2\mu)z+$$
$$+8b^4c_3\kappa\mu(\lambda+2\mu)z + 8a^3(c_2-bc_3)\kappa\mu(\lambda+2\mu)z - 12b^2c_3\kappa(\lambda+2\mu)^2r^2z-$$
$$-4a^2\kappa(\lambda+2\mu)(2bc_2\mu + 3c_3(\lambda+2\mu)R^2)z+$$
$$+(\lambda+2\mu)r^{-1-j}(r-R)R^{-1-j}(r+R)(4d_{11}\mu r^{1+j}R^jz + 4e_{11}\mu r^{1+j}R^jz+$$
$$+4\imath f_{11}\mu r^{1+j}R^jz + 4\imath\mu q_{11}r^{1+j}R^jz - 4d_{21}\mu r^jR^{1+j}z - 4e_{21}\mu r^jR^{1+j}z-$$
$$-4\imath f_{21}\mu r^jR^{1+j}z - 4\imath\mu q_{21}r^jR^{1+j}z - 8d_1\mu r^{1+j}R^{1+j}z + 8d_2\mu r^{1+j}R^{1+j}z-$$
$$-8\imath\mu q_1r^{1+j}R^{1+j}z + 8\imath\mu q_2r^{1+j}R^{1+j}z + 3c_3\lambda r^{3+j}R^{1+j}z + 3c_3\mu r^{3+j}R^{1+j}z-$$
$$-3c_3\lambda r^{1+j}R^{3+j}z - 3c_3\mu r^{1+j}R^{3+j}z + 4\mu rR((d_{1j}+e_{1j}+\imath(f_{1j}+q_{1j}))r^j-$$
$$-(d_{2j}+e_{2j}+\imath(f_{2j}+q_{2j}))R^j)z^j)+$$
$$+ar^{-1-j}R^{-1-j}z(-8\kappa\mu(\lambda+2\mu)rR((g_{1j}+\imath h_{1j})r^j - (g_{2j}+\imath h_{2j})R^j)z^j+$$
$$+r^jR^j(-8b^2c_2\kappa\mu(\lambda+2\mu)rR - 8b^3c_3\kappa\mu(\lambda+2\mu)rR + 12bc_3\kappa(\lambda+2\mu)^2rR(r^2+R^2)+$$
$$+4\kappa(\lambda+2\mu)rR(-2g_{10}\mu + c_2\lambda r^2 + 2\mu(g_{20} - \imath h_{10} + \imath h_{20} + c_2r^2) - c_2(\lambda+2\mu)R^2)+$$
$$+\mu(-8(g_{11}+\imath h_{11})\kappa(\lambda+2\mu)r + (8(g_{21}+\imath h_{21})\kappa(\lambda+2\mu)-$$
$$-3(a_3-\imath b_3)(1+\kappa)(\lambda+\mu)r^3)R + 3(a_3-\imath b_3)(1+\kappa)(\lambda+\mu)rR^3)z))+$$
$$+br^{-1-j}R^{-1-j}z(8(g_{1j}+\imath h_{1j})\kappa\mu(\lambda+2\mu)r^{1+j}Rz^j+$$
$$+R^j(-8(g_{2j}+ih_{2j})\kappa\mu(\lambda+2\mu)rRz^j + r^j(4\kappa(\lambda+2\mu)rR(2g_{10}\mu - c_2\lambda r^2-$$
$$-2\mu(g_{20} - \imath h_{10} + \imath h_{20} + c_2r^2) + c_2(\lambda+2\mu)R^2) + \mu(8(g_{11}+\imath h_{11})\kappa(\lambda+2\mu)r-$$
$$-(8(g_{21}+\imath h_{21})\kappa(\lambda+2\mu) - 3(a_3-\imath b_3)(1+\kappa)(\lambda+\mu)r^3)R-$$
$$-3(a_3-\imath b_3)(1+\kappa)(\lambda+\mu)rR^3)z)))),$$

$$\phi_0(z) = \frac{1}{4(a-b)\kappa(\lambda+2\mu)}(r^{-1-j}R^{-1-j}(4a^2b(a_2+\imath b_2)\mu(\lambda+2\mu)r^{1+j}R^{1+j}+$$

$$+4a^3b(a_3+\imath b_3)\mu(\lambda+2\mu)r^{1+j}R^{1+j}-br^{1+j}(4\imath f_{10}\lambda\mu R^{1+j}+8\imath f_{10}\mu^2R^{1+j}+$$

$$+4d_{10}\mu(\lambda+2\mu)R^{1+j}+4e_{10}\mu(\lambda+2\mu)R^{1+j}+4i\lambda\mu q_{10}R^{1+j}+8\imath\mu^2q_{10}R^{1+j}+$$

$$+a_2\lambda\mu R^{3+j}+\imath b_2\lambda\mu R^{3+j}+3a_2\mu^2R^{3+j}+3\imath b_2\mu^2R^{3+j}+4d_{11}\lambda\mu R^jz+4e_{11}\lambda\mu R^jz+$$

$$+4\imath f_{11}\lambda\mu R^jz+8d_{11}\mu^2R^jz+8e_{11}\mu^2R^jz+8\imath f_{11}\mu^2R^jz+4\imath\lambda\mu q_{11}R^jz+$$

$$+8\imath\mu^2q_{11}R^jz-8d_1\lambda\mu R^{1+j}z-16d_1\mu^2R^{1+j}z-8\imath\lambda\mu q_1R^{1+j}z-16\imath\mu^2q_1R^{1+j}z-$$

$$-3c_3\lambda^2R^{3+j}z-9c_3\lambda\mu R^{3+j}z-6c_3\mu^2R^{3+j}z-a_2\lambda\mu R^{1+j}z^2+\imath b_2\lambda\mu R^{1+j}z^2-$$

$$-a_2\mu^2R^{1+j}z^2+\imath b_2\mu^2R^{1+j}z^2+4\mu(\lambda+2\mu)(d_{1j}+e_{1j}+\imath(f_{1j}+q_{1j}))Rz^j)+$$

$$+aR^{1+j}(-4\imath b^2b_2\lambda\mu r^{1+j}-4\imath b^3b_3\lambda\mu r^{1+j}+4d_{20}\lambda\mu r^{1+j}+4e_{20}\lambda\mu r^{1+j}+$$

$$+4\imath f_{20}\lambda\mu r^{1+j}-8\imath b^2b_2\mu^2r^{1+j}-8\imath b^3b_3\mu^2r^{1+j}+8d_{20}\mu^2r^{1+j}+8e_{20}\mu^2r^{1+j}+$$

$$+8\imath f_{20}\mu^2r^{1+j}+4\imath\lambda\mu q_{20}r^{1+j}+8\imath\mu^2q_{20}r^{1+j}+\imath b_2\lambda\mu r^{3+j}+3\imath bb_3\kappa\lambda\mu r^{3+j}+$$

$$+3\imath b_2\mu^2r^{3+j}+3\imath bb_3\kappa\mu^2r^{3+j}-3\imath bb_3\kappa\lambda\mu r^{1+j}R^2-3\imath bb_3\kappa\mu^2r^{1+j}R^2-$$

$$-a_3b\mu r^{1+j}(4b^2(\lambda+2\mu)+3\kappa(\lambda+\mu)(-r^2+R^2))+4d_{21}\lambda\mu r^jz+4e_{21}\lambda\mu r^jz+$$

$$+4\imath f_{21}\lambda\mu r^jz+8d_{21}\mu^2r^jz+8e_{21}\mu^2r^jz+8\imath f_{21}\mu^2r^jz+4\imath\lambda\mu q_{21}r^jz+8\imath\mu^2q_{21}r^jz-$$

$$-8d_2\lambda\mu r^{1+j}z-16d_2\mu^2r^{1+j}z-8\imath\lambda\mu q_2r^{1+j}z-16\imath\mu^2q_2r^{1+j}z-3c_3\lambda^2r^{3+j}z-$$

$$-9c_3\lambda\mu r^{3+j}z-6c_3\mu^2r^{3+j}z+\imath b_2\lambda\mu r^{1+j}z^2+\imath b_2\mu^2r^{1+j}z^2+$$

$$+4\mu(\lambda+2\mu)(d_{2j}+e_{2j}+\imath(f_{2j}+q_{2j}))rz^j-$$

$$-a_2\mu r^{1+j}(4b^2(\lambda+2\mu)-(\lambda+3\mu)r^2+(\lambda+\mu)z^2))-(a-b)(\lambda+\mu)^2r^{1+j}R^{1+j}ih_1(z))).$$

Again analyticity of the functions ϕ and ψ implies

$$d_1=\frac{1}{64(a-b)^2(-1+\kappa)^2\mu^2(\lambda+2\mu)rR}(8b^3c_2(-1+\kappa)\mu(\lambda+\mu)((-1+\kappa)\lambda+$$

$$+(-1+3\kappa)\mu)rR+8a^4c_3(-1+\kappa)\mu(\lambda+\mu)((-1+\kappa)\lambda+(-1+3\kappa)\mu)rR+$$

$$+8b^4c_3(-1+\kappa)\mu(\lambda+\mu)((-1+\kappa)\lambda+(-1+3\kappa)\mu)rR+$$

$$+8a^3(c_2-bc_3)(-1+\kappa)\mu(\lambda+\mu)((-1+\kappa)\lambda+(-1+3\kappa)\mu)rR-$$

$$-4b^2(-1+\kappa)(\lambda+2\mu)r(8(d_{11}+e_{11})\mu^2+3c_3(\lambda+\mu)((-1+\kappa)\lambda+(-1+3\kappa)\mu)r^2R-$$
$$-6c_3\mu(\lambda+\mu)R^3)-4a^2(-1+\kappa)r(8(d_{11}+e_{11})\mu^2(\lambda+2\mu)+2bc_2\mu(\lambda+\mu)((-1+\kappa)\lambda+$$
$$+(-1+3\kappa)\mu)R+3c_3(-1+\kappa)(\lambda+\mu)(\lambda+2\mu)(\lambda+3\mu)R^3)-$$
$$-4a(-1+\kappa)r(-16b(d_{11}+e_{11})\mu^2(\lambda+2\mu)-$$
$$-(\lambda+\mu)((-1+\kappa)\lambda+(-1+3\kappa)\mu)(-2(b^2(c_2+bc_3)+g_{10}-g_{20})\mu+$$
$$+(c_2+3bc_3)(\lambda+2\mu)r^2)R+(\lambda+\mu)(\lambda+2\mu)((c_2-3bc_3)(-1+\kappa)\lambda+(3bc_3(5-3\kappa)+$$
$$+c_2(-1+3\kappa))\mu)R^3)+(\lambda+\mu)((-1+\kappa)\lambda+(-1+3\kappa)\mu)(r-R)(r+R)(4(d_{11}+$$
$$+e_{11})\mu r-(4(d_{21}+e_{21})\mu-3c_3(\lambda+\mu)r^3)R-3c_3(\lambda+\mu)rR^3)+$$
$$+4b(-1+\kappa)(\lambda+\mu)((-1+\kappa)\lambda+$$
$$+(-1+3\kappa)\mu)rR(2g_{10}\mu-2g_{20}\mu+c_2(\lambda+2\mu)(-r^2+R^2))),$$
$$q_1=-\frac{1}{16(a-b)^2(1+\kappa)^2\mu(\lambda+2\mu)rR}(-8a^2(1+\kappa)\mu(\lambda+2\mu)(f_{11}+q_{11})r-$$
$$-8b^2(1+\kappa)\mu(\lambda+2\mu)(f_{11}+q_{11})r+2b(h_{10}-h_{20})(1+\kappa)(\lambda+\mu)((-1+\kappa)\lambda+$$
$$+(-1+3\kappa)\mu)rR+2a(1+\kappa)r(8b\mu(\lambda+2\mu)(f_{11}+q_{11})-$$
$$-(h_{10}-h_{20})(\lambda+\mu)((-1+\kappa)\lambda+(-1+3\kappa)\mu)R)+$$
$$+(\lambda+\mu)((-1+\kappa)\lambda+(-1+3\kappa)\mu)(r-R)(r+R)((f_{11}+q_{11})r-(f_{21}+q_{21})R)),$$
$$d_2=\frac{1}{64(a-b)^2(-1+\kappa)^2\mu^2(\lambda+2\mu)rR}(4(d_{11}+e_{11})\mu(\lambda+\mu)((-1+\kappa)\lambda+$$
$$+(-1+3\kappa)\mu)r^3+(-32(a-b)^2(d_{21}+e_{21})(-1+\kappa)\mu^2(\lambda+2\mu)+$$
$$+8(a-b)(a^2c_2+a^3c_3-b^2(c_2+bc_3)-g_{10}+g_{20})(-1+\kappa)\mu(\lambda+\mu)((-1+\kappa)\lambda+$$
$$+(-1+3\kappa)\mu)r-4(d_{21}+e_{21})\mu(\lambda+\mu)((-1+\kappa)\lambda+(-1+3\kappa)\mu)r^2+$$
$$+4(a-b)(-1+\kappa)(\lambda+\mu)(\lambda+2\mu)(c_2(-1+\kappa)\lambda+c_2(-1+3\kappa)\mu+$$
$$+3c_3(2a\mu+b(-1+\kappa)(\lambda+3\mu)))r^3+3c_3(\lambda+\mu)^2((-1+\kappa)\lambda+(-1+3\kappa)\mu)r^5)R-$$
$$-4(d_{11}+e_{11})\mu(\lambda+\mu)((-1+\kappa)\lambda+(-1+3\kappa)\mu)rR^2+2(\lambda+\mu)((-1+\kappa)\lambda+$$

$$+(-1+3\kappa)\mu)(2(d_{21}+e_{21})\mu-2(a-b)(c_2+3ac_3)(-1+\kappa)(\lambda+2\mu)r-$$

$$-3c_3(\lambda+\mu)r^3)R^3+3c_3(\lambda+\mu)^2((-1+\kappa)\lambda+(-1+3\kappa)\mu)rR^5),$$

$$q_2=\frac{1}{16(a-b)^2(1+\kappa)^2\mu(\lambda+2\mu)rR}(-(\lambda+\mu)((-1+\kappa)\lambda+$$

$$+(-1+3\kappa)\mu)(f_{11}+q_{11})r^3+(8(a-b)^2(1+\kappa)\mu(\lambda+2\mu)(f_{21}+q_{21})+$$

$$+2(a-b)(h_{10}-h_{20})(1+\kappa)(\lambda+\mu)((-1+\kappa)\lambda+(-1+3\kappa)\mu)r+(\lambda+\mu)((-1+\kappa)\lambda+$$

$$+(-1+3\kappa)\mu)(f_{21}+q_{21})r^2)R+(\lambda+\mu)((-1+\kappa)\lambda+$$

$$+(-1+3\kappa)\mu)(f_{11}+q_{11})rR^2-(\lambda+\mu)((-1+\kappa)\lambda+(-1+3\kappa)\mu)(f_{21}+q_{21})R^3).$$

So

$$\psi_0(z)=\frac{1}{4(a-b)(\lambda+2\mu)z}(r^{-1-j}R^{-1-j}(-8a^2a_2b\mu(\lambda+2\mu)r^{1+j}R^{1+j}z-$$

$$-8a^3a_3b\mu(\lambda+2\mu)r^{1+j}R^{1+j}z+aR^{1+j}(3c_3\lambda^2r^{3+j}z^2-$$

$$-8\mu(\lambda+2\mu)(d_{2j}+\imath q_{2j})rz^{1+j}-8d_2\mu(\lambda+2\mu)r^{1+j}(r-z)(r+z)+$$

$$+\mu^2r^j(16\imath q_2r(r^2+z^2)+z(-2r(8(d_{20}+\imath q_{20})+a_2(-8b^2+3r^2)+$$

$$+a_3b(-8b^2+3\kappa(r-R)(r+R)))-16(d_{21}+\imath q_{21})z+6c_3r^3z+(a_2-\imath b_2)rz^2))+$$

$$+\lambda\mu r^j(8\imath q_2r(r^2+z^2)+z(-8(d_{21}+\imath q_{21})z+r(-8(d_{20}+\imath q_{20})+$$

$$+a_3(8b^3+6b\kappa(-r^2+R^2))+z(9c_3r^2-\imath b_2z)+a_2(8b^2-2r^2+z^2)))))+$$

$$+br^{1+j}(-3c_3\lambda^2R^{3+j}z^2+8d_1\mu(\lambda+2\mu)R^{1+j}(R-z)(R+z)+$$

$$+\mu^2(16(d_{1j}+\imath q_{1j})Rz^{1+j}+R^j(-16\imath q_1R(R^2+z^2)+z(2R(8d_{10}+8\imath q_{10}+3a_2R^2)+$$

$$+2(8d_{11}+8\imath q_{11}-3c_3R^3)z-(a_2-\imath b_2)Rz^2)))+\lambda\mu(8(d_{1j}+\imath q_{1j})Rz^{1+j}+$$

$$+R^j(-8\imath q_1R(R^2+z^2)+z(8(d_{11}+\imath q_{11})z+R(8d_{10}+8\imath q_{10}+2a_2R^2-9c_3R^2z-$$

$$-(a_2-\imath b_2)z^2)))))+r^{1+j}R^{1+j}(-4(\lambda+2\mu)(ar^2-bR^2)\phi_0'(z)-$$

$$-4ab(\lambda+2\mu)(r-R)(r+R)\phi_1'(z)+(\lambda+\mu)(-(\lambda+3\mu)(ar^2-bR^2)h_1(z)+$$

$$+(a-b)(\lambda+\mu)z\imath h_1(z))+4(a-b)\kappa(\lambda+2\mu)z\phi_0(z)))).$$

And

$$\psi_1(z) = \frac{1}{4(a-b)(\lambda+2\mu)z}(-8\lambda\mu(d_1 - iq_1)R^2 - 16\mu^2(d_1 - \imath q_1)R^2 +$$

$$+8\lambda\mu(a^2a_2 + a^3a_3 - b^2(a_2 + a_3b) - d_{10} + d_{20} - \imath(q_{10} - q_{20}))z + 16\mu^2(a^2a_2 + a^3a_3 -$$

$$-b^2(a_2 + a_3b) - d_{10} + d_{20} - \imath(q_{10} - q_{20}))z - 3c_3\lambda^2(r-R)(r+R)z^2 -$$

$$-8\mu(\lambda+2\mu)r^{-j}R^{-j}((d_{1j} + \imath q_{1j})r^j - (d_{2j} + \imath q_{2j})R^j)z^{1+j} +$$

$$+8d_2\mu(\lambda+2\mu)(r-z)(r+z) - 8i\lambda\mu q_2(r^2+z^2) - 16\imath\mu^2 q_2(r^2+z^2) +$$

$$+\frac{1}{rR}(\mu^2 z(6(a_2 + a_3b\kappa)r^3R - 6(a_2 + aa_3\kappa)rR^3 - 2(8d_{11}r + 8\imath q_{11}r -$$

$$-R(8d_{21} + 8\imath q_{21} + 8d_1r + 8\imath q_1r - 3c_3r^3 + 3c_3rR^2))z + 3(a-b)(a_3 - \imath b_3)rRz^2)) +$$

$$+\frac{1}{rR}(\lambda\mu z(2(a_2 + 3a_3b\kappa)r^3R - 2(a_2 + 3aa_3\kappa)rR^3 - (8d_{11}r + 8\imath q_{11}r -$$

$$-R(8d_{21} + 8\imath q_{21} + 8d_1r + 8\imath q_1r - 9c_3r^3 + 9c_3rR^2))z + 3(a-b)(a_3 - \imath b_3)rRz^2)) +$$

$$+4(\lambda+2\mu)(r-R)(r+R)\phi_0'(z) + 4(\lambda+2\mu)(br^2 - aR^2)\phi_1'(z) +$$

$$+(\lambda+\mu)(\lambda+3\mu)(r-R)(r+R)h_1(z) + 4(a-b)\kappa(\lambda+2\mu)z\phi_1(z)).$$

10.3 Solution pseudocode

Here we give the outline of the algorithm which can be used for finding of the displacement coordinates for the cylider and the spline of the first type.

We need five auxiliary functions

```
//complex conjugate
function CONJUGATE( α = a + i * A )
DO RETURN( a − i * A )
//differentiation
function D(a, k)
DO RETURN(k*a)
//integration
```

function T(a, k)

DO RETURN(a/(k+1))

// this function is constructed by the recurrent formula with the help of function T, the last three arguments are arrays

function Ψ (k, j, ϕ , ψ , ρ)

DO RETURN($\Psi_{k,j}$)

// this function is constructed by the recurrent formula with the help of function T, the last three arguments are arrays

function Φ (k, j, ϕ , ψ , ρ)

DO RETURN($\Phi_{k,j}$)

// we first need a solution in one segment

procedure SEGMENT(h, H)

BEGIN

// first we define the number of levels

INPUT(n)

//we introduce the curves levels

FOR k from n to 0 by -1

DO INPUT (h_k)

// we fix the degree of the Fourier polynomials for boundary conditions

INPUT(m)

// we input the boundary conditions

FOR j from n to 0 by -1

DO FOR k from -m to m by 1

DO INPUT ($A_{j,k}, a_{j,k}, B_{j,k}, b_{j,k}, C_{j,k}, c_{j,k}$)

// we find the values of Fourier coefficients for ϕ_k , ρ_k and ϕ_k using the coefficients with nonnegative degrees of z from the recurrent formulas and inverse to Vandermonde matrix. The functions expressions themselves are again rather cumbersome.

FOR l from 0 to m by 1

DO BEGIN

$\phi_{n+1,l} = 0$;

$\psi_{n+1,l} = 0$;

$\rho_{n+1,l} = 0$;

END

FOR k from n to 0 by -1

DO FOR l from 0 to m by 1

DO

$\phi_{k,l} = \phi_{k,l}(\phi_{k+1,l}, \psi_{k+1,l}, \rho_{k+1,l})$;

FOR k from n to 0 by -1

DO FOR l from 0 to m by 1

DO

$\rho_{k,l} = \rho_{k,l}(\phi_{k+1,l}, \psi_{k+1,l}, \rho_{k+1,l})$;

FOR k from n to 0 by -1

DO FOR l from 0 to m by 1

DO $\psi_{k,l} = \psi_{k,l}(\phi_{k+1,l}, \psi_{k+1,l}, \rho_{k+1,l})$;

// we find the auxiliary functions Φ_k , Ψ_k , the general form of these functions can be found in the appendices

FOR k from n to 0 by -1

DO FOR l from -m to m by 1

DO BEGIN

$\Phi_{k,l} = \Phi(k, l, \phi, \psi, \rho)$;

$\Psi_{k,l} = \Psi(k, l, \phi, \psi, \rho)$;

END

// we use the analyticity conditions or similarly the coefficients with the negative degrees of z from the recurrent formulas in order to find the auxiliary constants; again some of the formulas for the constants are given at the appedices

FOR k from n to 0 by -1

DO BEGIN

$D_k = D_k(\phi_{n,j})$;

$E_k = E_k(\phi_{n,j})$;

END

// we find $\phi_{n,l}$, $\psi_{n,l}$, $\rho_{n,l}$ from the the boundary conditions by solving the relevant systems of linear equations

FOR l from 0 to m by 1

DO

BEGIN

$\phi_{0,l} = \phi_{0,l}(A_{j,k}, a_{j,k}, B_{j,k}, b_{j,k}, C_{j,k}, c_{j,k})$;

$\psi_{0,l} = \psi_{0,l}(A_{j,k}, a_{j,k}, B_{j,k}, b_{j,k}, C_{j,k}, c_{j,k})$;

$\rho_{0,l} = \rho_{0,l}(A_{j,k}, a_{j,k}, B_{j,k}, b_{j,k}, C_{j,k}, c_{j,k})$;

END

// we find the displacement coordinates

FOR k from n to 0 by -1

DO FOR l from -m to m+1 by 1

DO

BEGIN

$u_{k,l} = 0$;

$v_{k,l} = 0$;

$w_{k,l} = 0$;

END

FOR l from -m to m by 1

DO

BEGIN

$u_{k,l} = u_{k,l} - \kappa/(2\mu)(\phi_{k,l} + CONJUGATE(\phi_{k,l}) + 1/(2\mu)(\psi_{k,l} + CONJUGATE(\psi_{k,l}) + 1/(2\mu)(D(\phi_{k,l+1}, l+1) + CONJUGATE(D(\phi_{k,l+1}, l+1)) + 1/(2\mu)(\Phi_{k,l} + CONJUGATE(\Phi_{k,l})$;

$v_{k,l} = v_{k,l} - \kappa/(2\mu)(\phi_{k,l} - CONJUGATE(\phi_{k,l}) - 1/(2\mu)(\psi_{k,l} - CONJUGATE(\psi_{k,l}) - 1/(2\mu)(D(\phi_{k,l+1}, l+1) - CONJUGATE(D(\phi_{k,l+1}, l+1)) + 1/(2\mu)(\Phi_{k,l} - CONJUGATE(\Phi_{k,l})$;

$w_{k,l} = w_{k,l} + 1/2(\rho_{k,l} + CONJUGATE(\rho_{k,l})) + 1/2(\Psi_{k,l} + CONJUGATE(\Psi_{k,l}))$

END

RETURN(u, v, w)

END //procedure SEGMENT

The program itself then is as follows

// we introduce the number of segments

INPUT(N)

FOR j from 0 to N by 1

DO

INPUT (h_j)

// then we solve the problem for each segment

FOR j from 0 to N-1 by 1

DO

SEGMENT (h_j, h_{j+1})

The procedure SEGMENT for the spline of the first type is similar to one given here for all the other types of solids with the following diferences: 1) $n = 2$, 2) the order of reconstruction is ϕ_1 , ρ_1 , ρ_0 , ϕ_0 , ψ_1 , ψ_0 . The formulas are given in Appendix 2.

For the spline of the second kind again this part of the procedure SEGMENT will be different, namely it will contain the representation of the additional variables (coefficients of ϕ_2 , ψ_2 , ρ_2 for end gluing solutions) through the given zero boundary and arbitrary end data. The formulas for gluing of the ends in the case of cylindrical solid are given in the first Appendix. All the formulas for the tube are given in the second part of chapter 5.

References

[1] A. Y. Alexandrov, "Solution of three-dimensional problems of the theory of elasticity for solids of revolution by means of analytical functions," *International Journal of Solids and Structures.* vol. 4, no. 7, pp. 701-721, July 1968.

[2] Bitsadze A.V. *Equations of mathematical physics*, Moscow: Mir, 1980.

[3] F. A. Bogashov, "Working out and application of the apparatus of complex space potentials in theory of elasticity: the dissertation," Doctor of physical and mathematical sciences dissertation: N.Novgorod State University, N.Novgorod, Russia, 1995.

[4] F.D. Gahov *Kraevye zadachi* (in russian), Moscow: Nauka, 1977.

[5] L. Komzsik, *Approximation techniques for engineers*, CRC Press, 2006.

[6] L.S. Lejbenzon, *Course of the theory of elasticity*, Moscow, 1947.

[7] N.I.Muskhelishvili. *Some Basic Problems of the Mathematical Theory of Elasticity*, Moscow: Nauka, 1966.

[8] I.I. Privalov, B.M. Pchelin, "Sur la théorie générale des fonctions polyharmoniques" *C.R. Acad. Sci. Paris*, Vol. 204, pp. 328-330, 1937.

[9] L. L. Schumaker, *Spline functions: Basic theory*, New York: Wiley, 1981.

Elena A. Shirokova and Pyotr N. Ivanshin (Eds)

[10] E.A.Shirokova, "On 3-d analog of the second basic problem of the theory of elasticity for a cylindrical solid," *Mech. Res. Com.* vol. 31, pp. 29-37, 2004.

[11] E.A.Shirokova, "Elastic solutions for a pressurised tube with given exterior displacements," *Int. J. Pressure Vessels and Piping*, vol. 81, pp. 731-738, 2004.

[12] E.A.Shirokova, "Reconstruction of the displacements and stresses at the points of pressurised tube via the given exterior displacements," *Pressure Vessels and Piping: Codes, Standards, Design and Analysis. Narosa Publishing House, New Delhi, India*, pp. 451-459, 2009.

[13] Ye.A.Shirokova, "The interpolation solution of the second basic plane problem of the dynamics of elastic solids," *J.Appl.Math and Mech.*, vol. 73, pp. 63-70, 2009.

[14] A. Tsalik, "Quaternionic representation of the 3D elastic and thermoelastic boundary problems," *Mathematical Methods in the Applied Sciences*, vol. 18 no. 9, pp 697 - 708, 1994.

[15] I.N. Vekua, "Solution of the basic boundary value problem for the equation $\Delta^{n+1}u = 0$," (Russian) *Soobshch. Akad. Nauk Gruz. SSR*, vol. 3, pp. 213-220, 1942.

INDEX

www.ingramcontent.com/pod-product-compliance
Lightning Source LLC
Chambersburg PA
CBHW050821220326
41598CB00006B/280
* 9 7 8 1 6 0 8 0 5 6 2 0 0 *